AF342051

IUTAM Symposium on Relations of Shell, Plate, Beam, and 3D Models

IUTAM BOOKSERIES
Volume 9

Series Editors

G.L.M. Gladwell, *University of Waterloo, Waterloo, Ontario, Canada*
R. Moreau, *INPG, Grenoble, France*

Editorial Board

J. Engelbrecht, *Institute of Cybernetics*, Tallinn, Estonia
L.B. Freund, *Brown University, Providence, USA*
A. Kluwick, *Technische Universität, Vienna, Austria*
H.K. Moffatt, *University of Cambridge, Cambridge, UK*
N. Olhoff *Aalborg University, Aalborg, Denmark*
K. Tsutomu, *IIDS, Tokyo, Japan*
D. van Campen, *Technical University Eindhoven, Eindhoven, The Netherlands*
Z. Zheng, *Chinese Academy of Sciences, Beijing, China*

Aims and Scope of the Series

The IUTAM Bookseries publishes the proceedings of IUTAM symposia
under the auspices of the IUTAM Board.

For other titles published in this series, go to
www.springer.com/series/7695

IUTAM Symposium on Relations of Shell, Plate, Beam, and 3D Models

Proceedings of the IUTAM Symposium on the Relations of Shell, Plate, Beam, and 3D Models, Dedicated to the Centenary of Ilia Vekua's Birth, held in Tbilisi, Georgia, April 23-27, 2007

Edited by

George Jaiani

I. Vekua Institute of Applied Mathematics,
Iv. Javakhishvili Tbilisi State
University, Tbilisi, Georgia

Paolo Podio-Guidugli

Department of Civil Engineering,
University of Rome,
"Tor Vergata", Rome,
Italy

 Springer

Editors

Prof. Dr. George Jaiani
Iv. Javakhishvili Tbilisi State University
I.Vekua Institute of Applied Mathematics
University Street 2
0186 Tbilisi
Georgia
george.jaiani@gmail.com

Prof. Dr. Paolo Podio-Guidugli
Universita' di Roma TorVergata
Dipartimento di Ingegneria Civile
Viale Politecnico, 1
00133 Roma
Italy
ppg@uniroma2.it

ISBN: 978-1-4020-8773-8   e-ISBN: 978-1-4020-8774-5

Library of Congress Control Number: 2008931595

Ilia Vekua
(April 23, 1907 – December 2, 1977)

Editorial Preface

During its 2004 meeting in Warsaw the General Assembly of the International Union of Theoretical and Applied Mechanics (IUTAM) decided to support a proposal of the Georgian National Committee to hold in Tbilisi (Georgia), on April 23–27, 2007, the IUTAM Symposium on the *Relation of Shell, Plate, Beam, and 3D Models*, dedicated to the Centenary of Ilia Vekua. The scientific organization was entrusted to an international committee consisting of Philipppe G. Ciarlet (Hong Kong), the late Anatoly Gerasimovich Gorshkov (Russia), Jorn Hansen (Canada), George V. Jaiani (Georgia, Chairman), Reinhold Kienzler (Germany), Herbert A. Mang (Austria), Paolo Podio-Guidugli (Italy), and Gangan Prathap (India).

The main topics to be included in the scientific programme were chosen to be: hierarchical, refined mathematical and technical models of shells, plates, and beams; relation of 2D and 1D models to 3D linear, non-linear and physical models; junction problems. The main aim of the symposium was to thoroughly discuss the relations of shell, plate, and beam models to the 3D physical models. In particular, peculiarities of cusped shells, plates, and beams were to be emphasized and special attention paid to junction, multibody and fluid-elastic shell (plate, beam) interaction problems, and their applications. The expected contributions of the invited participants were anticipated to be theoretical, practical, and numerical in character.

According to these premises, all the lecturers were invited personally for their active interest in the field covered by the symposium. In all there were 50 participant from 15 countries. The programme included an Opening Lecture, which was an impressive eulogy of Ilia Vekua, presented by B. Bojarsky (Poland), his former PhD student; the Closing Lecture was given by P. Podio-Guidugli (Italy), and there were 23 30min-lectures. The afternoon sessions were ended by general discussions (round tables). The Georgian National Science Foundation has established a "Best Scientific Paper Award" to be bestowed on three young participants to the IUTAM Symposium. According to the decision of the International Scientific Committee of IUTAM

Symposium, these awards went to Natalia Chinchaladze (Georgia), Lorenzo Freddi (Italy), and Rainer Schlebusch (Germany).

The local arrangements of the symposium were in the hands of a committee consisting of Gia Avalishvili (I.Vekua Institute of Applied Mathematics), Natalia Chinchaladze (I.Vekua Institute of Applied Mathematics, Secretary), David Gordeziani (I.Vekua Institute of Applied Mathematics), George Jaiani (I.Vekua Institute of Applied Mathematics, Chairman), Gela Kipiani (Georgian Technical University), Tengiz Meunargia (Iv. Javakhishvili Tbilisi State University), Nugzar Shavlakadze (A.Razmadze Mathematical Institute), Ilia Tavkhelidze (Iv. Javakhishvili Tbilisi State University), Tamaz Vashakmadze (Iv. Javakhishvili Tbilisi State University).

The working sessions of the symposium were held in lecture-halls at the I. Vekua Institute of Applied Mathematics of Iv. Javakhishvili Tbilisi State University. In the occasion of the opening session, the symposium was welcomed by G. Khubua, Rector of Iv. Javakhishvili Tbilisi State University, N. Jokhadze, Director of the Georgian National Science Foundation, and D. H. Van Campen, IUTAM Secretary-General. Excursions and an interesting ladies programme complemented the scientific activities. In addition, the participants visited the Pantheon, the Georgian national cemetery of statesmen, scientists, and writers, with graves of I. Vekua and N. Muskhelishvili, founder and first president of the Georgian Academy of Sciences and the USSR National Committee of Theoretical and Applied Mechanics.

The volume includes 18 peer-reviewed papers presented at the symposium. The editors are indebted to Springer-Verlag for their courteous and effective production of these Proceedings.

March, 2008

G. Jaiani (Tbilisi)
P. Podio-Guidugli (Rome)

Contents

Ilia Vekua
(April 23, 1907 - December 2, 1977)

April 23, 2007 is the day of 100th birth anniversary of Ilia Vekua, an eminent scholar in mathematics and mechanics.

Ilia Vekua's research works cover various fields of mathematics and mechanics. Many of them are devoted to the theory of partial differential equations, in which Ilia Vekua took a great interest. In the analytical theory of linear differential equations of elliptic type with two independent variables, an important part was played by formulas of general representation of solutions by means of analytic functions of one complex variable. These formulas made it possible to widen considerably the field of application of the methods of the classical theory of analytic functions of a complex variable. Based on these studies, Ilia Vekua developed new methods for solving boundary value problems, which enabled him to investigate a vast class of boundary value problems formulated in nonclassical sense. The method he proposed for reducing boundary value problems to singular integral equations is one of the most powerful means for studies in this field. Concerning the merits of I. Vekua in the theory of singular integral equations, one can read in a well-known monograph "Singular Integral Equations" by N. Muskhelishvili: "Under the influence of a number of results obtained by the participants of the seminar and mainly due to I. N. Vekua's fine works, the range of the problems I wished to study has considerably changed and I can note with a great and quite comprehensible pleasure that the most part of this book content should be considered as a result of joint work of a group of young scientific collaborators from the Tbilisi Mathematical Institute of the Georgian Academy of Sciences with I. Vekua and me". Special mention should be made of a general boundary value problem for elliptic equations, which Ilia Vekua formulated and studied most completely. The well known boundary value problems of Dirichlet, Neumann and Poincaré are particular cases of this problem. Ilia Vekua derived the formulas of integral representation of holomorphic functions, which in the mathematical literature are named after him, and used them as an important tool in investigating the problem. Ilia Vekua is one of the founders of the theory of generalized analytic functions.

Ilia Vekua worked out several versions of the mathematical theory of elastic shells. In general, it should be stressed that all his works in mathematics were aimed at applications to topical problems of mechanics.

In recognition of his many pioneering contributions, Ilia Vekua

- was elected a Corresponding member of the Georgian Academy of Sciences in 1944,
- was elected a Corresponding member of the USSR Academy of Sciences and an Academician of the Georgian Academy of Sciences in 1946,
- was elected an Academician of the USSR Academy of Sciences in 1958,
- was elected a Foreign member of German Academy of Sciences in 1968,

- was elected a Foreign member of the Academy of Natural Sciences "Leopoldina" (Halle) in 1969,
- was elected a Foreign member of the Academy of Sciences of Literature and Art (Sicilian Academy of Sciences) in 1976,
- received the USSR Stalin Prize of the second degree for his monograph "New methods of solution of elliptic equations" (in Russian, published in 1948) in 1950,
- received the USSR Lenin Prize for his monograph "Generalized analytic functions" (in Russian, published in 1959) in 1963,
- received (posthumously) the State Prize for his research work "Some general methods of constructing various versions of the shell theory" (Russian, published in 1982) in 1984.

Ilia Vekua was born on April 23, 1907, in Abkhazian village Shesheleti (West Georgia). After finishing a secondary school in the West Georgian town Zugdidi in 1925, he moved to Tbilisi, the capital of Georgia, where he studied at the Faculty for Physics and Mathematics of Tbilisi State University. He graduated with honors in 1930 and, on the recommendation of Academician Niko Muskhelishvili, left Tbilisi for Leningrad (now St. Petersburg) to continue his education there as a post-graduate student at the USSR Academy of Sciences. His initial research was conducted under the supervision of the well-known mathematician A. N. Krylov. In Leningrad, Ilia Vekua published his first papers on problems of torsion and bending of elastic bars. He also worked on the theory of propagation of electric waves in an infinite layer with parallel plane boundaries and obtained the results which subsequently formed the basis of his thesis for the Candidate of Science degree.

After finishing the post-graduate course in 1933, Ilia Vekua returned to Tbilisi to work at his alma mater. He wholly devoted himself to scientific, educational and organizational activities. Ilia Vekua became an active participant in the famous seminar run by Niko Muskhelishvili. He delivered lectures on mathematical physics, calculus of variations, differential and integral equations and was one of the founders of the Mathematical Institute of the Georgian Branch of the USSR Academy of Sciences (now A. Razmadze Mathematical Institute).

In 1951, Ilia Vekua moved to Moscow where he was officially invited for permanent residence and work. Together with his outstanding colleagues and friends M. A. Lavrent'ev, I. G. Petrovskii, and S. L. Sobolev, he directed the research seminars at V. A. Steklov Mathematical Institute and M. V. Lomonosov Moscow University.

Ilia Vekua was the founding Rector (1959–1964) of Novosibirsk University. When living in Siberia, Ilia Vekua simultaneously combined several duties: he headed the theoretical department at the Hydrodynamics Institute of the Siberian Branch of the USSR Academy of Sciences, held the mathematical physics chair of Novosibirsk University, and supervised the work of several scientific seminars.

After the USSR National Committee on Theoretical and Applied Mechanics was formed in 1956, Ilia Vekua became a permanent member. From 1963 he was also a member of the National Committee of Soviet Mathematicians.

At the end of 1964, Ilia Vekua returned to Tbilisi, where he was elected vice-president of the Georgian Academy of Sciences (1964–1965) and head of the mathematics chair at Tbilisi State University (1966-1972). On his initiative and under his guidance, the Department of Mechanics was organized (1964) at A. Razmadze Mathematical Institute of the Georgian Academy of Sciences, and the Problem Laboratory of Applied Mathematics was founded (1966) at Tbilisi State University and shortly reorganized as the Institute of Applied Mathematics (1968). The latter institute is named after Ilia Vekua, who was its founder and remained its director and scientific leader (1968–1977) till the last days of his life. From 1972 to 1977, Ilia Vekua served as the president of the Georgian Academy of Sciences.

Through the last years of his life, in spite of his grave illness, Ilia Vekua continued to pursue his scientific, teaching and organizational activities. His last monographs were published posthumously. In September 1976, on Ilia Vekua's suggestion, the IUTAM's General Assembly decided to organize the 3rd International Symposium on the Theory of Shells in Tbilisi, Georgia. Ilia Vekua was appointed chairman both of the international scientific committee and of the national organizing committee. Preparations for the symposium were underway when the whole scientific world was deeply saddened by the untimely demise of Ilia Vekua on December 2, 1977. Nevertheless, the symposium which the IUTAM Bureau decided to dedicate to the memory of Ilia Vekua, was held in Tbilisi in August 22–28, 1978.

In recognition of his special services to mechanics in the occasion of the centenary of his birth, the IUTAM Symposium on Relation of Shell, Plate, Beam, and 3D Models was dedicated to I. Vekua. The symposium was held in Tbilisi on 23–27 April, 2007. The works of well-known scientists presented at the symposium are collected in this issue.

G. Jaiani

Main Publications of Ilia Vekua

(i) monographs

1. New methods of solution of elliptic equations. (Russian) Gostekhizdat, Moscow-Leningrad, 1948, 296p.
2. Systeme von Differentialgleichungen erster Ordnung vom elliptischen Typus und Randwertaufgaben mit einer Anwendung in der Theorie der Schalen. Deutscher Verlag. Wiss., 1956, 107p.
3. Generalized analytic functions. (Russian) Fizmatgiz, Moscow, 1959, 628p.
4. Systems of first order differential equations of elliptic type and boundary value problems with an application to the shell theory. (Chinese) Peking, Gao den tsiya-o-yu chuban-she, 1960, VII , 204p.
5. Generalized analytic functions. Oxford-London-New York-Paris, 1962, 668p.
6. Verallgemeinerte analytische Funktionen. Berlin, Akad. Verlag, 1963, 538p.
7. On a version of the theory of shallow thin shells. (Russian) Izd. Novosib. Gos. Univ., Novosibirsk, 1964, 68p.
8. Theory of thin and shallow shells of varying thickness. (Russian) Izd. Novosib. Gos. Univ., Novosibirsk, 1964, 39p.
9. Fundamentals of tensor analysis. (Russian) Izd. Novosib. Gos. Univ., Novosibirsk, 1964, 138p.
10. On a version of the bending theory of elastic shells. Univ. Maryland, USA, 1964, 42p.
11. Fundamentals of tensor analysis. (Russian) Tbilisi University Press, Tbilisi, 1967, 137p.
12. New methods for solving elliptic equations. North-Holland Publ. Co., Amsterdam, 1967.
13. Variational principles of construction of the shell theory. (Russian) Tbilisi University Press, Tbilisi, 1970, 17p.
14. Fundamentals of tensor analysis and theory of covariants. (Russian) Nauka, Moscow, 1978, 296p.
15. Some general methods of constructing various versions of shell theory. (Russian) Nauka, Moscow, 1982, 286p.
16. Fundamentals of tensor analysis and theory of covariants. (Georgian), Metcniereba, Tbilisi, 1982, 365p.
17. Shell theory: general methods of construction. Pitman Advanced Publishing Program, Boston-London-Melbourne, 1985, 287p.
18. Generalized analytic functions.(Russian).2nd ed., revised, Nauka, Moscow, 1988, 509p.
19. Some general methods of construction of different versions of the theory of shells. (Georgian) Tbilisi University Press, Tbilisi, 2007, 287p.

(ii) papers

1. Problem of torsion of a circular cylinder reinforced with a longitudinal circular rod. (Russian) Izv. Akad. Nauk SSSR, Otd. Mat. Estestv. Nauk, Ser. 7(1933), No. 3, 373–386 (coauthor A. K. Rukhadze).

2. Torsion and bending by transverse force of a bar composed of two elastic materials bounded by confocal ellipses. (Russian) Prikl. Mat. Mekh. 1(1933), No. 2, 167–178 (coauthor A. K. Rukhadze).

3. Propagation of elastic waves in an infinite layer bounded by two parallel planes. (Russian) Proc. II All-Union Math. Congr. (Leningrad, June 24–30, 1934), vol. 2, 363–364, USSR Acad. Sci., Moscow-Leningrad, 1936.

4. Sur une représentation complèxe de la solution générale des équations du problème stationnaire plan de la théorie de l'elasticité. C. R. Acad. Sci. URSS, 16(1937), No. 3, 155–160.

5. Sur la représentationn générale des solutions de équations aux dérivées partielles du second ordre. C. R. Acad. Sci. URSS, 17(1937), No. 6, 295–299.

6. A general representation of solutions of partial differential equations of elliptic type which are linear with respect to the Laplace operator. (Russian) Trudy Tbilis. Mat. Inst. 2(1937), 227–240.

7. A boundary value problem of oscillation of an infinite layer. (Georgian) Trudy Tbilis. Mat. Inst. 1(1937), 141–164.

8. To the question of propagation of elastic waves in an infinite layer bounded by two parallel planes. (Russian) Trudy Tbilis. Geophys. Inst. 2(1937), 23–50.

9. Some remarks in connection with I. G. Kurdiani's paper "Some problems of stratification instability of air masses". (Russian) Trudy Tbilis. Geophys. Inst. 4(1939), 165–171.

10. A complex representation of solutions of elliptic differential equations and its application to boundary value problems. (Russian) Trudy Tbilis. Mat. Inst. 7(1939), 161–253.

11. Sur les équations intégrales linéaires singulières contenant des intégrales au sens de la valeur principale de Cauchy. C. R. Acad. Sci. URSS, 26(1940), No. 4, 327–330.

12. Boundary value problems of the theory of linear elliptic differential equations with two independent variables 1. (Russian) Soobshch, Gruz. Fil. Akad. Nauk SSSR, 1(1940), No. 1, 29–34.

13. Boundary value problems of the theory of linear elliptic differential equations with two independent variables 2. (Russian) Soobshch. Gruz. Fil. Akad. Nauk SSSR, 1(1940), No. 3, 181–186.

14. Boundary value problems of the theory of linear elliptic differential equations with two independent variables 3. (Russian) Soobshch. Gruz. Fil. Akad. Nauk SSSR, 1(1940), No. 7, 497–500.

15. Remarks in connection with the Fourier method. (Russian) Soobshch. Gruz. Fil. Akad. Nauk SSSR, 1(1940), No. 9, 647–650 (coauthor D. F. Kharazov).

16. An application of Academician N. Muskhelishvili's method to the solution of boundary value problems of the plane theory of elasticity of an anisotropic medium. (Russian) Soobshch. Gruz. Fil. Akad. Nauk SSSR, 1(1940), No. 10, 719–724.

17. Allgemeine Darstellung der Lösungen elliptischer Differentialgleichungen in einem mehrfach zusammenhängenden Gebiet. Soobshch. Fil. Akad. Nauk SSSR, 1(1940), No. 5, 329–335.

18. On one new integral representation of analytic functions and its application. (Russian) Soobshch. Akad. Nauk Gruz. SSR, 2(1941), No. 6, 477–484.

19. On one class of singular integral equations with an integral in the sense of the Cauchy principal value. Soobshch. Akad. Nauk Gruz. SSR, 2(1941), No. 7, 579–586.

20. On reducing singular integral equations to the Fredholm equation. (Russian) Soobshch. Akad. Nauk Gruz. SSR, 2(1941), No. 8, 697–700.

21. On harmonic and metaharmonic functions in a space. (Russian) Soobshch. Akad. Nauk Gruz. SSR, 2(1941), No. 1, 20–32.

22. Supplement to the paper "On one new integral representation of analytic functions and its application". (Russian) Soobshch. Akad. Nauk Gruz. SSR, 2(1941), No. 8, 701–706.

23. Integral equations with a singular kernel of the Cauchy type. (Russian) Trudy Tbilis. Mat. Inst. 10(1941), 45–72.

24. Über harmonische und metaharmonische Funktionen im Raum. Soobshch. Akad. Nauk Gruz. SSR, 2(1941), No. 1–2, 29–34.

25. On the approximation of solutions of elliptic differential equations. (Russian) Soobshch. Akad. Nauk. Gruz. SSR, 3(1942), No. 2, 97–102.

26. Solution of the basic boundary value problem for the equation $\Delta^{n+1}u = 0$. (Russian) Soobshch. Akad. Nauk Gruz. SSR, 3(1942), 213–220.

27. On solutions of equation $\Delta u + \lambda^2 u = 0$. (Georgian) Soobshch. Akad. Nauk Gruz. SSR, 3(1942), No. 4, 307–314.

28. On the bending of a plate with a free edge. (Russian) Soobshch. Akad. Nauk Gruz. SSR, 3(1942), No. 7, 641–648.

29. To the theory of singular integral equations. (Russian) Soobshch. Akad. Nauk Gruz. SSR, 3(1942), No. 9, 869–876.

30. On one linear boundary value problem of Riemann. (Russian) Trudy Tbilis. Mat. Inst. 11(1942), 109–139.

31. On the solution of a mixed boundary value problem of the theory of a Newtonian potential for a multiply connected domain. (Russian) Soobshch. Akad. Nauk Gruz. SSR, 3(1942), 753–758.

32. Green's function for a spherical layer. (Georgian) Trudy Tbilis. Gosud. Univ. 25(1942), 225–228.

33. On some basic properties of metaharmonic functions. (Russian) Soobshch. Akad. Nauk Gruz. SSR, 4(1943), No. 4, 281–288.

34. Remarks on a general representation of solutions of differential equations of elliptic type. (Russian) Soobshch. Akad. Nauk Gruz. SSR, 4(1943), No. 5, 385–392.

35. To a general diffraction problem. (Russian) Soobshch. Akad. Nauk Gruz. SSR, 4(1943), No. 6, 503–506.

36. On one integral representation of solutions of differential equations. (Russian, Georgian) Soobshch. Akad. Nauk Gruz. SSR, 4(1943), No. 9, 843–852.

37. On one new representation of solutions of differential equations. (Georgian) Soobshch. Akad. Nauk Gruz. SSR, 4(1943), No. 10, 941–950.

38. On metaharmonic functions. (Russian) Trudy Tbilis. Mat. Inst. 12(1943), 105–174.

39. Correction to Ilia Vekua's paper "On one linear boundary value problem of Riemann". (Russian) (see Trudy Tbil. Mat. Inst. 11(1942), 109–139). Trudy Tbilis. Math. Inst. 12(1943), 215.

40. Sur certain dévelopment des fonctions métaharmoniques. C. R. Acad. Sci. URSS, 48(1945), No. 1, 3–6.

41. Représentation genérale des solutions d'une équation différentielle des fonctions sphériques. C. R. Acad. Sci. URSS, 49(1945), No. 5, 311–314a.

42. Inversion of one integral transformation and some of its applications. (Russian) Soobshch. Akad. Nauk Gruz. SSR, 6(1945), No. 3, 179–183.

43. On the integrodifferential equation of Prandtl. (Russian) Prikl. Mat. Mekh. 9(1945), No. 2, 143–150.

44. Integration of equations of a spherical shell. (Russian) Prikl. Mat. Mekh. 9(1945), No. 5, 368–388.

45. To the theory of Legendre's functions. (Russian) Soobshch. Akad. Nauk Gruz. SSR, 7(1946), No. 1–2, 3–10.

46. To the theory of cylindrical functions. (Russian) Soobshch. Akad. Nauk Gruz. SSR, 7(1946), No. 3, 95–101.

47. Sur une généralisation de l'intégrale de Poisson pour le demi-plan. C. R. Acad. Sci. URSS, 56(1947), No. 3, 229–231.

48. Some basic problems of the theory of a thin spherical shell. (Russian) Prikl. Mat. Mekh. 11(1947), No. 5, 499–516.

49. Approximation of solutions of second order differential equations of elliptic type. (Georgian) Trudy Tbilis. Gos. Univ. 30a(1947), 1–21.

50. On one generalization of the Poisson integral for a half-plane. (Georgian) Trudy Tbilis. Mat. Inst. 15(1947), 149–154.

51. On one method of solution of boundary value problems of sinusoidal oscillation of an elastic cylinder. (Russian) Dokl. Akad. Nauk SSSR, 60(1948), No. 5, 779–782.

52. To the theory of shallow thin elastic shells. (Russian) Prikl. Mat. Mekh. 12(1948), No. 1, 69–74.

53. To the theory of elastic shells. (Russian) Dokl. Akad. Nauk SSSR, 68(1949), No. 3, 453–455.

54. On one representation of solutions of differential equations of elliptic type. (Russian) Soobshch. Akad. Nauk Gruz. SSR, 11(1950), No. 3, 137–141.

55. On the proof of some uniqueness theorem occurring in the stationary oscillation theory. (Russian) Dokl. Akad. Nauk SSSR, 80(1951), No. 3, 341–343.

56. Systems of first order differential equations of elliptic type and boundary value problems with an application in the shell theory. (Russian) Mat. Sb. 31(1952), No. 2, 217–314.

57. A general representation of functions of two independent variables admitting derivatives in the Sobolev sense and the problem of primitives. (Russian) Dokl. Akad. Nauk SSSR, 89(1953), No. 5, 773–775.

58. On the completeness of a system of harmonic polynomials in a space. (Russian) Dokl. Akad. Nauk SSSR, 90(1953), No. 4, 495–498.

59. On the completeness of a system of metaharmonic functions. (Russian) Dokl. Akad. Nauk SSSR, 90(1953), No. 5, 715–718.

60. A boundary value problem with an oblique derivative for an elliptic type equation. (Russian) Dokl. Akad. Nauk SSSR, 92(1953), No. 6, 1113–1116.

61. On one property of the solution of a generalized system of Cauchy-Riemann equations. (Russian) Soobshch. Akad. Nauk Gruz. SSR, 14 (1953), No. 8, 449–453.

62. On some properties of solutions of a system of elliptic type equations. (Russian) Dokl. Akad. Nauk SSSR, 98(1954), No. 2, 181–184.

63. On the solution of boundary value problems of the shell theory. (Russian) Soobshch. Akad. Nauk GSSR, 15(1954), No. 1,3–6.

64. Problem of reducing differential equations of elliptic type to the canonical form and the generalized Cauchy-Riemann system. (Russian) Dokl. Akad. Nauk SSSR, 100(1955), No. 2, 197–200.

65. On one method of solution of boundary value problems of partial differential equations. (Russian) Dokl. Akad. Nauk SSSR, 101(1955), No. 4, 593–596.

66. On one method of calculating of prismatic shells. (Russian) Trudy Tbilis. Mat. Inst. 21(1955), 191–259.

67. Theory of generalized analytic functions and its applications in geometry and mechanics. (Russian) III All-Union Math. Congr., (Moscow, June-July, 1956), Abstracts of survey and section reports, 9–11, Moscow, Izd. Akad. Nauk SSSR, 1956.

68. On some rigidity conditions for surfaces of positive curvature. (Russian) Delivered at the IV Congress of Czechoslovak mathematicians in Prague, 6.IX. 1955. Czech. Math. J. 6(1956), No. 2, 143–160.

69. Some problems of infinitesimal bendings of surfaces. (Russian) Dokl. Akad. Nauk SSSR, 112(1957), No. 3, 377–380.

70. Civeta probleme ale teoriei functilor analitice generalizate si ale aplicatülor ei in geometrie i mecanica. Bull. Math. Soc. Sci. Mat. Fiz. R.P.Roumanie, 1(1957), No. 2, 229–243.

71. Theory of generalized analytic functions and some of its applications in geometry and mechanics. (Russian) Proc. III All-Union Math. Congr. (Moscow, June-July, 1956), v. 3, Survey reports, 42–64, Izd. Akad. Nauk SSSR, Moscow, 1958.

72. Über die korrekte Stellung der Riemann - Hilbertschen Aufgabe. Proc. Intern. Colloq. on Theory of Functions. Ann. Acad, Sci. Fenn., ser. A. 1(1958), p. 14.

73. Proof of the rigidity of piecewise-regular closed convex surfaces of non-negative curvature. (Russian) Izv. Akad. Nauk SSSR. Ser. Math. 22(1958), No. 2, 165–176 (coauthor B. V. Boyarski).

74. On the conditions providing a momentless stressed state of equilibrium of the convex shell. (Russian) Soobshch. Akad. Nauk Gruz. SSR, 20(1958), No. 5, 525–532.

75. On the conditions of a momentless stressed state of convex shells. (Russian) Soobshch. Akad. Nauk Gruz. SSR, 21(1958), No. 6, 649–652.

76. On the conditions for realizing a momentless stressed equilitrium state of convex shells. (Russian) Steklov Mat. Inst. Akad. Nauk SSSR, Moscow, 1958, 23 p.

77. On the conditions of momentless stressed equilibrium state of a convex shell. (Russian) Proc. All-Union Conf. on Diff. Equations (Yerevan, November, 1958), 32–44, Izd. Akad. Nauk Arm. SSR, Yerevan, 1960.

78. Über die Bedingungen der Verwirklichung des momentenfreien Spannungsleichgewichtes von Schalen positiver Krümmung, Proceedings of the symposium on the theory of thin elastic shells. North-Holland Publ. Co., Amsterdam, 1960, 270–280.

79. A remark on the properties of solutions of equations $\Delta u = -2ke^u$. (Russian) Sibirsk. Mat. Zh. 1(1960), No. 3, 331–342.

80. Projective properties of force and deflection fields. (Russian) Problems of Mechanics of Continua. To the 70th birthday anniversary of Academician N. I. Muskhelishvili, 83–91, Izd. Akad. Nauk SSSR, Moscow, 1961.

81. A projective property of force and deflection fields. Problems of continuum mechanics, 582–591, Philadelphia, 1961.

82. On some properties of solutions of the Gauss equation. (Russian) Trudy Mat. Inst. Steklov. 64(1961), 5–8.

83. To the theory of quasiconformal mappings. (Russian) Some problems of mathematics and mechanics, 57–64, Izd. Akad. Nauk SSSR, Novosibirsk, 1961.

84. On the compactness of a family of generalized analytic functions. (Russian) Trudy Tbilis. Univ. Ser. mech.-math. sci., 1961, 17–21.

85. Methods of the theory of analytic functions in the theory of elasticity. (Russian) Proc. All-Union Congr. on Theoretical and Applied Mechanics (January 27-February 3, 1960). Survey Reports, Izd. Akad. Nauk SSSR, Moscow-Leningrad, 1962, 310–338 (coauthor N. I. Muskhelishvili).

86. Fixed singular points of generalized analytic functions. (Russian) Dokl. Akad. Nauk SSSR, 145(1962), No. 1, 24–26.

87. Equations and systems of equations of elliptic type. (Russian) Proc. IV All-Union Math. Congr. v. 1, 29–48, 1963.

88. New methods in mathematical shell theory. Proc. XI Intern. Congr. Appl. Mech. Munich, 47–58, Springer-Verlag, 1964.

89. Theory of thin and shallow elastic shells with variable thickness. (Russian) Appplications of the theory of functions in continuum mechanics, v. 1, 410–431, Nauka, Moscow, 1965.

90. Theory of thin shallow shells of variable thickness. (Russian) Trudy Tbilis. Mat. Inst. Razmadze, 30(1965), 5–103.

91. Über eine Verallgemeinerung der Biegetheorie der Schalen. Intern. Kongr. Anwendungen der Mathematik in Ingenieurwiss. mit Rahmenthema, Anwendungen elektronischer Rechenanlagen im Bauwesen, Bd. 1(1967), 260–280.

92. On construction of an approximate solution of the equation of a shallow spherical shell. Intern. J. Solids Struct. 10(1968), 991–1003.

93. On conformal invariant differential forms in shell theory. Functional theoretical methods in partial differential equations, 303–311, Acad. Press, 1968.

94. On one version of the consistent theory of elastic shells. IUTAM Symp. (Copenhagen, 1967), 59–84, Springer-Verlag, Berlin-Heidelberg-New York, 1969.

95. On the integration of a system of equations of an elastic equilibrium of a plate. (Russian) Dokl. Akad. Nauk SSSR, 186(1969), No. 3, 541–544.

96. On the integration of equilibrium equations of a cylindrical shell. (Russian) Dokl. Akad. Nauk SSSR, 186(1969), No. 4, 787–790.

97. On one class of an irregular elliptic system of first order equations. (Russian) Abstracts of reports of the Seminar of Inst. Appl. Math. Tbilis. State Univ. 1(1969), 5–9.

98. On one class of elliptic systems with singularities. Proc. Intern. Conf. on Functional Analysis and Related Topics, Tokyo, 1969, 142–147.

99. On one method of solving of the basic biharmonic boundary value problem and the Dirichlet problem. (Russian) Some problems of mathematics and mechanics, 120–127, Leningrad, 1970.

100. Equations of thin elastic shells. (Russian) Abstracts of reports Inst. Appl. Math. Tbilis. State Univ. 5(1971) 69–75.

101. On one trend of construction of shell theory. (Russian) Mechanics in the USSR for 50 years, 3, 267–290, Nauka, Moscow, 1972.

102. On two ways of constructing the theory of elastic shells. Proc. XIII Intern. Congr. Theoret. and Appl. Mech., (Moscow), 1972, 322–339, Springer-Verlag, Berlin-Heidelberg-New York, 1973.

103. On a functional equation of the theory of minimal surfaces. (Russian) Dokl. Akad. Nauk SSSR, 217(1974), No. 5, 997–1000.

104. On two ways of constructing a noncontradictory theory of elastic shells. (Russian) Proc. I All-Union School on the Theory and Numerical Methods

of Calculation of Shells and Plates, (Gegechkori), 5–10, Metsniereba, Tbilisi, 1975.

105. On one method of solving the first biharmonic boundary value problem and the Dirichlet problem. J. Amer. Math. Soc. 104(1976), 104–111.

106. On one class of statically definable problems of shell theory (Russian). Soobshch. Akad. Nauk Gruz. SSR, 83(1976), No. 2, 273–276; No. 3, 529–532.

An Asymptotic Method for Solving Three-Dimensional Boundary Value Problems of Statics and Dynamics of Thin Bodies

Lenser A. Aghalovyan

Institute of Mechanics of NAS of Armenia, Marshal Baghramyan Ave. 24B, Yerevan, 375019, Armenia, `aghal@mechins. sci.am`

Abstract The equations of the three-dimensional problem of elasticity for thin bodies (bars, beams, plates, shells) in dimensionless coordinates are singularly perturbed by a small geometrical parameter. The general solution of such a system of equations is a combination of the solutions of an internal problem and a boundary-layer problem.

The asymptotic orders of the stress tensor components and of the displacement vector in the second and mixed boundary value problems for thin bodies are established; the inapplicability of classical theory hypothesis for the solution of these problems is proved.

In the case of a plane first boundary value problem for a rectangular strip a connection of the asymptotic solution with the Saint-Venant principle is established and its correctness is proved.

Free and forced vibrations of beams, strips and possibly anisotropic and layered plates are considered by an asymptotic method. The connection of free-vibration frequency values with the propagation velocities of seismic shear and longitudinal waves is established. In a three-dimensional setting forced vibrations of two-layered, three-layered and multi-layered plates under the action of seismic and other dynamic loadings are considered and the resonance conditions are established.

At theoretical justification for the expediency of using seismoisolators in an aseismic construction is given.

Keywords: elasticity, space problem, plates, shells, asymptotic method, vibrations

1 Introduction

One of the basic methods for finding the stress and strain states in thin bodies (beams, bars, plates, shells) is the reduction of a three-dimensional problem to the solution of comparably simpler two-dimensional or one-dimensional problems of mathematical physic. This kind of reduction has been based on the method of decomposition by alternating coordinate characterizing the thickness and on the method of hypotheses. Each of these methods has its branches. The method of decomposition in power series by alternating parameter of thickness, originating in the works of Cauchy and Poisson, was developed by N.A. Kilchevsky, who see main contribution was the formulation of noncontradictory boundary conditions for the reduced two-dimensional equations [1]. The method of representation of the solution in the form of a series featuring certain special functions of the transversal coordinate (for example, Legendre polynomials) is also a method of decomposition with respect to the transversal coordinate. Such an approach was suggested by I.N. Vekua [2]. A characteristic feature of this approach is the growth in the exponent of the reduced partial differential equations, a fact that gives rise with the increase of the approximations to significant mathematical difficulties. Method hypothesis originates from Bernoulli-Caulomb-Euler beam theory, based on the plane cross-section hypothesis. For plates, this method was developed by Kirchhoff, and for shells by Love, on the assumption that normal fibers remain orthogonal to the deformed mid-surface (classic theory of plates and shells) [3]. The theories by E. Reissner, S.A. Ambartsumyan and S.P. Timoshenko are based on softened assumptions [4–8]. In all the above mentioned theories, only the first boundary value problem of elasticity theory was considered, i.e. the case when on the end faces of the thin body the values of the appropriate components of the stress tensor are specified. Under other boundary conditions, the assumptions of the classic theory bring to contradictory equations. That is why up to the last decades, when the asymptotic method began to be used, the question of solution of such boundary value problems remained open, though these problems are basic in many areas, for example, in base constructions and seismology. A lot of important results in the theory of beams, plates and shells have been obtained by means of the asymptotic method. Since one of the geometrical measures of a thin body essentially differs from the two others, when the equations of elasticity theory are posed in terms of dimensionless coordinates and displacement vector components, the transformed system contains a small geometrical parameter. For solving such systems, it is natural to use asymptotic methods. But, the system turns out to be singularly perturbed; besides, the small parameter is the coefficient not of the whole senior operator, but of the part, of it. This required a new approach. The construction of the mathematical theory of singularly perturbed differential equations has not been completed, although it has been intensively developed [9–13]. For singularly perturbed differential equations, the general solution is a combination of the solutions of an internal problem (basic solution) and of a boundary-layer problem. For plate and shell theories, the first papers on asymptotic integration of the three-dimensional equations of elasticity theory were authored by Friedrichs, Dressler, Goldenweiser and Green [14–16]. Below we shall reduce to asymptotic solutions of plane and three-dimensional problems of elasticity theory for anisotropic strips, plates and shells, being either one-layered, or multi-layered. We consider static and dynamic problems. We discuss the relationships of the obtained solutions with the results from classic and precise theories, and we give a mathematical justification of the Saint-Venant principle. We also show the effectiveness of the asymptotic method for the solution of new classes of thin-body problems (nonclassical boundary value problems), and delineat the possible applications of the solutions.

2 Asymptotic Solution of the First Boundary Plane Problem of Elasticity Theory for Anisotropic Thermoelastic Strip

At first we consider a plane problem, the solution of which by the asymptotic method permits to represent the essence and possibilities of the method more availably. State the problem: Find the solution of the equations of the plane stress state of the anisotropic thermoelastic strip-rectangular $D = \{(x, y): 0 \le x \le l, |y| \le h, h << l\}$ under boundary conditions

$$\sigma_{yy}(y = \pm h) = \pm Y^{\pm}(x), \qquad \sigma_{xy}(y = \pm h) = \pm X^{\pm}(x), \tag{1}$$

and yet arbitrary conditions on torsional cross-sections $x = 0, l$. In the equations of equilibrium and elasticity correlations, taking into account the temperature summands according Duhamel-Neuman model, passing to dimensionless displacements vector components $U = u/l, V = v/l$ and dimensionless coordinates $\xi = x/l, \zeta = y/h$, we get singularly perturbed by small geometrical parameter $\varepsilon = h/l$ system [17, 18]

$$\frac{\partial \sigma_{xx}}{\partial \xi} + \varepsilon^{-1} \frac{\partial \sigma_{xy}}{\partial \zeta} + lF_x(\xi, \zeta) = 0$$

$$\frac{\partial \sigma_{xy}}{\partial \xi} + \varepsilon^{-1} \frac{\partial \sigma_{yy}}{\partial \zeta} + lF_y(\xi, \zeta) = 0$$

$$\frac{\partial U}{\partial \xi} = a_{11}\sigma_{xx} + a_{12}\sigma_{yy} + a_{16}\sigma_{xy} + \alpha_{11}\Theta(\xi, \zeta) \tag{2}$$

$$\varepsilon^{-1} \frac{\partial V}{\partial \zeta} = a_{12}\sigma_{xx} + a_{22}\sigma_{yy} + a_{26}\sigma_{xy} + \alpha_{22}\Theta(\xi, \zeta)$$

$$\varepsilon^{-1} \frac{\partial U}{\partial \zeta} + \frac{\partial V}{\partial \xi} = a_{16}\sigma_{xx} + a_{26}\sigma_{yy} + a_{66}\sigma_{xy} + \alpha_{12}\Theta(\xi, \zeta)$$

where σ_{ij} are the stresses tensor components, a_{ik} are the elasticity constants, α_{ik} are the coefficients of the heat extension, F_x, F_y are the components of the volume forces, in the capacity of which weight, reduced seismic forces and other may appear, $\Theta(\xi, \zeta) = T(\xi, \zeta) - T_0(\xi, \zeta)$ is the temperature field change of which is considered to be known.

The solution of the singularly perturbed system (2) is combined from the internal problem and boundary layers solutions [9–13, 18]

$$I = I^{int} + I_b^{I} + I_b^{II} \tag{3}$$

4   L.A. Aghalovyan

The solution of the internal problem has the shape

$$I^{int} = \varepsilon^{q_Q+s} Q^{(s)}(\xi, \zeta), \qquad s = \overline{0, N} \tag{4}$$

where Q is any of the required values, the integers q_Q should be chosen so,that after the substitution (4) into (2) and setting equal the coefficients in each equation under the same degrees ε, to get a noncontradictory system for determining $Q^{(s)}$. This stage is the most responsible in any physical problem, considered in the restricted area. In our case it is established that

$$q_{\sigma_{xx}} = -2, \ q_{\sigma_{xy}} = -1, \ q_{\sigma_{yy}} = 0, \ q_u = -2, \ q_v = -3 \tag{5}$$

Under any other value of q the system relative to $Q^{(s)}$ will turn to be contradictory. The volume forces and the temperature field will exhibit themselves from the input approximation if we represent them in the shape of

$$lF_x = \varepsilon^{-2+s} F_x^{(s)}, \ lF_y = \varepsilon^{-1+s} F_y^{(s)}, \ \Theta = \varepsilon^{-2+s}\Theta^{(s)}, \ F_x^{(0)} = \varepsilon^2 lF_x$$

$$F_y^{(0)} = \varepsilon l F_y, \ \Theta^{(0)} = \varepsilon^2 \Theta, \ F_x^{(s)} = F_y^{(0)} = 0, \ \Theta^{(s)} = 0, \ s \neq 0 \tag{6}$$

Substituting (4) into (2) and solving the obtained system we get

$$V^{(s)} = v^{(s)}(\xi) + v_*^{(s)}$$

$$U^{(s)} = -\frac{dv^{(s)}}{d\xi}\zeta + u^{(s)}(\xi) + u_*^{(s)}$$

$$\sigma_{xx}^{(s)} = -\frac{1}{a_{11}}\frac{d^2 v^{(s)}}{d\xi^2}\zeta + \frac{1}{a_{11}}\frac{du^{(s)}}{d\xi} + \sigma_{xx*}^{(s)} \tag{7}$$

$$\sigma_{xy}^{(s)} = \frac{1}{a_{11}}\frac{d^3 v^{(s)}}{d\xi^3}\frac{\zeta^2}{2} - \frac{1}{a_{11}}\frac{d^2 u^{(s)}}{d\xi^2}\zeta + \sigma_{xy0}^{(s)}(\xi) + \sigma_{xy*}^{(s)}(\xi,\zeta)$$

$$\sigma_{yy}^{(s)} = -\frac{1}{a_{11}}\frac{d^4 v^{(s)}}{d\xi^4}\frac{\zeta^3}{6} + \frac{1}{a_{11}}\frac{d^3 u^{(s)}}{d\xi^3}\frac{\zeta^2}{2} - \frac{d\sigma_{xy0}^{(s)}}{d\xi}\zeta + \sigma_{yy0}^{(s)}(\xi) + \sigma_{yy*}^{(s)}(\xi,\zeta)$$

where $Q_*^{(s)}$ are known functions, if $Q^{(m)}, (m < s)$ are known, $Q^{(m)} \equiv 0, Q_*^{(m)} \equiv 0$ when $m < 0$. We have

$$v_*^{(s)} = \int_0^\zeta \left[a_{12}\sigma_{xx}^{(s-2)} + a_{22}\sigma_{yy}^{(s-4)} + a_{26}\sigma_{xy}^{(s-3)} + \alpha_{22}\Theta^{(s-2)} \right] d\zeta$$

$$u_*^{(s)} = \int_0^\zeta \left[a_{16}\sigma_{xx}^{(s-1)} + a_{26}\sigma_{yy}^{(s-3)} + a_{66}\sigma_{xy}^{(s-2)} + \alpha_{12}\Theta^{(s-1)} - \frac{\partial v_*^{(s)}}{\partial \xi} \right] d\zeta \tag{8}$$

$$\sigma_{xx*}^{(s)} = \frac{1}{a_{11}} \left[\frac{\partial u_*^{(s)}}{\partial \xi} - a_{12}\sigma_{yy}^{(s-2)} - a_{16}\sigma_{xy}^{(s-1)} - \alpha_{11}\Theta^{(s)} \right]$$

$$\sigma_{xy*}^{(s)} = -\int_0^\zeta \left[F_x^{(s)} + \frac{\partial \sigma_{xx*}^{(s)}}{\partial \xi} \right] d\zeta, \qquad \sigma_{yy*}^{(s)} = -\int_0^\zeta \left[F_y^{(s)} + \frac{\partial \sigma_{xy*}^{(s)}}{\partial \xi} \right] d\zeta$$

Solution (7) contains unknown functions $v^{(s)}, u^{(s)}, \sigma_{xy0}^{(s)}, \sigma_{yy0}^{(s)}$, which should be determined from the boundary conditions. Satisfying conditions (1), $\sigma_{xy0}^{(s)}$, $\sigma_{yy0}^{(s)}$ will be expressed through $u^{(s)}, v^{(s)}$

$$\sigma_{xy0}^{(s)} = -\frac{1}{2a_{11}}\frac{d^3 v^{(s)}}{d\xi^3} + \frac{1}{2}\left[X^{+(s)} - X^{-(s)} - \sigma_{xy*}^{(s)}(\xi,1) - \sigma_{xy*}^{(s)}(\xi,-1)\right]$$

$$\sigma_{yy0}^{(s)} = \frac{1}{2}\left[Y^{+(s)} - Y^{-(s)} - \frac{1}{2}\frac{dq_x^{(s)}}{d\xi} - \sigma_{yy*}^{(s)}(\xi,1) - \sigma_{yy*}^{(s)}(\xi,-1)\right] \qquad (9)$$

$$X^{\pm(0)} = \varepsilon X^\pm, \quad Y^{\pm(0)} = Y^\pm, \quad X^{\pm(s)} = Y^{\pm(s)} = 0, \quad s \neq 0$$

$v^{(s)}, u^{(s)}$ are obtained from the equations

$$\frac{2}{a_{11}}\frac{d^2 u^{(s)}}{d\xi^2} = q_x^{(s)}, \quad q_x^{(s)} = -\left[X^{+(s)} + X^{-(s)} - \sigma_{xy*}^{(s)}(\xi,1) + \sigma_{xy*}^{(s)}(\xi,-1)\right]$$

$$\frac{2}{3a_{11}}\frac{d^4 v^{(s)}}{d\xi^4} = q^{(s)}, \qquad (10)$$

$$q^{(s)} = Y^{+(s)} + Y^{-(s)} - \sigma_{yy*}^{(s)}(\xi,1) + \sigma_{yy*}^{(s)}(\xi,-1) +$$

$$+\frac{d}{d\xi}\left[X^{+(s)} - X^{-(s)} - \sigma_{xy*}^{(s)}(\xi,1) - \sigma_{xy*}^{(s)}(\xi,-1)\right]$$

The solutions of Equations (10) in totality contain six arbitrary constants $C_i^{(s)}$, three of them characterize rigid displacement, and three enter the expressions for stresses according formulae (7). These constants should be determined from the conditions when $x = 0, l$. It is not possible to satisfy these conditions at each point of the end-walls. For example, if the conditions are given at $x = 0$

$$\sigma_{xx}(x = 0, \zeta) = \varphi(\zeta), \qquad \sigma_{xy}(x = 0, \zeta) = \psi(\zeta) \qquad (11)$$

by the solution of the internal problem the conditions (11) may be satisfied only in isolated point, which once more points to singularly perturbance of the input boundary problem. For the satisfaction of conditions (11) or analogous conditions it is important to have a new solution, the solution of the boundary layer is like that.

3 The Solution of the Boundary Layer

The boundary layer is localized not far from the torsion sections $x = 0, l$. In order to find the solution of the boundary layer, the corresponding section $x = 0$, in the Equation (2) without taking into account the volume forces and the temperature (they are considered in the solution of the internal problem) the change of variable $\gamma = \xi/\varepsilon$ is introduced and the solution of the transformed system is sought in the form of [18]

$$I_b^I = \varepsilon^{\chi_b + s} Q_b^{(s)}(\zeta) \exp(-\lambda\gamma), \qquad s = \overline{0, N} \qquad (12)$$

It is possible to establish that $\chi_{\sigma_{ik}} = \chi - 1, \chi_{u,v} = \chi$, the integer χ is determined during the conjunction of the internal problem and the boundary

6   L.A. Aghalovyan

layer solutions. As inhomogeneous conditions at $\zeta = \pm 1$ are already satisfied, the solution of the boundary layer should satisfy the conditions

$$\sigma_{xyb}(\zeta = \pm 1) = 0, \qquad \sigma_{yyb}(\zeta = \pm 1) = 0 \tag{13}$$

Substituting (12) into transformed Equations (2), we get

$$\sigma_{yyb}^{(s)} = A_n^{(s)} F_n(\zeta), \quad \sigma_{xyb}^{(s)} = A_n^{(s)} \frac{F_n'(\zeta)}{\lambda_n} \quad \sigma_{xxb}^{(s)} = A_n^{(s)} \frac{F_n''(\zeta)}{\lambda_n^2}$$

$$u_b^{(s)} = -A_n^{(s)}(a_{11}\lambda_n^{-3} F_n'' + a_{16}\lambda_n^{-2} F_n' + a_{12}\lambda_n^{-1} F_n) \tag{14}$$

$$v_b^{(s)} = -A_n^{(s)}(a_{11}\lambda_n^{-4} F_n''' + 2a_{16}\lambda_n^{-3} F_n'' + (a_{12} + a_{66})\lambda_n^{-2} F_n' + a_{26}\lambda_n^{-1} F_n)$$

Functions F_n satisfy equation

$$a_{11}F_n^{IV} + 2a_{16}\lambda_n F_n''' + (2a_{12} + a_{66})\lambda_n^2 F_n'' + 2a_{26}\lambda_n^3 F_n' + a_{22}\lambda_n^4 F_n = 0 \tag{15}$$

Having solved Equation (15), satisfied conditions (13), i.e.

$$F_n(\zeta = \pm 1) = 0, \qquad F_n'(\zeta = \pm 1) = 0 \tag{16}$$

we get characteristic equation relative to λ_n and final solution, where $A_n^{(s)}$ are the integration constants. Depending to the roots type of the characteristic equation corresponding to (16) various variants are possible [18]. For example, in the problem of bending isotropic strip we have

$$F_n(\zeta) = \sin \lambda_n \zeta - \zeta tg\lambda_n \cos \lambda_n \zeta, \quad \sin 2\lambda_n - 2\lambda_n = 0 \tag{17}$$

for the orthotropic strip

$$F_n(\zeta) = \sin \beta_2 \lambda_n \sin \beta_1 \lambda_n \zeta - \sin \beta_1 \lambda_n \sin \beta_2 \lambda_n \zeta$$

$$\omega \sin z_n - \sin \omega z_n = 0, \ z_n = (\beta_1 + \beta_2)\lambda_n, \ \omega = \frac{\beta_2 - \beta_1}{\beta_1 + \beta_2}, \ 0 < \omega < 1 \tag{18}$$

where β_1, β_2 are some parameters, depending to the values of the elasticity constants. The roots with $Re\lambda_n > 0$ will interest us, as then the solution will be fading. Each λ_n corresponds $\overline{\lambda_n}$, as a result solution (12), (14) will be real. The magnitudes of the boundary layer diminish from the end-wall $x = 0$ as $O\left[\exp(-Re\lambda_n\gamma)\right]$.

The stresses of the boundary layer have very important property – they are self-balanced in the arbitrary cross-section $\gamma = \gamma_k$

$$\int_{-1}^{1} \sigma_{xxb}(\gamma, \zeta)d\zeta = 0, \ \int_{-1}^{1} \zeta\sigma_{xxb}(\gamma, \zeta)d\zeta = 0, \ \int_{-1}^{1} \sigma_{xyb}(\gamma, \zeta)d\zeta = 0, \ \forall\gamma \tag{19}$$

In which it is easy to get sure of using formulae (12), (14), (16). The displacements don't have this property, i.e.

$$\int_{-1}^{1} u_b d\zeta \neq 0, \qquad \int_{-1}^{1} \zeta u_b d\zeta \neq 0, \qquad \int_{-1}^{1} v_b d\zeta \neq 0 \qquad (20)$$

The data for the boundary layer at $x = l$ can be obtained from the above cited formal changing γ with $\gamma_1 = 1/\varepsilon - \gamma = (l - x)/h$. Admitting $A_n^{(s)} = \frac{1}{2}(A_{1n}^{(s)} - iA_{2n}^{(s)})$, formulae (12), (14) will have the form

$$I_b^I = \varepsilon^{\chi_b + s} \left[A_{1n}^{(s)} Re\tilde{Q}_{bn}(\zeta) + A_{2n}^{(s)} Im\tilde{Q}_{bn}(\zeta) \right], \quad s = \overline{0, N}, \, n = \overline{0, k} \quad (21)$$

where $A_{1n}^{(s)}, A_{2n}^{(s)}$ are real constants, $\tilde{Q}_{bn}(\zeta) = Q_{bn}(\zeta) \exp(-\lambda_n \gamma)$, Q_{bn} is calculated by formula (14) as coefficient at $A_n^{(s)}$ for the corresponding quantity, k is the number of chosen boundary functions. Note that solution (21) for each s is exact, the classic theory of beams does not consider it, it is impossible to get it on the base of known hypotheses. This solution practically coincides with well known in elasticity theory Shiff-Papkovich-Lourie homogeneous solution.

4 Conjugation of the Internal Problem and Boundary Layer Solutions, Mathematical Justification of Saint-Venant Principle

The solutions of the internal problem and boundary layers in total contain sufficient number of arbitrary constants to satisfy the boundary conditions at $x = 0, l$. When satisfying the conditions at $x = 0$ we usually neglect the influence of the boundary layer R_b^{II} and on the contrary. It is possible if $1 + \exp\left(-Re\lambda_1 \frac{l}{h}\right) \approx 1$. In the problem of bending isotropic strip $Re\lambda_1 \approx 3.748, Re\lambda_2 \approx 6.95, Re\lambda_3 \approx 10.119$, and for the strip of an anisotropic material SVAM $Re\lambda_1 \approx 1.783, Re\lambda_2 \approx 3.076, Re\lambda_3 \approx 4.401$. Taking into account, that usually $l \geq 10h$, this condition is always fulfilled.

Let conditions (11) be given at $x = 0$. According to (3), (4), (5), (12), we have

$$\sigma_{xx} = \varepsilon^{-2+s}\sigma_{xx}^{(s)}(\xi, \zeta) + \varepsilon^{\chi-1+s}\sigma_{xxb}^{(s)} \exp(-\lambda\gamma)$$
$$\sigma_{xy} = \varepsilon^{-1+s}\sigma_{xy}^{(s)}(\xi, \zeta) + \varepsilon^{\chi-1+s}\sigma_{xyb}^{(s)} \exp(-\lambda\gamma) \qquad (22)$$

Substituting (22) into (11) we get noncontradictory conditions only at $\chi = -1$, as a result we have

$$\sigma_{xx}^{(s)}(0, \zeta) + \sigma_{xxb}^{(s)}(\zeta) = \varphi^{(s-2)}(\zeta)$$
$$\sigma_{xy}^{(s-1)}(0, \zeta) + \sigma_{xyb}^{(s)}(\zeta) = \psi^{(s-2)}(\zeta) \qquad (23)$$
$$\varphi^{(0)} = \varphi, \quad \varphi^{(m)} = 0, \quad m \neq 0, \qquad (\varphi, \psi)$$

As $\sigma_{xxb}, \sigma_{xyb}$ satisfy conditions (19), from (23), (19)

$$\int_{-1}^{1} \sigma_{xx}^{(s)}(0,\zeta)d\zeta = \int_{-1}^{1} \varphi^{(s-2)}d\zeta, \quad \int_{-1}^{1} \zeta\sigma_{xx}^{(s)}(0,\zeta)d\zeta = \int_{-1}^{1} \zeta\varphi^{(s-2)}d\zeta,$$

$$\int_{-1}^{1} \sigma_{xy}^{(s)}(0,\zeta)d\zeta = \int_{-1}^{1} \psi^{(s-1)}d\zeta \tag{24}$$

follow. According to formulae (7), (10) the expressions for the stresses in the internal problem contain exactly three unknown constants which uniquely are determined from conditions (24). This fact is very interesting and testifies the existence of the inner harmony in elasticity theory. Returning to conditions (23), taking into account (21) we get the conditions

$$A_{1n}^{(s)}Re\tilde{\sigma}_{xxbn}(\zeta) + A_{2n}^{(s)}Im\tilde{\sigma}_{xxbn}(\zeta) = \varphi^{(s-2)}(\zeta) - \sigma_{xx}^{(s)}(0,\zeta), \quad n = \overline{1,k}$$

$$A_{1n}^{(s)}Re\tilde{\sigma}_{xybn}(\zeta) + A_{2n}^{(s)}Im\tilde{\sigma}_{xybn}(\zeta) = \psi^{(s-2)}(\zeta) - \sigma_{xy}^{(s-1)}(0,\zeta) \tag{25}$$

From conditions (25), the right parts of which are already known functions, are determined $A_{1n}^{(s)}$, $A_{2n}^{(s)}$. They can be determined using the collocation, least squares or Fourie methods. From (24), (25) follows that the boundary layer takes the self-balanced part of the load on itself, i.e. the same inner stress state corresponds to statically equivalent end-wall loads, which expresses Saint-Venant principle. Thus, in case of the first boundary problem of elasticity theory for a strip, Saint-Venant principle is mathematically exactly fulfilled. The corresponding illustrative examples are brought in [18].

If at $x = 0, l$ displacements values are given, particularly, the conditions of the rigid sealing, the conjugation of the solutions must not be done in the above described way taking into account the character of (20) which means that Saint-Venant principle for displacements is not true. The conjugation here may be realized determining the solutions constants of the internal problem and boundary layer simultaneously, using, for example, (3) and the method of least squares.

5 The Connection with the Classical Beam Theory of Bernoulli-Coulomb-Euler

The solution of the internal problem will be written in the initial dimensional coordinates, assuming

$$u = \varepsilon^{s}u_{s}, \quad v = \varepsilon^{s}v_{s}, \quad s = \overline{0,N} \tag{26}$$

Using (4), (5), we have

$$u^{(s)} = l^{-1}\varepsilon^{2}u_{s}, \quad v^{(s)} = l^{-1}\varepsilon^{3}v_{s} \tag{27}$$

Substituting (27) into the Equations (10) we shall have

$$E_1 F \frac{d^2 u_s}{dx^2} = q_{x0}^{(s)}, \quad E_1 J \frac{d^4 v_s}{dx^4} = q^{(s)} \tag{28}$$

where $q_{x0}^{(s)} = \varepsilon^{-1} q_x^{(s)}$, $E_1 F$ is the stiffness of the bar on tension, $E_1 J$ is the stiffness of the beam on bending, $F = 2h \cdot 1$ is the area of the bar cross-section, $J = \frac{2}{3} h^3 \cdot 1$ is the moment of inertia of the beam cross-section. At $s = 0$

$$q_{x0}^{(0)} = -(X^+ + X^-), \quad q^{(0)} = Y^+ + Y^- + h \frac{d}{dx}(X^+ - X^-) \tag{29}$$

Equations (28) coincide with the equations of the classic theory of Bernoulli-Coulomb-Euler beams, based on the plane sections hypothesis. Approximations $s \geq 1$ make more precise the classical theory. Even at $s = 0$ the asymptotic theory gives more information, than the classic one, as by formulae (4), (5), (7) are calculated stresses σ_{xy}, σ_{yy}, too, the last one in classical theory is neglected at all. Admitting the hypothesis of plane sections, a whole solution – Shiff-Papkovich-Lourie homogeneous solution (the solution of the boundary layer) is lost, which cannot be obtained on the base of "softened hypothesis" of Reissner, Ambartsumyan, of Timoshenko type theories as well. Therefore, in "more precise theories" of thin bodies it should be made more precise – what kind of new solutions they consider from the position of three-dimensional problem solution.

Note another property of the asymptotic solution. If functions $X^{\pm}(\xi), Y^{\pm}(\xi)$ are polynomials of power m, iteration process in the internal problem terminates on the approximation $s = m + 1$ and we get mathematically exact solution of the problem in the sense of Saint-Venant. For isotropic beams this solution coincides with the well known in elasticity theory Menage-Timoshenko solution.

The asymptotic method is applied in more complicated objects, too. Asymptotics (3)–(4) is true for layered beams, too. If the beam consists of $n + m$ layers, from them n layers are situated above, m layers are lower from the surface of the reference, by the above cited procedure all the quantities are expressed through the components of the displacements $u^{(n)}, v^{(n)}$ of the n-type layer, which are determined from the equations [19]

$$C \frac{d^2 u^{(n,s)}}{d\xi^2} + K \frac{d^3 v^{(n,s)}}{d\xi^3} = p^{(s)}$$

$$D \frac{d^4 v^{(n,s)}}{d\xi^4} + K \frac{d^3 u^{(n,s)}}{d\xi^3} = q^{(s)} \tag{30}$$

where

$$C = \sum_{k=1}^{n} \frac{1}{a_{11}^{(k)}}(\zeta_k - \zeta_{k-1}) - \sum_{k=1}^{m} \frac{1}{a_{11}^{(-k)}}(\zeta_{-k} - \zeta_{-k+1})$$

$$D = \frac{1}{3}\sum_{k=1}^{n} \frac{1}{a_{11}^{(k)}}(\zeta_k^3 - \zeta_{k-1}^3) - \frac{1}{3}\sum_{k=1}^{m} \frac{1}{a_{11}^{(-k)}}(\zeta_{-k}^3 - \zeta_{-k+1}^3) \qquad (31)$$

$$K = -\frac{1}{2}\sum_{k=1}^{n} \frac{1}{a_{11}^{(k)}}(\zeta_k^2 - \zeta_{k-1}^2) + \frac{1}{3}\sum_{k=1}^{m} \frac{1}{a_{11}^{(-k)}}(\zeta_{-k}^2 - \zeta_{-k+1}^2)$$

$$\zeta_k = \frac{1}{h}\sum_{j=1}^{k} h_j \quad (k=1,2,...,n), \quad \zeta_{-k} = -\frac{1}{h}\sum_{j=1}^{k} h_{-j} \quad (k=1,2,...,m)$$

$$h = \sum_{j=1}^{n} h_j + \sum_{j=1}^{m} h_{-j}$$

At $s = 0$ Equations (30) coincide with the equations on the base of classical hypothesis of plane sections. It is always possible to choose the reference line so, that $K = 0$, then the problems of tension-compression and bending will be separated.

6 Space Problems of Anisotropic Plates and Shells

In the equations of elasticity theory three-dimensional problems of the anisotropic body passing to dimensionless coordinates $\xi = x/l, \eta = y/l, \zeta = z/h$ and dimensionless displacement vector components $U = u/l, V = v/l, W = w/l$, the equations will again turn out to be singularly perturbed of the relative small parameter $\varepsilon = h/l$, where $2h$ is the thickness, l is the characteristic tangential dimension of the plate.

The solution has the form

$$I = I^{int} + I_b \qquad (32)$$

where I^{int} is the solution of the internal problem, I_b is the solution of the boundary layer localized not far from the lateral surface of the plate. I^{int} has the form (4). In case when on the facial surfaces $z = \pm h$ of the plate the values of $\sigma_{xz}, \sigma_{yz}, \sigma_{zz}$ are given (the first boundary problem), noncontradictory iteration process is obtained at

$$q = -2 \text{ for } \sigma_{xx}, \sigma_{xy}, \sigma_{yy}; \qquad q = -1 \text{ for } \sigma_{xz}, \sigma_{yz}; \qquad q = 0 \text{ for } \sigma_{zz}$$
$$q = -2 \text{ for } u, v; \qquad q = -3 \text{ for } w \qquad (33)$$

Making the procedure analogous done in the plane problem, the solution of three-dimensional problem in case of general anisotropy (21 constants of elasticity) is reduced in dimensional coordinates to the solution of the following system relative to three functions $u^{(s)}(x,y)$, $v^{(s)}(x,y)$, $w^{(s)}(x,y)$

$$l_{11}u^{(s)} + l_{12}v^{(s)} = p_1^{(s)}, \qquad l_{12}u^{(s)} + l_{22}v^{(s)} = p_2^{(s)} \tag{34}$$

$$D_{11}\frac{\partial^4 w^{(s)}}{\partial x^4} + 4D_{16}\frac{\partial^4 w^{(s)}}{\partial x^3 \partial y} + 2(D_{12} + 2D_{66})\frac{\partial^4 w^{(s)}}{\partial x^2 \partial y^2}$$

$$+4D_{26}\frac{\partial^4 w^{(s)}}{\partial x \partial y^3} + D_{22}\frac{\partial^4 w^{(s)}}{\partial y^4} = q^{(s)} \tag{35}$$

where

$$l_{11} = C_{11}\frac{\partial^2}{\partial x^2} + 2C_{16}\frac{\partial^2}{\partial x \partial y} + C_{66}\frac{\partial^2}{\partial y^2} \quad (1,2;x,y)$$

$$l_{12} = C_{16}\frac{\partial^2}{\partial x^2} + (C_{12} + C_{66})\frac{\partial^2}{\partial x \partial y} + C_{26}\frac{\partial^2}{\partial y^2} \tag{36}$$

$$C_{ik} = 2hB_{ik}$$

At $s = 0$ Equations (34) coincide with the equations of generalized plane problem, and Equation (35) with classical equation of the plate bending, when there is plane of an elastic symmetry [6, 18]. For approximations $s > 0$ only the right parts of the equations change, i.e. the shapes of the reduced loadings, where the coefficients of the elasticity characterizing general anisotropy enter. The solution of the boundary layer has got the structure, analogous to (12). For orthotropic plates it is divided into antiplane (lateral torsion) and plane boundary layers. The values of the antiplane boundary layer when removing from the lateral surface $x = 0$ diminish into the plate as

$$\exp\left(-\sqrt{\frac{G_{23}}{G_{12}}}\pi\frac{x}{h}\right) \quad \text{and} \quad \exp\left(-\sqrt{\frac{G_{23}}{G_{12}}}\frac{\pi}{2}\frac{x}{h}\right) \tag{37}$$

correspondingly in symmetric and skew-symmetric (bending) problems. At $G_{23} << G_{12}$ the antiplane boundary layer penetrates deeply enough. The plane boundary layer diminishes as

$$\exp\left(-Re\lambda_1\frac{x}{h}\right) \tag{38}$$

where λ_1 is the first root with $Re\lambda_1 > 0$ of the corresponding characteristic equation. In the problem of the bending at $2G_{13} < \sqrt{E_1 E_3}$ (18) is this kind of equation, and at $2G_{13} > \sqrt{E_1 E_3}$

$$\omega \sin z_n - sh\omega z_n = 0, \quad z_n = 2\beta\lambda_n, \quad \omega = \alpha/\beta, \quad 0 < \omega < 1 \tag{39}$$

where α, β are material parameters, depending on constants of elasticity [18]. In the symmetric problem the equations differ from (18) to (39) with sight. The classic theory of the plates and shells neglects the boundary layers. Reissner,

Ambartsumyan, of Timoshenko type theories take into account one of the boundary layers – antiplane layer. In case of general anisotropy the boundary layers do not decompose and the determining equation is an ordinary differential equation of the sixth order. The solutions of the internal problem and boundary layers for plates and shells are built independently and their conjugation is fulfilled during satisfying boundary conditions on the lateral surface. The preference of this or the other applied theory is estimated from the position of three-dimensional problem how much exactly all the equations and boundary conditions on the lateral surface are satisfied.

Kirchhoff-Love theory is the original approximation in asymptotic theory of plates and shells and in the sense of the brought two-dimensional equations and in the sense of boundary conditions on the lateral surface. If restricted to this or the other number of approximations of the internal problem and the boundary layers, the connection of this or the other applied theory will be revealed. For example, the deflection equation of the plate corresponding to the first three approximations of the internal problem, practically coincides with Ambartsumyan iteration theory equation [20, 21]. It is also important to note that in space problems of plates and shells the boundary layer through the boundary conditions on the lateral surface influences on the solution of the internal problem, that is why its integral construction must be considered in applied theories claiming to make more precise the classic theory.

7 Asymptotic Solutions of Nonclassic Space Boundary Problems of Thin Bodies

Classic theory of beams, plates and shells, existing precise theories, are devoted to the case when on the facial surfaces of the thin body the values of the corresponding stresses tensor component are given. It is interesting to consider other boundary conditions as well, for example, the cases of giving the values of the displacement vector components or mixed conditions. These problems are called nonclassical in order to distinguish them from the mixed problems of classical theory, though in the elasticity theory they are classical, too (basic). This kind of problems are basic in base constructions, seismology, in calculations of conjunctions of pliable bodies with more rigid, during the study of the stamping process, etc. In direct test it is easy to be convinced that classical theory hypotheses here are not applicable, therefore, asymptotics is not applicable either (33). It is necessary to find another asymptotics, which is found in [22] for beams and bars, and for plates and shells – in [23, 24].

Let us have a plate, occupying area $\Omega = \{(x, y, z) : 0 \leq x \leq a, 0 \leq y \leq b, |z| \leq h, h << l, l = min(a, b)\}$. It is required to find the solution of space problem equations of elasticity theory of an anisotropic body in area Ω under boundary conditions

$$u(x, y, -h) = u^-(x, y), \ v(x, y, -h) = v^-(x, y), \ w(x, y, -h) = w^-(x, y) \ (40)$$

and conditions

$$u(x,y,h) = u^+(x,y), \ v(x,y,h) = v^+(x,y), \ w(x,y,h) = w^+(x,y) \quad (41)$$

or

$$\sigma_{xz}(x,y,h) = \sigma_{xz}^+(x,y), \ \sigma_{yz}(x,y,h) = \sigma_{yz}^+(x,y), \ \sigma_{zz}(x,y,h) = \sigma_{zz}^+(x,y) \quad (42)$$

and under arbitrary yet conditions on the lateral surface of the plate. Passing in the equations of equilibrium and Hooke generalized law, taking into account volume forces and temperature field by Duhamel-Neuman model:

$$\frac{\partial \sigma_{xx}}{\partial x} + \frac{\partial \sigma_{xy}}{\partial y} + \frac{\partial \sigma_{xz}}{\partial z} + F_x(x,y,z) = 0, \quad (x,y,z)$$

$$\frac{\partial u}{\partial x} = a_{11}\sigma_{xx} + a_{12}\sigma_{yy} + a_{13}\sigma_{zz} + a_{14}\sigma_{yz} + a_{15}\sigma_{xz} + a_{16}\sigma_{xy} + \alpha_{11}\theta$$

$$(x,y,z;u,v,w;a_{1i},a_{2i},a_{3i};\alpha_{11},\alpha_{22},\alpha_{33}) \quad (43)$$

$$\frac{\partial w}{\partial y} + \frac{\partial v}{\partial z} = a_{14}\sigma_{xx} + a_{24}\sigma_{yy} + a_{34}\sigma_{zz} + a_{44}\sigma_{yz} + a_{45}\sigma_{xz} + a_{46}\sigma_{xy} + \alpha_{23}\theta$$

$$(x,y,z;u,v,w;a_{i4},a_{i5},a_{i6};\alpha_{23},\alpha_{13},\alpha_{12})$$

to dimensionless coordinates and displacements $\xi = x/l, \eta = y/l, \zeta = y/h; U = u/l, V = v/l, W = w/l$, again we get singularly perturbed by small parameter $\varepsilon = h/l$ system, the solution of which has the form of (32). The solution I^{int} of he internal problem is sought in the form

$$I^{int} = \varepsilon^{qQ+s}Q^{(s)}(\xi,\eta,\zeta) \quad (44)$$

Meanwhile the asymptotics of classical theory (33) brings to contradictory system. For this class of problems principally new asymptotics is established

$$q_{\sigma_{ij}} = -1, \qquad q_{u,v,w} = 0 \quad (45)$$

Substituting (44), (45) into the transformed system (43), arising from there the systems are determined $Q^{(s)}(\xi,\eta,\zeta)$. The arbitrary functions of integration are uniquely determined from boundary conditions (40), (41) or (40), (42). Unlike classic theory, the solution of the internal problem is fully determined by the conditions on the facial surfaces, i.e. the conditions on the lateral surfaces do not influence on the solution.

The found solution, as a rule will not satisfy the conditions on the lateral surfaces of the plate. The appeared residual is removed by the solution of the boundary layer. The boundary layer is built as in item 3, and the conjugation of the internal problem and the boundary layer solutions is realized as in item 4.

If functions $u^\pm, v^\pm, w^\pm, \sigma_{jk}^\pm$ are polynomials from coordinates ξ,η, the iteration process terminates and we get a mathematically exact solution of the internal problem. For the illustration we bring this solution for an orthotropic plate when $u^- = v^- = w^- = 0, u^+ = const, v^+ = const, w^+ = const$:

$$\sigma_{xz} = G_{13}\frac{u^+}{2h}, \quad \sigma_{yz} = G_{23}\frac{v^+}{2h} \quad \sigma_{zz} = \frac{1}{A_{33}}\frac{w^+}{2h}$$

$$\sigma_{xx} = \frac{A_{13}}{A_{33}}\frac{w^+}{2h}, \quad \sigma_{yy} = \frac{A_{23}}{A_{33}}\frac{w^+}{2h}, \quad \sigma_{xy} = 0$$

$$u = \frac{u^+}{2h}(h + z), \quad v = \frac{v^+}{2h}(h + z), \quad w = \frac{w^+}{2h}(h + z) \tag{46}$$

$$A_{13} = (a_{12}a_{23} - a_{13}a_{22})/\Delta_1, \quad A_{23} = (a_{12}a_{13} - a_{23}a_{11})/\Delta_1$$

$$A_{33} = a_{13}A_{13} + a_{23}A_{23} + a_{33}, \quad \Delta_1 = a_{11}a_{22} - a_{12}^2$$

Conditions $u^- = v^- = w^- = 0, \sigma_{xz}^+ = const, \sigma_{yz}^+ = const, \sigma_{zz}^+ = const$ corresponds solution

$$\sigma_{xz} = \sigma_{xz}^+, \quad \sigma_{yz} = \sigma_{yz}^+, \quad \sigma_{zz} = \sigma_{zz}^+,$$

$$\sigma_{xx} = A_{13}\sigma_{zz}^+, \quad \sigma_{yy} = A_{23}\sigma_{yy}^+, \quad \sigma_{xy} = 0 \tag{47}$$

$$u = \frac{\sigma_{xz}^+}{G_{13}}(h + z), \quad v = \frac{\sigma_{yz}^+}{G_{23}}(h + z) \quad w = A_{33}\sigma_{zz}^+(h + z)$$

Asymptotics (44), (45) is just for multi-layered plates and shells too, when between the layers full contact takes place. From the exact solution of the problem for the two-layered plate with the rigidly fastened lower bound $z = -h_2$ of the second layer of thickness $h_2 - u_{II}(-h_2) = v_{II}(-h_2) = w_{II}(-h_2) = 0$ and the bound $z = h_1$ of the upper layer bearing normal loading of constant intensity $\sigma_{zzI}(h_1) = \sigma_{\gamma\gamma}^+, \sigma_{xz}^+ = \sigma_{yz}^+ = 0$, Winkler's modulus of foundation of the elastic orthotropic foundation follows

$$K = \frac{1}{h_2}A_{33}^{II} \tag{48}$$

This formula also follows from the last formula (47) as a reaction of foundation of thickness $2h$ under normal loading of intensity $\sigma_{zz}^+(K = 1/(2hA_{33}))$. Having calculated A_{33} by formula (46), we shall have

$$K = \frac{(1 - \nu_{12}\nu_{21})E_3}{h_2(1 - \nu_{12}\nu_{21} - \nu_{13}\nu_{31} - \nu_{23}\nu_{32} - 2\nu_{12}\nu_{23}\nu_{31})} \tag{49}$$

For isotropic foundation

$$K = \frac{(1 - \nu)E}{h_2(1 + \nu)(1 - 2\nu)} \tag{50}$$

where ν is Poisson's ratio, E is Young's modulus. The asymptotic solution also permits to reduce the formula for modulus of foundation for n-layered foundation

$$K_n = \left(\sum_{k=1}^{n} h_k A_{33}^{(k)} \right)^{-1} \tag{51}$$

Note that with general anisotropy (21 constants of elasticity) the concept of modulus of foundation becomes not applicable as the normal deflection of the plate is possible under tangential boundary effects as well.

By the same method in [25] solutions of a lot of applied problems are obtained.

8 Asymptotics of Free and Forced Vibrations of Thin Bodies

A lot of questions of seismology and seismosteady engineering, study of wave propagation in layered media, contact interactions of thin and massive bodies, reduce to the solution of dynamic problems of thin bodies. Among these the problems on free vibrations of beams and plates on absolutely rigid foundation (including layered) and the problems of forced vibrations caused by harmonically changing in time vector of displacement applied to the foot of the thin body, distinguish. The latter, particularly, simulates the effect of seismic wave on the base of the constructions and airport covering of the flying strip. The problem on free vibrations of orthotropic plates on rigid foundation is formulated – to find non-trivial solution of dynamic equations of three-dimensional problem of elasticity theory of anisotropic body in the domain $D = \{(x,y,z) :, x, y \in D_0, |z| \leq h, h << l\}$ occupied by the plate, where D_0 is the domain of the middle surface, l is the characteristic dimension, $2h$ is the thickness of the plate, under boundary conditions on the facial surfaces $z = \pm h$ of the plate

$$u(x,y,-h) = 0, \quad v(x,y,-h) = 0, \quad w(x,y,-h) = 0 \tag{52}$$
$$u(x,y,h) = 0, \quad v(x,y,h) = 0, \quad w(x,y,h) = 0 \tag{53}$$

or

$$\sigma_{xz}(x,y,h) = \sigma_{yz}(x,y,h) = \sigma_{zz}(x,y,h) = 0 \tag{54}$$

Here the conditions on the lateral surface do not influence on the frequency of free vibrations. The solution of the stated problems is sought in the form of

$$\sigma_{\alpha\beta} = \sigma_{jk}(x,y,z)\exp(i\omega t), \quad (u,v,w) = (u_x(x,y,z), u_y, u_z)\exp(i\omega t) \tag{55}$$
$$\alpha, \beta = x, y, z; \quad j, k = 1, 2, 3$$

where ω is the unknown frequency of free vibrations. Substitute (55) into the equations of the three-dimensional problem, then passing to coordinates

16   L.A. Aghalovyan

$\xi = x/l, \eta = y/l, \zeta = z/h$ and displacements $U = u_x/l, V = u_y/l, W = u_z/l,$ the solution of the obtained singularly perturbed by small parameter system is represented in the form of [26, 27]

$$\sigma_{jk} = \varepsilon^{-1+s} \sigma_{jk}^{(s)}(\xi, \eta, \zeta), \ (U, V, W) = \varepsilon^s (U^{(s)}(\xi, \eta, \zeta), V^{(s)}, W^{(s)}) \quad (56)$$

$$\omega^2 = \varepsilon^s \omega_s^2, \quad s = \overline{0, N}$$

Substituting (56) into the above noted equations we succeed to express $\sigma_{jk}^{(s)}$ through $U^{(s)}, V^{(s)}, W^{(s)}$, which are determined from ordinary by ζ, differential equations. Solving these equations, satisfying conditions (52), (53) we get at $s = 0$ three independent systems of algebraic homogeneous equations. From the condition of existence of these systems non-trivial solutions three dispersion equations follow, from where the following principal values of frequencies of free vibrations follow

$$\omega_n^{xz} = \frac{\pi n}{2h} \sqrt{\frac{G_{13}}{\rho}} = \frac{\pi n}{2h} v^{xz}, \ \omega_n^{yz} = \frac{\pi n}{2h} \sqrt{\frac{G_{23}}{\rho}} = \frac{\pi n}{2h} v^{yz}, \ n \in N, \ \omega_n^p = \frac{\pi n}{2h} v_p$$

$$v_p = \sqrt{\frac{E}{\rho} \cdot \frac{1 - \nu_{12}\nu_{21}}{1 - \nu_{12}\nu_{21} - \nu_{13}\nu_{31} - \nu_{23}\nu_{32} - \nu_{12}\nu_{23}\nu_{31} - \nu_{21}\nu_{13}\nu_{32}}} \quad (57)$$

where v^{xz}, v^{yz} are the propagation of shear waves in an orthotropic medium, v_p is the velocity of longitudinal waves. The boundary conditions (52), (54) correspond such kind of principal values of frequencies

$$\omega_n^{xz} = \frac{\pi}{4h}(2n + 1)v^{xz}, \quad \omega_n^{yz} = \frac{\pi}{4h}(2n + 1)v^{yz}, \quad \omega_n^p = \frac{\pi}{4h}(2n + 1)v_p \quad (58)$$

Thus, in orthotropic plates two shear and one longitudinal free vibrations arise. When $s > 0$ the correction to the values of frequencies is of order $O(\varepsilon^2)$. To each free frequency ω free vibrations of the boundary layer not far from the lateral surface correspond [28]. Analogously free vibrations of layered beams, plates and shells are considered [29].

Consider forced vibrations. Let to the surface $z = -h$ of an orthotropic plate be applied displacement vector

$$(u(-h), v(-h), w(-h)) = (u^-(\xi, \eta), v, w) \exp(i\Omega t) \quad (59)$$

where Ω is the frequency of the forcing action, and at $z = h$ conditions (53) or (54) are given. The problem simulates the action of the seismic wave on the foot of the base construction and on the covering of the air-strips. The solution of problems (53), (59) and (54), (59) we seek in the form of (55), (56) replacing ω with Ω. Doing the same operations we get the equations relative to $U^{(s)}, V^{(s)}, W^{(s)}$, solving them and satisfying conditions (53), (54), (59) we get a final asymptotic solution of the internal dynamic problem. We bring the solution of problem (54), (59) at $u^-, v^-, w^- = const$

$$(u, v, w) = (\tilde{u}, \tilde{v}, \tilde{w}) \exp i\Omega t, \quad \sigma_{jk} = \tilde{\sigma}_{jk} \exp i\Omega t, \quad j, k = 1, 2, 3$$

$$\tilde{u} = \frac{u^-}{\cos 2\Omega_* \sqrt{a_{55}}} \cos \Omega_* \sqrt{a_{55}}(1 - \zeta), \quad \Omega_*^2 = \rho h^2 \Omega^2, \quad (u, v, w; a_{55}, a_{44}, 1/B_{11})$$

$$\tilde{\sigma}_{13} = \frac{1}{ha_{55}} \frac{\partial \tilde{u}}{\partial \zeta}, \quad \tilde{\sigma}_{23} = \frac{1}{ha_{44}} \frac{\partial \tilde{v}}{\partial \zeta}, \quad \tilde{\sigma}_{33} = \frac{B_{11}}{h} \frac{\partial \tilde{w}}{\partial \zeta}, \quad \sigma_{12} = 0 \qquad (60)$$

$$\tilde{\sigma}_{11} = -\frac{B_{23}}{h} \frac{\partial \tilde{w}}{\partial \zeta}, \quad \tilde{\sigma}_{22} = -\frac{B_{13}}{h} \frac{\partial \tilde{w}}{\partial \zeta}, \quad B_{11} = (a_{11}a_{22} - a_{12}^2)/\Delta$$

$$B_{13} = (a_{11}a_{23} - a_{12}a_{13})/\Delta, \quad B_{23} = (a_{22}a_{13} - a_{12}a_{23})/\Delta$$

$$\Delta = a_{11}a_{22}a_{33} + 2a_{12}a_{23}a_{13} - a_{11}a_{23}^2 - a_{22}a_{13}^2 - a_{33}a_{12}^2$$

To the boundary conditions on the lateral surface a dynamic boundary layer corresponds, i.e. solution rapidly damping when removing from the lateral surface into the inside of the plate. By asymptotic method dynamic problems of thin multi-layered bodies are also solved [30, 31].

We bring the solution for two-layered orthotropic plate $D = \{(x, y, z) : x, y \in D_0, -h_2 \leq z \leq h_1)\}$ corresponding to boundary conditions (59) at $z = -h_2$ and (53) at $z = h_1$ when $u^-, v^-, w^- = const$

$$u_m = \tilde{u}_m \exp i\Omega t, \quad (u, v, w), \quad \sigma_{jkm} = \tilde{\sigma}_{jkm} \exp i\Omega t, \quad m = I, II \quad (61)$$

the quantities of the first layer $(0 \leq \zeta \leq \zeta_1, \zeta_1 = h_1/h, h = h_1 + h_2)$

$$\tilde{u}_I = u^- \sqrt{\frac{\rho_{II}}{a_{55}^{II}}} \frac{\Omega_*}{\Delta_1} \sin a_1^I \Omega_*(\zeta_1 - \zeta), \qquad \Omega_* = h\Omega$$

$$\tilde{v}_I = v^- \sqrt{\frac{\rho_{II}}{a_{44}^{II}}} \frac{\Omega_*}{\Delta_2} \sin a_2^I \Omega_*(\zeta_1 - \zeta)$$

$$\tilde{w}_I = w^- \sqrt{B_{11}^{II} \rho_{II}} \frac{\Omega_*}{\Delta_3} \sin a_3^I \Omega_*(\zeta_1 - \zeta) \qquad (62)$$

$$\tilde{\sigma}_{13I} = \frac{1}{ha_{55}^I} \frac{\partial \tilde{u}_I}{\partial \zeta}, \quad \tilde{\sigma}_{23I} = \frac{1}{ha_{44}^I} \frac{\partial \tilde{v}_I}{\partial \zeta}, \quad \tilde{\sigma}_{33I} = \frac{B_{11}^I}{h} \frac{\partial \tilde{w}_I}{\partial \zeta}$$

$$\tilde{\sigma}_{12I} = 0, \quad \tilde{\sigma}_{11I} = -\frac{B_{23}^I}{h} \frac{\partial \tilde{w}_I}{\partial \zeta}, \quad \tilde{\sigma}_{22I} = -\frac{B_{13}^I}{h} \frac{\partial \tilde{w}_I}{\partial \zeta}$$

the quantities of the second layer $(-\zeta_2 \leq \zeta \leq 0, \zeta_2 = h_2/h)$

$$\tilde{u}_{II} = \left[\sqrt{\frac{\rho_{II}}{a_{55}^{II}}} \sin a_1^I \Omega_* \zeta_1 \cos a_1^{II} \Omega_* \zeta - \sqrt{\frac{\rho_I}{a_{55}^I}} \cos a_1^I \Omega_* \zeta_1 \sin a_1^{II} \Omega_* \zeta \right] u^- \frac{\Omega_*}{\Delta_1}$$

$$\tilde{v}_{II} = \left[\sqrt{\frac{\rho_{II}}{a_{44}^{II}}} \sin a_2^I \Omega_* \zeta_1 \cos a_2^{II} \Omega_* \zeta - \sqrt{\frac{\rho_I}{a_{44}^I}} \cos a_2^I \Omega_* \zeta_1 \sin a_2^{II} \Omega_* \zeta \right] v^- \frac{\Omega_*}{\Delta_2}$$

$$\tilde{w}_{II} = \left[\sqrt{B_{11}^{II} \rho_{II}} \sin a_3^I \Omega_* \zeta_1 \cos a_3^{II} \Omega_* \zeta \right.$$

$$\left. - \sqrt{B_{11}^I \rho_I} \cos a_3^I \Omega_* \zeta_1 \sin a_3^{II} \Omega_* \zeta \right] w^- \frac{\Omega_*}{\Delta_3} \tag{63}$$

$$\tilde{\sigma}_{13II} = \frac{1}{h a_{55}^{II}} \frac{\partial \tilde{u}_{II}}{\partial \zeta}, \quad \tilde{\sigma}_{23II} = \frac{1}{h a_{44}^{II}} \frac{\partial \tilde{v}_{II}}{\partial \zeta}, \quad \tilde{\sigma}_{33II} = \frac{B_{11}^{II}}{h} \frac{\partial \tilde{w}_{II}}{\partial \zeta}$$

$$\tilde{\sigma}_{12II} = 0, \quad \tilde{\sigma}_{11II} = -\frac{B_{23}^{II}}{h} \frac{\partial \tilde{w}_{II}}{\partial \zeta}, \quad \tilde{\sigma}_{22II} = -\frac{B_{13}^{II}}{h} \frac{\partial \tilde{w}_{II}}{\partial \zeta}$$

$$a_1^m = \sqrt{a_{55}^m \rho_m}, \quad a_2^m = \sqrt{a_{44}^m \rho_m}, \quad a_3^m = \sqrt{\frac{\rho_m}{B_{11}^m}}, \quad m = I, II$$

$$\Delta_1 = \left(\sqrt{\frac{\rho_I}{a_{55}^I}} \cos a_1^I \Omega_* \zeta_1 \sin a_1^{II} \Omega_* \zeta_2 + \sqrt{\frac{\rho_{II}}{a_{55}^{II}}} \sin a_1^I \Omega_* \zeta_1 \cos a_1^{II} \Omega_* \zeta_2 \right) \Omega_*$$

$$\Delta_2 = \left(\sqrt{\frac{\rho_I}{a_{44}^I}} \cos a_2^I \Omega_* \zeta_1 \sin a_2^{II} \Omega_* \zeta_2 + \sqrt{\frac{\rho_{II}}{a_{44}^{II}}} \sin a_2^I \Omega_* \zeta_1 \cos a_2^{II} \Omega_* \zeta_2 \right) \Omega_*$$

$$\Delta_3 = \left(\sqrt{B_{11}^I \rho_I} \cos a_3^I \Omega_* \zeta_1 \sin a_3^{II} \Omega_* \zeta_2 \right.$$

$$\left. + \sqrt{B_{11}^{II} \rho_{II}} \sin a_3^I \Omega_* \zeta_1 \cos a_3^{II} \Omega_* \zeta_2 \right) \Omega_*$$

It is not difficult to extract the solution for three-layered plate. The analysis of the solution for three-layered plate shows, that if the middle layer is soft (rubber-like material), then tangential displacements applied to the lower layer, practically are not transmitted to the above situated layer. Taking into account that when earthquakes take place, great danger produce the very applied tangential displacements, this effect is of great importance and proves the necessity of using seismoisolators in seismosteady construction.

An additional information on applying the asymptotic method for the solution of three-dimensional problems may be obtained from review article [32].

9 Conclusion

An asymptotic method for solution of two-dimensional and three-dimensional static and dynamic problems of elasticity theory for isotropic and anisotropic beams, plates and shells is expounded. The relationships of the asymptotic solutions with results from classic and refined theories and with the Saint-Venant principle, is established.

The effectiveness of the method for the solution of new classes of thin-body problems – in particular, for those problems that cannot be solved by the existing theories, is illustrated.

Acknowledgment

The author expresses his gratitude to INTAS, grant 06-100017-8886 which made this investigation possible.

References

1. Kilchevsky N.A. Grounds of analytical mechanics of shells. Kiev. Publish. House AS USSR, 1963. 354p.
2. Vekua I.N. Some general methods of constructing various variants of shell theory. M.: Nauka, 1982. 288p.
3. Timoshenko S.P., Voinovski-Kriger S. Plates and shells. M.: Physmathgiz, 1963. 636p.
4. Reissner E. On the theory of bending of elastic plates. J. Math. Phys. 1944. V. 23. pp. 184–191.
5. Ambartsumyan S.A. Theory of anisotropic shells. M.: Physmathgiz. 1961. 344p.
6. Ambartsumyan S.A. Theory of anisotropic plates. M.: Nauka. 1967. 266p.
7. Pelekh B.L. Theory of shells with finite shear stiffness. Kiev. Naukova Dumka. 1973. 248p.
8. Galimov K.Z. To non-linear theory of thin shells of Timoshenko type. Proc AS USSR. MTT. 1976. V. 4. pp. 155–166.
9. Vazov V. Asymptotic decompositions of solutions of ordinary differential equations. M.: Mir. 1968. 464p.
10. Nayfeh A.H. Perturbation methods. John Wiley and Sons, 1973. 455p.
11. Vasiljeva A.B., Boutuzov V.F. Asymptotic decompositions of solutions of singularly perturbed equations. M.: Nauka. 1973. 272p.
12. Lomov S.A. Introduction into the general theory of singular perturbation. M.: Nauka. 1981. 398p.
13. Iljin A.M. Concordance of asymptotic decompositions of solutions of boundary value problems. M.: Nauka. 1989. 336p.
14. Friedrichs K.O., Dressler R.F. A boundary-layer theory for elastic plates. Comm. Pure Appl. Math. 1961. V. 14. N 1.
15. Goldenveiser A.L. Construction of approximation theory of plate bending with the asymptotic integration method of equations of elasticity theory. J. Appl. Math. Mech. 1962. V. 26 Edition 4. pp. 668–686.
16. Green A.E. On the linear theory of thin elastic shells. Proc. Roy. Soc Ser. A. 1962. V. 266. N 1325.
17. Lekhnitsky S.G. Elasticity theory of anisotropic body. M.: Nauka. 1972. 416p.
18. Aghalovyan L.A. Asymptotic theory of anisotropic plates and shells. M.: Nauka. Fizmatlit, Moscow. 1997. 414p.
19. Aghalovyan L.A., Khachatryan A.M. Asymptotic analysis of stress-strain state of anisotropic layered beams. Proc. AS ARM SSR. Mech. 1986. V. 39. N 2. pp. 3–14.

20. Aghalovyan L.A. On bending equations of anisotropic plates. Proceeding of VII All-Union conference on theory of shells and plates. M.: Nauka 1970. pp. 17–21.

21. Aghalovyan L.A. On reduction of space problem of elasticity theory to two-dimensional for orthotropic shells and errors of some applied theories. Rep. AS ARMSSR. 1979. V. 69. N 3. pp. 151–156.

22. Aghalovyan L.A. On the structure of solution of one class of plane problems of elasticity theory of anisotropic body. Mechanics: Interuniversity Transactions: Yerevan University Publishing House. 1982. Edition 2. pp. 7–12.

23. Aghalovyan L.A., Gevorgyan R.S. On the asymptotic solution of mixed three-dimentional problems for two-layered anisotropic plates. Appl. Math Mech. 1986. V. 50. N 2. pp. 271–278.

24. Aghalovyan L.A., Gevorgyan R.S. On asymptotic solution of nonclassical boundary value problems for two-layered anisotropic thermoelastic shells. Proc. AS ARMSSR. Mech. 1989. V. 42. N 3. pp. 28–36.

25. Aghalovyan L.A., Gevorgyan R.S. Nonclassical boundary-value problems of anisotropic layered beams, plates and shells. Yerevan, Publish. House "Gitutjun" NAS of Armenia. 2005. 468p.

26. Aghalovyan L.A. To the asymptotic method of solution of dynamic mixed problems of anisotropic strips and plates. Publish. House of IHE of Russia. North-Caucasus region. Nat. Sci. 2000. N 3(111). pp. 8–11.

27. Aghalovyan L.A., Aghalovyan M.L. To the determination of frequencies and forms of orthotropic strip free vibrations. reports NAS RA. 2003. V. 103. N 4. pp. 296–301.

28. Aghalovyan M.L. On solution of the boundary layer in the problem on free vibrations of the strip. In collect. of conf.: Contemporary questions of optimal control of vibrations and stability of systems. Yerevan: Publish. House of Yerevan University. 1997. pp. 132–135.

29. Aghalovyan L.A., Gulgazaryan L.G. Asymptotic solutions of non-classical boundary-value problems of the natural vibrations of orthotropic shells. J. App. Math. Mech. 2006. 70. pp. 102–115.

30. Aghalovyan L.A. On one class of the problems on forced vibrations of anisotropic plates. Problems of mechanics of thin deformable bodies. Yerevan. Armenia. 2002. pp. 9–19.

31. Hovhannisyan R. Sh. The asymptotic form of forced vibrations of three-layered orthotropic plate in case of full contact conditions between layers. Elasticity, plastisity and creep selected topics. Yerevan, Armenia. 2006. pp. 242–248.

32. Aghalovyan L.A. Asymptotic of solution of classical and nonclassical boundary value problems of statics and dynamics of thin bodies. Int. Appl. Mech. 2002. V. 38. N 7. pp. 3–24.

Multiscale Assessment of Low-Temperature Performance of Flexible Pavements

E. Aigner[1], R. Lackner[1,2], M. Spiegl[3], M. Wistuba[3], R. Blab[3], and H. Mang[1]

[1] Christian-Doppler-Laboratory for "Performance-based Optimization of Flexible Pavements", Institute for Mechanics of Materials and Structures, Vienna University of Technology, Karlsplatz 13, 1040 Vienna, Austria, {elisabeth.aigner,roman.lackner,herbert.mang}@tuwien.ac.at

[2] Computational Mechanics, Technical University of Munich, Arcisstraße 21, 80333 Munich, Germany, lackner@bv.tum.de

[3] Christian-Doppler-Laboratory for "Performance-based Optimization of Flexible Pavements", Institute for Road Construction and Maintenance, Vienna University of Technology, Gußhausstraße 28, 1040 Vienna, Austria, {mspiegl,mwistuba,rblab}@istu.tuwien.ac.at

Abstract The increase of heavy-load traffic within Europe requires the development of appropriate tools for the assessment of existing and new road infrastructure. In this paper, such a tool is presented, combining multiscale material modeling of asphalt with structural analysis of flexible pavements representing plate-like structures at low temperatures. At this temperature regime, rapid cooling of the road surface in consequence of temperature drops may result in so-called top-down cracking. These cracks, when propagating further into the base layer, significantly reduce the service life of road infrastructure. Within the proposed multiscale model for asphalt, the temperature-dependent viscoelastic properties of asphalt are related to the constituent bitumen, exhibiting the thermorheological behavior, accounting for

- the large variability of asphalt mixtures, resulting from different mix design, different constituents (e.g. bitumen, filler, aggregate), and the allowance of additives, and
- changing material behavior in consequence of thermal, chemical, and mechanical loading.

The parameters of the employed viscoelastic material model for asphalt are obtained from upscaling of parameters identified at the bitumen-scale up towards the macroscale i.e., the scale of structural analysis, with the viscoelastic behavior of bitumen serving as input. Upscaling is performed in the framework of continuum micromechanics using the elastic-viscoelastic correspondence principle. The presented multiscale model is applied to asphalts typically used for surface and base layers of flexible pavements. The so-obtained macroscopic model parameters are employed in the numerical analysis of flexible pavements, giving access to stresses resulting from (i) a sudden drop of the surface temperature in consequence of changing weather conditions and (ii) traffic loading. Comparison of the so-obtained surface stresses with the tensile strength of asphalt at the respective surface temperature allows an assessment of the risk of top-down cracking in flexible pavements.

Keywords: bitumen, asphalt, pavement, viscoelasticity, multiscale model, upscaling, identification, validation, finite element method

1 Motivation

Since the recent enlargement of the European Union (EU), in 2004, Austria is almost completely surrounded by member states of the EU. As a consequence of the single market program of the EU, in addition to North-South transit between Germany and Italy, a pronounced increase of East-West transit is observed and expected to increase further. The increased loading of the Austrian road infrastructure by the number of vehicles and the use of so-called single tires requires (i) a rigorous assessment of the existing road infrastructure and (ii) a reliable performance prediction for future projects. So far, however, this assessment is performed exclusively by means of empirical methods and/or experiments.

As a remedy, numerical analysis tools (already employed in structural engineering and soil mechanics) represent a promising alternative for the assessment of pavements. In contrast to material models that have been adopted so far for the simulation of pavement structures (see e.g. [4,8,9]), the incorporation of finer-scale information is proposed in this paper.

According to [5], four additional observation scales may be introduced below the *macro*-scale (see Fig. 1). Changes in the material behavior in consequence of thermal, chemical, and mechanical loading can be considered at the respective scale of observation and, via upscaling, their effect on the macroscopic material behavior is obtained. Within this paper, upscaling of viscoelastic properties from the *bitumen-* to the *macro*-scale is carried out, allowing to relate the rheological behavior of asphalt to the behavior of bitumen, i.e., to the only constituent in asphalt exhibiting viscoelastic behavior.

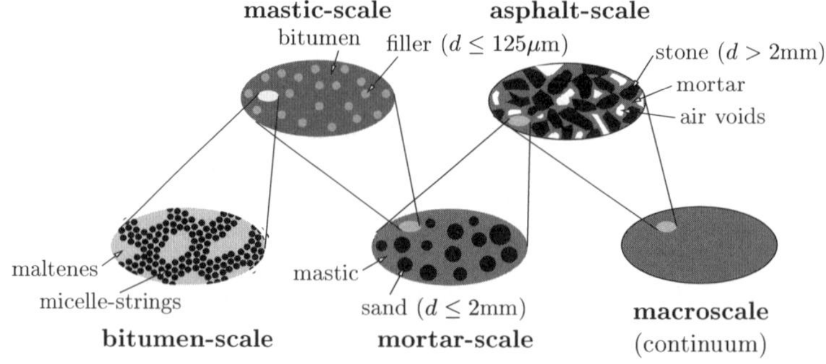

Fig. 1 Multiscale model with four additional observation scales below the *macro*-scale

In the following section, the theoretical basis for upscaling of viscoelastic properties within continuum micromechanics will be presented. Thereafter, identification of viscoelastic properties of bitumen and verification of the proposed upscaling scheme will be carried out for different types of asphalt. In Section 4, the obtained macroscopic model parameters will be employed in the numerical assessment of flexible pavements.

2 Upscaling of Creep Properties

Upscaling techniques are used to shift finer-scale information from an observation scale to the next-higher one, considering the respective composite microstructure. For upscaling of viscoelastic properties of asphalt, use of continuum micromechanics is proposed in this paper.

2.1 Homogenization of elastic properties – continuum micromechanics

Continuum micromechanics is an analytical technique for determination of the effective behavior of composites, taking the arrangement and properties of the material phases into account. The constitutive law for the r-th material phase at the position $\mathbf{x}$ is given by

$$\boldsymbol{\sigma}_r(\mathbf{x}) = \mathbb{c}_r \; : \; \boldsymbol{\varepsilon}_r(\mathbf{x}) \; . \tag{1}$$

Within continuum micromechanics, a localization tensor $\mathbb{A}$ is introduced, relating the homogenized strain tensor $\mathbf{E}$ to the local strain tensor $\boldsymbol{\varepsilon}$ at $\mathbf{x}$:

$$\boldsymbol{\varepsilon}(\mathbf{x}) = \mathbb{A}(\mathbf{x}) \; : \; \mathbf{E} \; , \tag{2}$$

with

$$\mathbf{E} = \langle \boldsymbol{\varepsilon}(\mathbf{x}) \rangle_V = \frac{1}{V} \int_V \boldsymbol{\varepsilon}(\mathbf{x}) dV \; . \tag{3}$$

The homogenized stress tensor Σ is obtained from volume averaging of the local stress tensor $\boldsymbol{\sigma}(\mathbf{x})$, reading

$$\Sigma = \langle \boldsymbol{\sigma}(\mathbf{x}) \rangle_V = \frac{1}{V} \int_V \boldsymbol{\sigma}(\mathbf{x}) dV. \tag{4}$$

Considering Equations (1) and (2) in Equation (4) and comparing the so-obtained result with the macroscopic stress-strain law

$$\Sigma = \mathbb{C}_{eff} \; : \; \mathbf{E} \tag{5}$$

gives the effective material tensor $\mathbb{C}_{eff}$ as

$$\mathbb{C}_{eff} = \langle \mathbb{c} \; : \; \mathbb{A} \rangle_V \; , \tag{6}$$

with $\mathbb{c}$ as the material tensor of the single phase. Consideration of an idealized microstructure, allows estimating the unknown localization tensor $\mathbb{A}$. For a microstructure showing a clear matrix-inclusion morphology, the Mori-Tanaka scheme [7] is used. Hereby, the localization tensor for the I-th ellipsoidal inclusion is given by

$$\mathbb{A}_I = \left[\mathbb{I} + \mathbb{S}_I : \left(\mathbb{c}_M^{-1} : \mathbb{c}_I - \mathbb{I}\right)\right]^{-1} : \left\{f_M \mathbb{I} + f_I \left[\mathbb{I} + \mathbb{S}_I : \left(\mathbb{c}_M^{-1} : \mathbb{c}_I - \mathbb{I}\right)\right]^{-1}\right\}^{-1} ,\tag{7}$$

where the indices "I" and "M" refer to the inclusion and the matrix phase, respectively. In Equation (7), f_I and f_M refer to the volume fraction of the inclusions and the matrix, respectively, and $\mathbb{S}$ denotes the so-called fourth-order Eshelby tensor. From $\langle\mathbb{A}\rangle_V = \mathbb{I}$, $\langle\mathbb{A}\rangle_{V_M}$ is obtained as

$$\langle\mathbb{A}\rangle_{V_M} = \frac{1}{f_M}\left(\mathbb{I} - f_I \mathbb{A}_I\right) .\tag{8}$$

Considering Equations (7) and (8) in Equation (6), the effective shear modulus μ_{eff} is obtained for the case of spherical inclusions:

$$\mu_{eff} = \frac{f_M \mu_M + f_I \mu_I \left[1 + \beta\left(\mu_I/\mu_M - 1\right)\right]^{-1}}{f_M + f_I \left[1 + \beta\left(\mu_I/\mu_M - 1\right)\right]^{-1}} ,\tag{9}$$

where μ_I and μ_M denote the shear moduli of the inclusions and the matrix, respectively. $\beta = 6(k_M + 2\mu_M)/[5(3k_M + 4\mu_M)]$, with k_M as the bulk modulus of the matrix material, represents the Eshelby tensor for spherical inclusions.

2.2 Homogenization of viscoelastic properties – the elastic-viscoelastic correspondence principle

Employing this correspondence principle, the material parameters in the solution of the respective elastic problem are replaced by the respective Laplace-Carson transformed viscoelastic parameters. For example, the elastic shear compliance $1/\mu$ is replaced by the Laplace-Carson transformed creep-compliance function for deviatoric creep, $J_{dev}^*(p)$. The inverse Laplace-Carson

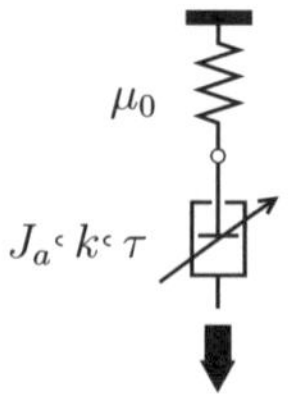

$$J(t) = \frac{1}{\mu_0} + J_a\left(\frac{t}{\bar{\tau}}\right)^k$$

$$J^*(p) = \mathcal{LC}\P J(t)\Diamond = \frac{1}{\mu_0} + J_a\left(\frac{1}{p\bar{\tau}}\right)^k \Gamma[k+1]$$

Fig. 2 Power-law (rheological) model for the description of deviatoric creep of bitumen/asphalt: model parameters, creep-compliance function, and Lacaplace-Carson transform of creep compliance

transformation delivers the corresponding solution in the time space. Figure 2 shows the Power-law (PL) model used to describe the deviatoric creep of bitumen, where μ_0 [MPa] is the shear modulus, J_a [MPa^{-1}] is the viscous part of the creep compliance at $t = \bar{\tau}$, and Γ denotes the Gamma function. With the expression for the effective elastic properties at hand (Equation (9)), the correspondence principle gives access to the effective (homogenized) creep compliance of a matrix/inclusion-composite as

$$J^*_{eff}(p) = \frac{f_M + f_I\left[1 + \beta^*\left(J^*_M/J^*_I - 1\right)\right]^{-1}}{f_M/J^*_M + f_I/J^*_I\left[1 + \beta^*\left(J^*_M/J^*_I - 1\right)\right]^{-1}} \ , \tag{10}$$

with $\beta^* = 6(k^*_M + 2/J^*_M)/[5(3k^*_M + 4/J^*_M)]$. For the case of deviatoric creep of the matrix and elastic behavior of the inclusions, $k^*_M = k_M$ and $1/J^*_I = \mu_I$. Application of the inverse Laplace-Carson transformation,

$$J_{eff}(t) = \mathcal{LC}^{-1}\left[J^*_{eff}(p)\right] \ , \tag{11}$$

leads to the effective creep compliance of the composite material in the time space, giving access to the effective viscoelastic model parameters.

3 Application to Asphalt

3.1 Identification of bitumen creep

In the following, experimental results from identification of viscous properties of bitumen will be presented. For this purpose, two types of experiments, the bending-beam rheometer (BBR) test [10] in the low temperature regime ($-30 < T < 0°C$) and the dynamic-shear rheometer (DSR) test [10] in the elevated temperature regime ($0 < T < 80°C$) are conducted:

- *Bending beam rheometer (BBR):* Based on the monitored displacement history $u(t)$, the creep compliance J [MPa^{-1}] is given by (see [6])

$$J(t) = \frac{\varepsilon(t)}{\sigma} = \frac{4bh^3}{Fl^3}u(t) \ . \tag{12}$$

The experimental results are well described by the aforementioned PL model, with the creep compliance J and the creep-compliance rate dJ/dt (see Fig. 3b) reading

$$J = \frac{1}{\mu_0} + J_a\left(\frac{t}{\bar{\tau}}\right)^k \quad \text{and} \quad dJ/dt = \frac{J_a\,k}{\bar{\tau}}\left(\frac{t}{\bar{\tau}}\right)^{k-1} = Ht^p \ , \tag{13}$$

where H [MPa^{-1}s^{-k}] represents the creep-compliance rate at $t = \bar{\tau}$ and p [-] is the slope of the linear model response in the log $dJ/dt - \log t$ diagram. BBR experiments on pure bitumen were carried out and the model parameters p and H were extracted from the creep-compliance rate.

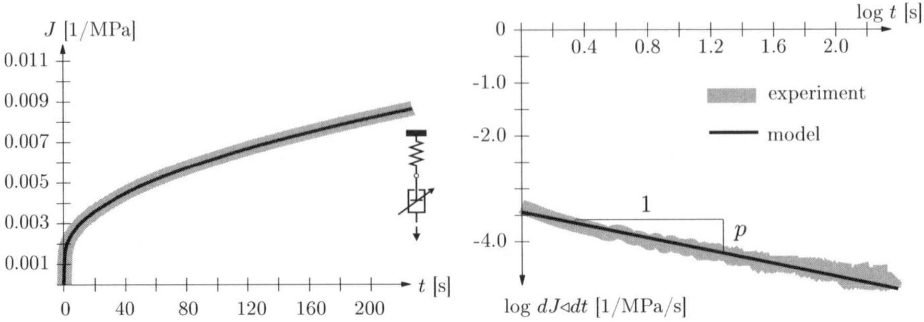

Fig. 3 Comparison of PL-model response with experimental data from BBR: (**a**) creep compliance and (**b**) creep-compliance rate

Since BBR tests are performed at different temperatures, the temperature-dependence of the model parameter p is obtained as

$$p(T) = p_0 + c(T - \bar{T}) \,, \tag{14}$$

with T as the actual temperature and $\bar{T}$ as the reference temperature. The temperature-dependence of the model parameter H is described by the Arrhenus-type law

$$H(T) = H_0 \exp\left[-\frac{E_a}{R}\left(\frac{1}{T} - \frac{1}{\bar{T}}\right)\right] \,. \tag{15}$$

Hereby, H_0 is the creep-compliance rate at $t = \bar{\tau}$ for $T = \bar{T}$, E_a [J/mol] is the activation energy, and R is the gas constant.

- *Dynamic shear rheometer (DSR):* DSR tests are conducted under cyclic loading undergoing a temperature and frequency sweep from -20 to $+46°$C and 0.1–50 Hz, respectively. The DSR delivers the phase angle φ and the complex shear modulus μ^*, giving the storage and loss modulus as $\mu' = \mathrm{Re}(\mu^*) = \mu^*\cos\varphi$ and $\mu'' = \mathrm{Im}(\mu^*) = \mu^*\sin\varphi$. The storage and loss modulus in viscoelastic solids measure the stored energy, representing the elastic portion, and the energy dissipated as heat, representing the viscous portion. The experimentally-obtained values for μ' and μ'' obtained for two types of bitumen are plotted in the so-called Cole-Cole diagram (see Fig. 4). The parameters of the PL model describing the material response of pure bitumen are obtained from curve fitting, aiming at the best fit between the model response and the experimental data in the Cole-Cole diagram. This approach is illustrated in Fig. 4 for bitumen B50/70 and a polymer-modified bitumen B50/90S.

3.2 Validation of upscaling result

In order to validate the model parameters for asphalt obtained from upscaling, static creep tests are used for determination of parameters representing

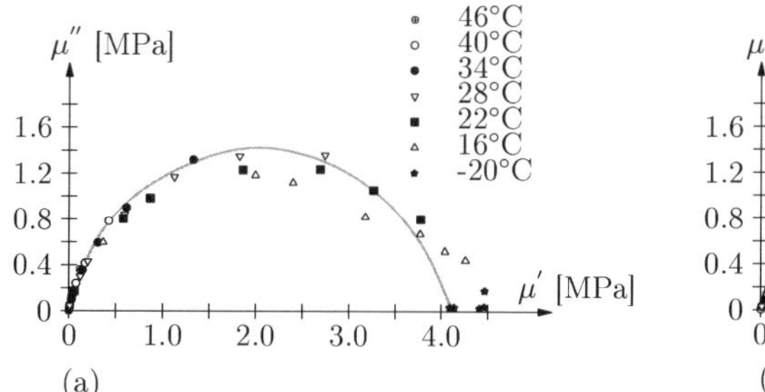

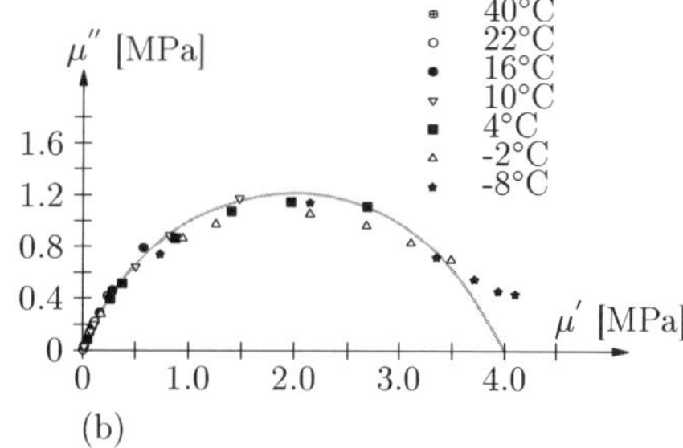

Fig. 4 DSR results and PL-model response for (**a**) bitumen B50/70 (model parameters: $1/\mu_0 = 0.24$ MPa^{-1}, $J_a = 1.2$ MPa^{-1}, $k = 0.8$, $\bar{\tau} = 1$ s) and (**b**) polymer-modified bitumen B50/90S (model parameters: $1/\mu_0 = 0.25$ MPa^{-1}, $J_a = 1.0$ MPa^{-1}, $k = 0.7$, $\bar{\tau} = 1$ s)

the long-term response and cyclic tests are performed at different frequency and temperature regimes in order to specify elastic and short-term viscous properties. The presented upscaling scheme was applied to four types of asphalt, with the mix design given in Table 1. Hereby, limestone dust was used as filler for all considered types of asphalt. Two types of aggregates were used: Diabas for SMA11 and Hollitzer for BT22.

Table 1 Composition of types of asphalt

Asphalt type	Bitumen [vol%]	Filler [vol%]	Aggregate 0–4mm [vol%]	Aggregate 4–22mm [vol%]	Air [vol%]
SMA11-B70/100-D	15.7	3.7	23.1	54.4	3.1
SMA11-B50/90S-D	15.5	3.7	22.9	54.1	3.7
BT22-B50/70-H	11.1	0.4	33.3	49.9	5.2
BT22-B50/90S-H	11.1	0.4	33.4	50.2	4.8

Static tests

Both $J_{a,eff}$ and k_{eff} obtained from upscaling, describing the long-term response of the material, were validated by static uniaxial creep tests using prismatic asphalt specimens with cross sections of 50 mm $\times$ 50 mm for testing of SMA11 and 60 mm $\times$ 60 mm for testing BT22. Within the conducted experimental program, the load level is adapted to the tensile strength of asphalt at the respective testing temperature. Figure 5 shows the creep compliance and creep-compliance rate as a function of time for BT22-B50/70-H for different temperatures. With the input parameters for bitumen given in Table 2 at hand, Equation (10) is used to compute the effective creep parameters for the considered types of asphalt. Hereby, upscaling is performed in

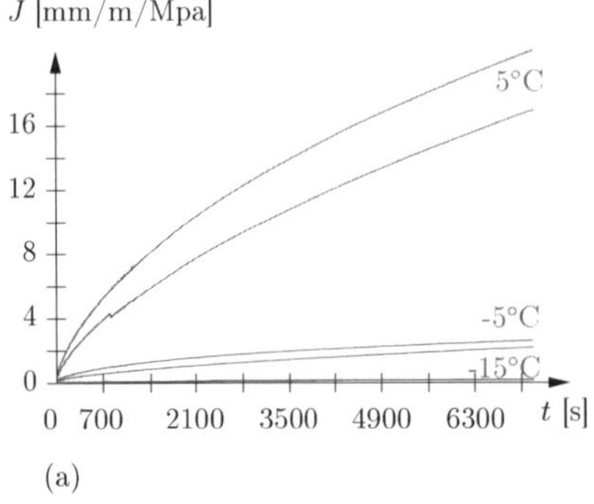

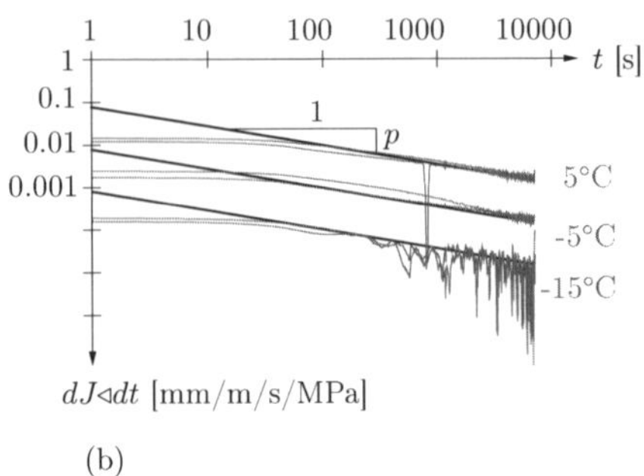

Fig. 5 Experimental results from static creep tests of BT22-B50/70-H: (a) creep compliance and (b) creep-compliance rate (*lines* refer to PL-model, with $H = H(T)$ and $p = -0.4$)

Table 2 Model parameters for considered types of bitumen obtained from BBR and DSR tests ($\bar{T} = -12°C$)

Bitumen	H_0 [mm/m/MPa/s]	p_0 [−]	c [−]	E_a/R [K]
B70/100	0.376	−0.41	0.01	9500
B50/90S	0.611	−0.46	0.011	8600
B50/70	0.254	−0.51	0.013	10600

a step-wise manner, following the multiscale model shown in Figure 1. Thus, only bitumen and filler are considered in the first step. In the second step, this bitumen-filler composite (mastic) becomes the new matrix material where additional aggregates, ranging from 0 to 4 mm, are added. In a third step, aggregates, ranging from 4 to 22 mm, and air voids are added to this new matrix material. When considering two inclusion phases (e.g., aggregate and air within the third step), Equation (10) is extended to

$$J^*_{eff} = \frac{f_M + f_s \left[1 + \beta^* \left(\frac{J^*_M}{J^*_s} - 1\right)\right]^{-1} + f_a \left[1 + \beta^* \left(\frac{J^*_M}{J^*_a} - 1\right)\right]^{-1}}{f_M/J^*_M + f_s/J^*_s \left[1 + \beta^* \left(\frac{J^*_M}{J^*_s} - 1\right)\right]^{-1} + f_a/J^*_a \left[1 + \beta^* \left(\frac{J^*_M}{J^*_a} - 1\right)\right]^{-1}},$$

(16)

with $\beta^* = 6(k^*_M + 2/J^*_M)/[5(3k^*_M + 4/J^*_M)]$ and the indices M, s, and a referring to matrix, stone, and air phase, respectively.

For all three upscaling steps, the matrix material is associated with viscoelastic behavior described by the PL model, whereas elastic behavior is assigned to the inclusions. Under hydrostatic loading, the matrix is assumed to behave elastically, giving $k^*_M = k_M$. Inserting the Laplace-Carson transformed creep compliance, $k^*_M = k_M$, $1/J^*_s = \mu_s = 2 \times 10^5$ MPa^{-1} [3] and

$1/J_a^* = \mu_a = 0$ into Equations (10) and (16), respectively, and performing the inverse Laplace-Carson transformation gives access to the effective creep compliance $J_{eff}(t)$ and, thus, to the effective creep parameters H_{eff} and p_{eff} of asphalt concrete.

Hereby, upscaling was performed using (i) the original, (ii) the effective filler, and (iii) the effective aggregate content.[1] The results obtained from upscaling from the bitumen-scale to the macroscale are depicted in Figs. 6–9 and compared with the respective values obtained from the uniaxial creep experiments. The slope of the $\ln(H)-1/T$ curve obtained from upscaling gives access to the activation energy which remains unchanged during scale transition from the bitumen scale to the macroscale. The so-obtained temperature dependence of the creep parameters H and p, which was already introduced at the bitumen-scale, captures the experimentally-obtained results for asphalt well.

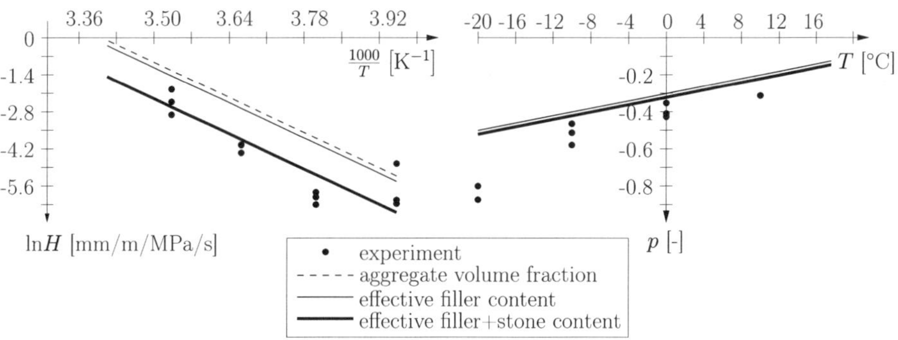

Fig. 6 Experimental results for H and p and multiscale predictions for SMA11-B70/100-D

Consideration of the effective filler/aggregate content in the upscaling procedure leads to a shift of the model parameters obtained from upscaling towards less viscous and more pronounced elastic response, yielding a better agreement between model predictions and experimental results.

---

[1] The *effective* volume content of particles in a particle-matrix composite, $f_{p,eff}$, is obtained as [5]

$$f_{p,eff} = \frac{f_p}{1 - f_a} \, , \tag{17}$$

where f_p is the volume content of the particles and f_a is the volume of air voids in case of maximum particle compaction. The latter is obtained for filler and sand/stone using standard test methods [1,2].

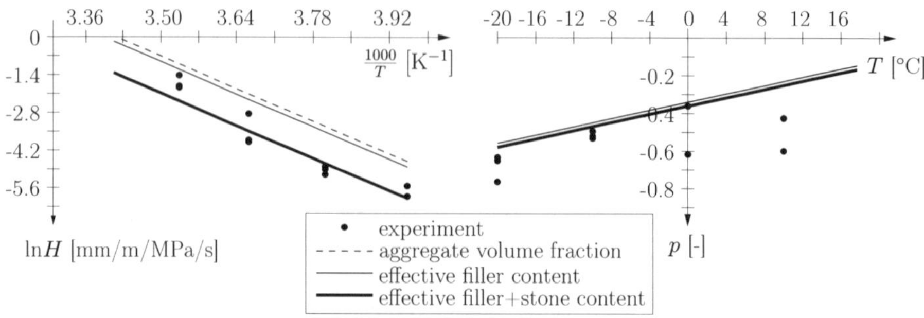

Fig. 7 Experimental results for H and p and multiscale predictions for SMA11-B50/90S-D

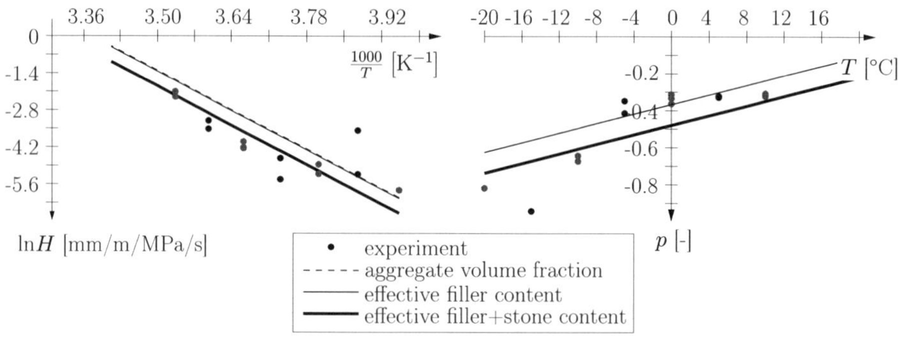

Fig. 8 Experimental results for H and p and multiscale predictions for BT22-B50/70-H

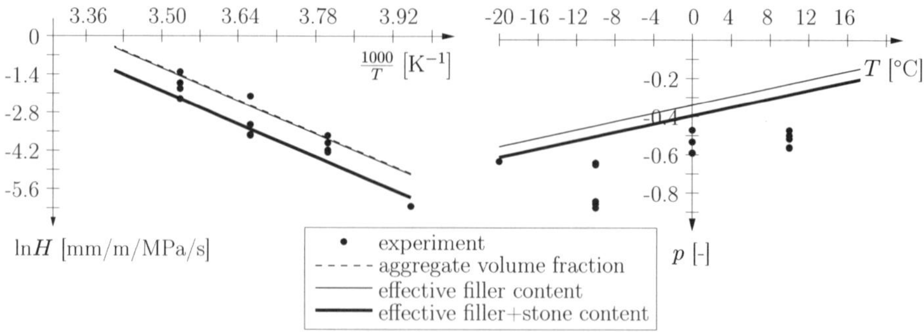

Fig. 9 Experimental results for H and p and multiscale predictions for BT22-B50/90S-H

Cyclic tests

Cyclic tests are performed at different temperatures (ranging from, e.g., $-20°$C to $45°$C) and different frequencies, ranging from 0.1 to 20 Hz, focusing on the short-term material response of asphalt. Hereby, either a uniaxial loading condition using prismatic specimens or a bending-beam experimental setup is employed. From the measured load (stress) and displacement (strain) history, the complex modulus E^* and the phase angle φ are determined and plotted in the Cole-Cole diagram (see Figs. 10 and 11). The experimental results are compared with the material response of asphalt predicted by the multiscale model. This comparison is shown in Figs. 10 and 11 for the four types of asphalt listed in Table 2. The model predictions fit well the experimentally-obtained results in the low-temperature regime. In fact, the PL model will be used in the following section for assessment of the risk of low-temperature cracking in flexible pavements only.

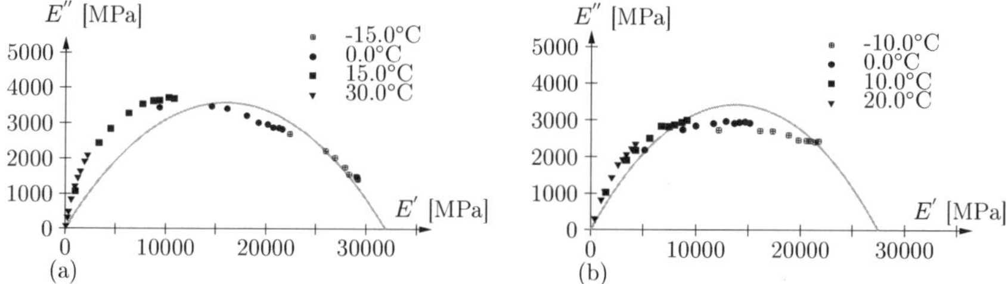

Fig. 10 Experimental results from cyclic tests and multiscale predictions for (**a**) SMA11-70/100-D (model parameters from upscaling: $1/\mu_{0,eff} = 3.0\ 10^{-5}$ MPa^{-1}, $J_{a,eff} = 1.7\ 10^{-5}$ MPa^{-1}, $k_{eff} = 0.3$, $\bar{\tau} = 1$ s) and (**b**) SMA11-50/90S-D (model parameters from upscaling: $1/\mu_{0,eff} = 3.03\ 10^{-5}$ MPa^{-1}, $J_{a,eff} = 3.5\ 10^{-5}$ MPa^{-1}, $k_{eff} = 0.3$, $\bar{\tau} = 1$ s)

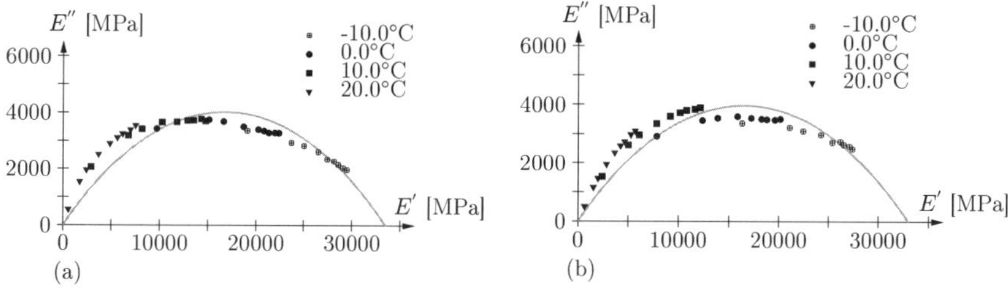

Fig. 11 Experimental results from cyclic tests and multiscale predictions for (**a**) BT22-50/70-H (model parameters from upscaling: $1/\mu_{0,eff} = 3.125\ 10^{-5}$ MPa^{-1}, $J_{a,eff} = 2.93\ 10^{-5}$ MPa^{-1}, $k_{eff} = 0.28$, $\bar{\tau} = 1$ s) and (**b**) BT22-50/90S-H (model parameters from upscaling: $1/\mu_{0,eff} = 3.57\ 10^{-5}$ MPa^{-1}, $J_{a,eff} = 5.4\ 10^{-5}$ MPa^{-1}, $k_{eff} = 0.3$, $\bar{\tau} = 1$ s)

4 Application to Low-Temperature Assessment of Flexible Pavements

So far, only the uniaxial stress situation was considered, however, the lateral confinement of asphalt in pavement structures induces three-dimensional stress states. For the extension of the one-dimensional viscoelastic models to three dimensions, the (fourth-order) normalized compliance tensor $\mathbb{G}$ is introduced, reading in matrix notation

$$\mathbb{G} = E\,\mathbb{C}^{-1} = \begin{bmatrix} 1 & -\nu & -\nu & 0 & 0 & 0 \\ -\nu & 1 & -\nu & 0 & 0 & 0 \\ -\nu & -\nu & 1 & 0 & 0 & 0 \\ 0 & 0 & 0 & 1+\nu & 0 & 0 \\ 0 & 0 & 0 & 0 & 1+\nu & 0 \\ 0 & 0 & 0 & 0 & 0 & 1+\nu \end{bmatrix}, \tag{18}$$

where $\mathbb{C}$ is the material matrix and ν is Poisson's ratio. Accordingly, the viscoelastic-strain tensor in consequence of three-dimensional loading represented by the history of the stress tensor $\boldsymbol{\sigma}(t)$ is given by the convolution integral as

$$\boldsymbol{\varepsilon}^{ve}(t) = \int_0^t J(t-\tau)\mathbb{G} : \frac{\partial \boldsymbol{\sigma}}{\partial \tau}\,d\tau \ . \tag{19}$$

In addition to the influence on model parameters, temperature changes result in thermal strains which are considered in the constitutive law as

$$\boldsymbol{\sigma}_{n+1} = \mathbb{C} : \left(\boldsymbol{\varepsilon}_{n+1} - \boldsymbol{\varepsilon}^{ve}_{n+1} - \boldsymbol{\varepsilon}^{T}_{n+1}\right) \ , \tag{20}$$

where the temperature-strain tensor is given by

$$\boldsymbol{\varepsilon}^{T}_{n+1} = \mathbf{1}\alpha_T(T_{n+1} - T_0) \ . \tag{21}$$

In Equation (21), α_T is the (constant) thermal dilatation coefficient, which is assumed to be temperature independent and T_0 represents the initial temperature at the respective point of the pavement cross-section.

4.1 Analysis of standard road sections

Surface-initiated top-down cracking in consequence of stresses caused by a temperature drop superposed by traffic-initiated stresses is a major mode of deterioration of asphalt pavements. In order to assess the risk of damage of flexible pavements (see Fig. 12), the proposed multiscale approach to modeling of asphalt is combined with computational mechanics.

The cross section considered in the numerical analysis (see Fig. 13) consists of a 40 mm bituminous top layer of stone-mastic asphalt SMA11-B50/90S-D, a 210 mm base layer of high modulus asphalt concrete BT22-B50/90S-H,

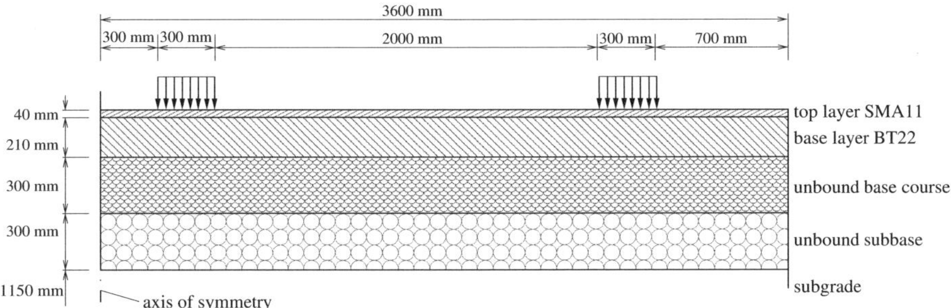

Fig. 12 Geometric dimensions of the considered pavement structure

a 600 mm unbound base course/sub-base of coarse gravel, and a sub-grade. The behavior of the unbound layers is described by an elastic material model. The time span assigned to loading in consequence of traffic is set to 0.02 s, corresponding to a vehicle velocity of 80 km/h. The axle load P [kN] is distributed over a circular contact area for each tire, given by the radius

$$a = \sqrt{\frac{P}{2p\pi}}, \tag{22}$$

where p is the internal tire pressure (0.7 MPa in the present study) and P denotes the axle load. In order to reduce calculation time, the numerical model was restricted to a plane-strain model, considering only a cross section of the pavement structure. Moreover, only one cooling scenario and one type of traffic load was considered (decrease of surface temperature from $-20°$C to $-25°$C within 1 h and a traffic load of 100 kN).[2] The temperature in a depth of 2 m is assumed to remain constant at $7.6°$C.

The solution of the coupled thermo-mechanical problem is performed in two steps: First, the temperature distribution in the road section is determined on the basis of the prescribed temperature cooling scenario. The so-obtained temperature profiles serve as input for the second step – the mechanical analysis, considering thermal shrinkage, the change of material parameters with temperature and, finally, the traffic load. This analysis is performed by using the finite element program FEAP [11]. Figure 13a shows the so-obtained distribution of the surface stress perpendicular to the road axis for different time instants. After simulation of cooling in the analysis, the traffic load is applied, resulting in a significant increase of tensile stresses between the two tires (see Fig. 13b). By comparing the obtained maximum stress with the tensile strength of asphalt at the respective surface temperature, the risk of top-down cracking is evaluated.

---

[2] Results obtained for different cooling scenarios and unbound-layer conditions can be found in [12].

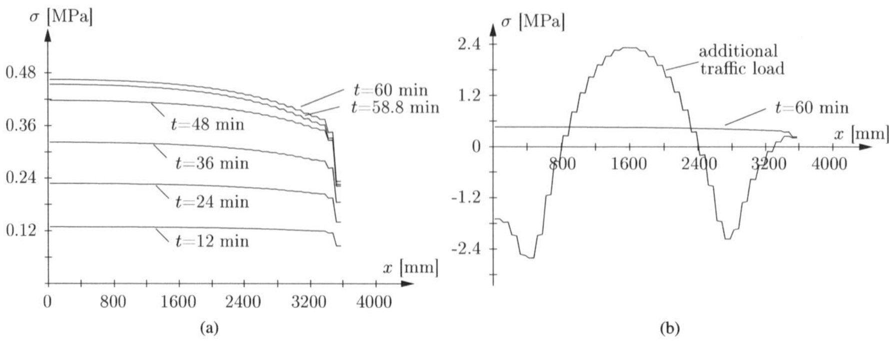

Fig. 13 Distribution of surface stress perpendicular to the road axis in consequence of (**a**) temperature changes and (**b**) additional traffic load

5 Conclusions

In this paper, the risk of low-temperature cracking of flexible pavements was assessed by combining multiscale modeling of asphalt with computational mechanics. Hereby, the thermorheological properties were assigned to the bitumen only. Via upscaling, the viscoelastic properties of asphalt were determined. For upscaling, continuum micromechanics was employed, using the so-called "correspondence principle". Hereby, the typical matrix-inclusion morphology of asphalt at different scales was considered by the Mori-Tanaka scheme used for upscaling of information from one observation scale to the next higher one. The viscous behavior of bitumen was described by the Power-law model, with the respective model parameters serving as input for the multiscale model. The predictive capabilities of the multiscale model were validated by asphalt testing, showing good agreement with the viscous properties of asphalt obtained from static tests conducted at different temperatures. The obtained model parameter representing the slope of the creep-compliance rate dJ/dt in the log $dJ/dt - \log t$ diagram showed almost no dependence on the amount of aggregates of the different observation scales. In fact, this parameter remained nearly constant during the upscaling procedure. The initial creep compliance rate, on the other hand, was decreasing with every upscaling step, as the amount of aggregates in the bituminous mixture increased. Finally, the combination of multiscale modeling of the viscoelastic behavior of asphalt and computational mechanics allowed realistic determination of the tensile loading of the pavement structure. Based on the obtained stresses resulting from (i) a sudden decrease in the temperature in consequence of changing weather conditions and (ii) traffic loading, the following conclusions can be drawn:

- Stress induced by temperature loading increases with increasing cooling rate and decreasing initial surface temperature. This effect is explained, on the one hand, by the reduced time available for relaxation of cryogenic stresses and, on the other hand, by the higher viscosity of the bounded layers as the temperature decreases.

- Traffic loading causes bending of the top layers of the pavement structure. Thus, the properties of the unbound layers, providing the support of the top layers become crucially important. The reduction of the elastic properties in consequence of thawing during spring time results in a significant increase of loading of the top layers.

Thus, based on the presented multiscale analysis framework, the stress state in the pavement structure can be computed for a given asphalt mix design and weather scenario. The risk of low-temperature top-down cracking is estimated by comparing the obtained surface stress with the respective material strength, i.e., the tensile strength of asphalt at the respective temperature. As regards the latter, future work will aim at extending the presented multiscale approach towards upscaling of strength properties, with the developed multiscale model finally giving access to both the viscoelastic and strength properties of asphalt.

Acknowledgements

The first author is grateful for financial support by the Austrian Academy of Sciences via the DOC-FFORTE program. Financial support by the Christian Doppler Gesellschaft (Vienna, Austria) is gratefully acknowledged.

References

1. DIN-18126 (1996). *Baugrund, Untersuchung von Bodenproben – Bestimmung der Dichte nichtbindiger Böden bei lockester und dichtester Lagerung [Soil, investigation and testing – Determination of density of granular soils for minimum and maximum compactness]*. Deutsches Institut für Normung e. V., Berlin, Germany. In German.
2. Ewers, J. and Heukelom, W. (1964). *Die Erhöhung der Viskosität von Bitumen durch die Zugabe von Füller [The increase of bitumen viscosity by allowance of filler]*. Straße und Autobahn, 15(2):31–39. In German.
3. Holl, A. (1971). *Bituminöse Straßen: Technologie und Bauweisen [Flexible pavements: Technology and design]*. Bauverlag, Wiesbaden. In German.
4. Huet, C. (1963). *Étude par une méthode d'impédance du comportement viscoélastique des matériaux hydrocarbonés [Study of the viscoelastic behavior of bituminous mixes by method of impedance]*. PhD thesis, Faculte des Sciences de Paris, Paris. In French.
5. Lackner, R., Blab, R., Jäger, A., Spiegl, M., Kappl, K., Wistuba, M., Gagliano, B., and Eberhardstcincr, J. (2004). *Multiscale modeling as the basis for reliable predictions of the behavior of multi-composed materials.* In Topping, B. and Mota Soares, C., editors, *Progress in Computational Structures Technology*, pages 153 187. Saxc-Coburg Publications, Stirling.

6. Lackner, R., Spiegl, M., Eberhardsteiner, J., and Blab, R. (2005). *Is low-temperature creep of asphalt mastic independent of filler shape and mineralogy? – Arguments from multiscale analysis. Journal of Materials in Civil Engineering (ASCE)*, 17(5):485–491.

7. Mori, T. and Tanaka, K. (1973). *Average stress in matrix and average elastic energy of materials with misfitting inclusions. Acta Metallurgica*, 21:571–574.

8. Olard, F. (2003). *Comportement thermoméchanique des Enrobés bitumineux à basses températures. Relations entre les propriétés du liant et de l'enrobé [Thermomechanical behavior of bituminous mixtures at low temperatures. Relations between characteristics of binder and properties of bituminous mixtures]*. PhD thesis, Ecole Nationale des TPE, Lyon. In French.

9. Sayegh, G. (1965). *Contribution à l'étude des propriétés viscoélastique des bitumes purs at des bétons bitumineux [Contribution of viscoelastic properties of pure bitumen on asphalt concrete]*. PhD thesis, Sorbonne, Paris. In French.

10. SHRP-A-370 (1994). *Binder Characterization and Evaluation. Volume 4: Test Methods*. Technical report, Strategic Highway Research Program, National Research Council, Washington, DC, USA.

11. Taylor, R. (2004). *FEAP – A Finite Element Analysis Program (Version 7.5 User Manual)*. Department of Civil and Environmental Engineering, University of California at Berkeley, California, USA.

12. Wistuba, M., Lackner, R., Spiegl, M., and Blab, B. (2006). *Risk evaluation of surface-initiated cracking in asphalt pavements by means of fundamental laboratory tests and numerical modeling. In International Conference on Asphalt Pavements, 12–17 August 2006, Québec, Canada*.

On the Different Possibilities to Derive Plate and Shell Theories

Holm Altenbach[1] and Johannes Meenen[2]

[1] Lehrstuhl für Technische Mechanik, Zentrum für Ingenieurwissenschaften,
  Martin-Luther-Universität Halle-Wittenberg, D-06099 Halle (Saale), Germany,
  `holm.altenbach@iw.uni-halle.de`
[2] Brabantstraße 10 -18, D-52070 Aachen, Germany,
  `johannes.meenen@t-online.de`

Abstract The plate theory is an old branch of solid mechanics – the first development of a general plate theory was made by Kirchhoff more than 150 years ago. After that many improvements were suggested, at the same time some research was focussed on the establishment of a consistent plate theory. Limiting ourself by the linear elastic case, it will be demonstrated that the von Kármán theory can be deduced from the three-dimensional continuum mechanics equation.

Keywords: homogeneous and inhomogeneous plates, von Kármán theory

1 Introductional Remarks

Thin-walled structures are used for various applications. Their main advantages are a high load bearing capacity, combined with low weight and excellent specific stiffness properties. Modern plate and shell structures are made from different materials – common structural materials like steel or concrete, but also modern composite materials like laminates, or sandwiches.

Increasing safety requirements or the need for optimization in an early development stage has lead again to a strong interest in the mechanical analysis of thin-walled structures. A common modeling approach is to use shell and plate theories, which can be derived by different procedures. In this presentation, a clear classification of structural models will be given and an overview over the modeling approaches for plate theories will be presented. Finally, it is shown how the von Kármán plate theory can be deduced from three-dimensional continuum mechanics in a consistent way.

2 Classification of Two-dimensional Theories for Thin-walled Load Bearing Structures

2.1 Structural models

One of the basic problems in engineering mechanics is the analysis of the strength, the vibration behavior and the stability of structural elements with the help of a structural model. In this context, structural models are special cases or approximations of a general continuum theory, and are intended for the analysis of a certain type of structural element. The structural models can be classified, for example,

- by their suitability for bodies with certain geometrical (spatial) dimensions,
- by their suitability for certain applied loads,
- by the use of kinematical and/or statical hypotheses approximating its mechanical behavior

Structural elements and the structural models for their analysis can be categorized into three classes, depending on the ratio of their characteristic dimensions. The first class is the class of three-dimensional structural elements, which can be defined as follows:

> *A three-dimensional structural element has three spatial dimensions of the same order, no predominant dimension exists.*

Typical examples of geometrically simple, compact structural elements in the theory of elasticity are *cube, prism, cylinder, sphere,* etc.

The second basic class is the class of two-dimensional structural elements which can be defined as follows:

> *Two-dimensional structural elements are bodies, which have two spatial dimensions of comparable size, and a third spatial dimension, the so-called thickness, which is at least one order of magnitude smaller.*

Typical examples of two-dimensional structural elements in civil engineering and structural mechanics are *membrane, disc, plate, shell, folded structure,* etc.

The last class is related to the one-dimensional structural elements which can be defined as follows:

> *Two spatial dimensions, which can be related to the cross-section, have a comparable size. The third dimension, which is related to the length of the structural element, is at least one order of magnitude larger than the size of the cross-section dimensions.*

Typical examples in engineering mechanics are *truss, beam, torsion beam,* etc.

In general, it is possible to introduce other classes. For example, in shipbuilding, thin-walled structural elements are often used. These are thin-walled

light-weight structures with a special profile and they require an extension of
the classical one-dimensional structural models:

> *If the spatial dimensions are of significantly different order and the
> thickness of the profile is small in comparison to the other cross-
> section dimensions, and the cross-section dimensions are much smaller
> in comparison to the length of the structure one can introduce quasi-
> one dimensional structural elements.*

Suitable theories for the analysis of quasi-onedimensional structural elements
are the thin-walled beam theory (Vlasov-Theory) and the semi-membrane
theory or generalized beam theory [5]. Typical thin-walled cross-section profiles are closed cross-section profiles, open cross-section profiles, open-closed
cross-section profiles, etc.

2.2 Theories for two-dimensional structures

Since the characteristic length in thickness direction is much smaller than the
characteristic length in the surface direction, for two-dimensional structures it
is tempting to look for procedures that eliminate the thickness dimension. The
mathematical consequence is obvious – instead of a three-dimensional problem, which is represented by a system of coupled partial differential equations
with respect to three spatial coordinates, one can analyze a two-dimensional
problem, which is described with respect to two spatial coordinates. These coordinates represent a surface in three-dimensional space, and a procedure has
to be developed that maps the real behavior in thickness direction onto the
mechanical behavior of the surface. The transition from the three-dimensional
to the two-dimensional problem is non-trivial, but once a two-dimensional
theory has been obtained, the solution effort decreases significantly and the
possibilities to solve problems analytically are increased.

During the last years many scientific papers, textbooks, monographs and
proceedings on the state of the art and recent developments in the plate
theories were published. Without any comments some of the most important
publications will be listed here:

- Reviews articles [9, 14, 15, 18, 31–36, 43, 48] among others
- Monographs and textbooks [2, 6, 11, 12, 19, 38, 44, 52, 54, 56, 60], etc.
- Proceedings: EUROMECH 444 [23], Shell Structures Theory & Applica-
  tions [39]

All of these approaches can be classified into two basic techniques: The reduction technique, which starts from the equations of 3D continuum and de-
velops approximate 2D continuum theories; and the direct approach, which
starts from a rigorous 2D continuum theory and maps the behavior of a three-
dimensional body onto this 2D continuum theory [3, 57]. If one starts from
the 3D continuum theory, common techniques are: the use of hypotheses to

approximate the three-dimensional equations (e.g. by introducing these hypotheses into the principle of virtual displacements), or the use of mathematical approaches, such as series expansions, special functions or asymptotic expansions.

All these approaches have their own advantages and disadvantages, and it is difficult to argue what is the best method for deriving a plate or shell theory. Additionally, in many cases different derivation methods result in identical sets of governing equations.

Theories which are based on hypotheses are preferred by engineers. For example, there is a huge number of theories which are based on displacement approximations. The three displacements in the classical three-dimensional continuum are split into in-plane displacements and transverse deflection. The first theory of plates, which is based on displacement assumptions, was presented by Kirchhoff [24], and was seriously improved about 100 years later (see, for example, Hencky [17] and Mindlin [30]). Kirchhoff's plate theory does not account for transverse shear strain and thickness change. In the improved theories, additional degrees of freedom (cross-section rotations) were introduced, so that transverse shear is considered in an approximate sense.

The introduction of independent rotations is in some cases not enough, since it is assumed that any cross-section will be plane before and after deformation. For example, in the case of plates made from rubber-like materials the assumption of plane cross-section is not valid. A weaker assumption was proposed by [26,42] among others. These refined theories can be understood as theories that introduce additional degrees of freedom, or as some part of a power series expansion. The first suggestion of this type was done by [27]. A generalization of the power series approach was given in [29,51].

The method of hypotheses considering assumptions for the stress and/or the strain (displacement) states was also applied in [7,8,17,25,26,30,42,45–47]. It can be shown that, for example, Mindlin's and Reissner's theories contain partly identical equations, but that the coefficients take slightly different values and that their physical interpretation is not the same.

Pure mathematical approaches are mostly based on power series, trigonometric functions, on special functions, etc. (see, e.g., [1,21,27,28,41,53,55]). The mathematical approaches are very helpful if one wants to check the accuracy of the given approximation. A nice comparison of the different approximations in the series approach is given in [22].

The direct approach is based on the a priori introduction of a two-dimensional deformable surface. This approach was applied by [13,16,31,37,49,50,58–60], etc. The main advantage of these theories is that their derivation does not rely on assumptions or series expansions and is mathematically and physically as strong and exact as the three-dimensional continuum mechanics. This approach is still under discussion, since the application is not trivial, and a relationship between the constitutive laws of the two-dimensional surface and the corresponding three-dimensional body has to be found.

3 Derivation from Three-dimensional Continuum Mechanics

Below let us derive a plate theory from the equations of three-dimensional nonlinear continuum mechanics. As an example, we discuss the von Kármán plate equations because they are, from the engineering point of view, "in between the actual and the reference configuration". In contrast to the classical Kirchhoff theory the equilibrium equations of the von Kármán theory are formulated for the actual configuration. The basics of the von Kármán theory were presented in an original paper of von Kármán [20], some discussions on the possibilities of this theory are included in [10] or [40]. For the sake of simplicity we are introducing a kinematical assumption which is adequate to the Kirchhoff's kinematics.

3.1 Kinematics

For moderate rotations, classical and refined plate theories can be derived from the nonlinear equations of 3D continuum mechanics in a consistent way. A special case of the plate theory (see [29]) can be developed with the kinematic assumption

$$\bar{u}_\alpha = \bar{u}_\alpha^0(X_\alpha) - \bar{u}_{3,\alpha}^0(X_\alpha)X_3$$
$$\bar{u}_3 = \bar{u}_3^0(X_\alpha)$$

which corresponds to the displacement field of the von Kirchhoff plate theory. To show this, we start from a given midsurface, which is described in the reference configuration by its position vector $\mathbf{X}^0$. The corresponding normal $\mathbf{N}$ can be defined by the conditions $\mathbf{N} \cdot \mathbf{N} = 1$ and $\mathbf{N} \cdot d\mathbf{X}^0 = 0$. We furthermore assume that the deformation of the midsurface into the current configuration can be described by a deformation gradient $\mathbf{F}^0 = (\partial \mathbf{x}^0/\partial \mathbf{X}^0)^{\mathrm{T}}$ and its polar decomposition $\mathbf{F}^0 = \mathbf{R}^0 \cdot \mathbf{U}^0$, where $\mathbf{R}^0$ is the rotation of the midplane and $\mathbf{U}^0$ its right stretch tensor

$$dx^0 = \mathbf{F}^0 \cdot d\mathbf{X}^0 = \mathbf{F}^0 \cdot (\mathbf{I} - \mathbf{NN}) \cdot d\mathbf{X}$$

From the second part of this equation, it can be seen that the components $\mathbf{F}^0 \cdot (\mathbf{NN})$ of the deformation gradient are not determined by $dx^0/d\mathbf{X}$, and we are free to introduce an additional vector $\mathbf{n} = \mathbf{F}^0 \cdot \mathbf{N}$. It can be shown that this vector is normal to the deformed midplane ($\mathbf{n} \cdot dx^0 = 0$) if the condition

$$\mathbf{N} \cdot (\mathbf{F}^0)^{\mathrm{T}} \cdot \mathbf{F}^0 \cdot (\mathbf{I} - \mathbf{NN}) = \mathbf{N} \cdot (\mathbf{U}^0)^2 \cdot (\mathbf{I} - \mathbf{NN}) = 0$$

is fulfilled, and $\mathbf{n} \cdot \mathbf{n} = 1$ leads to

$$\mathbf{N} \cdot (\mathbf{F}^0)^{\mathrm{T}} \cdot \mathbf{F}^0 \cdot \mathbf{N} = \mathbf{N} \cdot (\mathbf{U}^0)^2 \cdot \mathbf{N} = 1$$

A straightforward calculation shows that with these conditions, the polar decomposition becomes

$$\mathbf{U}^0 = \{[\mathbf{F}^0 \cdot (\mathbf{I} - \mathbf{N}\mathbf{N})]^{\mathrm{T}} \cdot [\mathbf{F}^0 \cdot (\mathbf{I} - \mathbf{N}\mathbf{N})]\}^{(1/2)} + (\mathbf{N}\mathbf{N})$$
$$\mathbf{R}^0 = \mathbf{F}^0 \cdot (\mathbf{I} - \mathbf{N}\mathbf{N}) \cdot \{[\mathbf{F}^0 \cdot (\mathbf{I} - \mathbf{N}\mathbf{N})]^{\mathrm{T}} \cdot [\mathbf{F}^0 \cdot (\mathbf{I} - \mathbf{N}\mathbf{N})]\}^{-(1/2)} + \mathbf{n}\mathbf{N}$$

This deformation gradient can used to describe a von Kirchhoff deformation of a body with

$$\mathbf{X} = \mathbf{X}^0(X_1, X_2) + \mathbf{N}(X_1, X_2)X_3$$
$$\mathbf{x} = \mathbf{x}^0(X_1, X_2) + \mathbf{n}(X_1, X_2)X_3$$

For small rotations and small membrane deformations, we have

$$R_{11} = R_{22} = R_{33} \approx 1, R_{13} = -R_{31} = \phi_2, R_{23} = -R_{32} = -\phi_1,$$
$$U_{11}, U_{22} \approx 1, U_{12}, U_{21} \ll 1$$

It is straightforward to show that the deformation gradient of the body is

$$F_{i\alpha}(X_1, X_2, X_3) = R_{ik}U_{k\alpha} + (R_{ik}U_{kj}N_j)_{,\alpha}X_3,$$
$$F_{i3}(X_1, X_2, X_3) = n_i(X_1, X_2)$$

Introducing these approximations into $\mathbf{F}^0 = \mathbf{R}^0 \cdot \mathbf{U}^0$ leads to a deformation gradient, which can be integrated to give the assumed kinematic assumption.

3.2 Stresses and stress resultants

In all what follows, we assume that the displacements are small and that the rotations of the plate normal are moderate. The first Piola-Kirchhoff-tensor Σ_{ij}^I is approximated by

$$\Sigma_{i\alpha}^I = \Sigma_{i\alpha} + \Sigma_{ik}\frac{\partial u_\alpha}{\partial X_k} \approx \Sigma_{i\alpha},$$

$$\Sigma_{i3}^I = \Sigma_{i3} + \Sigma_{ik}\frac{\partial u_3}{\partial X_k} \approx \Sigma_{i3} + \Sigma_{i\alpha}\frac{\partial u_3}{\partial X_\alpha},$$

and the Green-Lagrange strain tensor G_{ij} can be approximated by

$$G_{\alpha\beta} = \frac{1}{2}(u_{\alpha,\beta} + u_{\beta,\alpha} + u_{k,\alpha}u_{k,\beta}) \approx \frac{1}{2}(u_{\alpha,\beta} + u_{\beta,\alpha} + u_{3,\alpha}u_{3,\beta}),$$

$$G_{\alpha 3} = \frac{1}{2}(u_{\alpha,3} + u_{3,\alpha} + u_{k,\alpha}u_{k,3}) \approx \frac{1}{2}(u_{\alpha,3} + u_{3,\alpha}),$$

$$G_{33} = u_{3,3} + \frac{1}{2}u_{k,3}u_{k,3} \approx u_{3,3}$$

these approximations will be termed as the von Kármán stress and strain tensors. In all equations, Einstein's summation convention is applied. Greek indices take the values 1 and 2 and represent the in-plane directions of the plate, whereas lower arabic indices run from 1 to 3.

These kinematical assumptions are introduced into the principle of virtual displacements in the reference configuration. The stress resultants are calculated from the second Piola-Kirchhoff stress tensor by

$$N_{\alpha\beta}^q := \int_{-h/2}^{h/2} \Sigma_{\alpha\beta} X_3^q \mathrm{d}X_3, \qquad Q_\alpha := \int_{-h/2}^{h/2} \Sigma_{\alpha 3} \mathrm{d}X_3$$

which leads to the equations of equilibrium

$$
\begin{aligned}
N_{\alpha\beta,\beta}^0 &= -F_\alpha^0 - P_\alpha^0, \\
N_{\alpha\beta,\alpha\beta}^1 - (N_{\beta\alpha}^0 \bar{u}_{3,\alpha}^0)_{,\beta} &= -F_3^0 + F_{\alpha,\alpha}^1 - P_3^0 + P_{\alpha,\alpha}^1, \\
N_{\alpha\beta,\beta}^1 + Q_\alpha &= -F_\alpha^1 - P_\alpha^1
\end{aligned}
$$

In these equations, P_α^0, P_α^1 and P_3^0 are calculated from the loads p_i on top and bottom of the plate

$$P_\alpha^0 := \Big[p_\alpha\Big]_{X_3=-h/2}^{h/2}, \qquad P_\alpha^1 := \Big[p_\alpha X_3\Big]_{X_3=-h/2}^{h/2}, \qquad P_3^0 := \Big[p_3\Big]_{X_3=-h/2}^{h/2}$$

and F_α^0, F_α^1 and F_3^0 are resultants of the volume loads ρf_i

$$F_\alpha^0 := \int_{-h/2}^{h/2} \rho f_\alpha \mathrm{d}X_3, \qquad F_\alpha^1 := \int_{-h/2}^{h/2} \rho f_\alpha X_3 \mathrm{d}X_3, \qquad F_3^0 := \int_{-h/2}^{h/2} \rho f_3 \mathrm{d}X_3$$

3.3 Constitutive hypothesis

For small elastic deformations, a linear relationship between the second Piola-Kirchhoff stress tensor and the Green-Lagrange strain tensor can be formulated

$$\Sigma_{ij} = C_{ijkl} G_{kl}$$

Approximating the Green-Lagrange strain by the von Kármán strain tensor leads to

$$
\begin{aligned}
G_{\alpha\beta} &= \epsilon_{\alpha\beta} - \kappa_{\alpha\beta} x_3, \\
G_{\alpha 3} &= 0, \\
G_{33} &= 0
\end{aligned}
$$

with the membrane strain and bending curvature

$$
\begin{aligned}
\epsilon_{\alpha\beta} &= \frac{1}{2}(\bar{u}_{\alpha,\beta}^0 + \bar{u}_{\beta,\alpha}^0) + \bar{u}_{3,\alpha}^0 \bar{u}_{3,\beta}^0, \\
\kappa_{\alpha\beta} &= \bar{u}_{3,\alpha\beta}^0
\end{aligned}
$$

This strain field leads to the constitutive plate equations

$$N^0_{\alpha\beta} = \mathcal{A}_{\alpha\beta\gamma\delta}\epsilon_{\gamma\delta} + \mathcal{B}_{\alpha\beta\gamma\delta}\kappa_{\gamma\delta},$$
$$N^1_{\alpha\beta} = \mathcal{B}_{\alpha\beta\gamma\delta}\epsilon_{\gamma\delta} + \mathcal{D}_{\alpha\beta\gamma\delta}\kappa_{\gamma\delta}$$

with the stiffness tensors $\mathcal{A}_{\alpha\beta\gamma\delta}$ (in-plane or membrane stiffness tensor), $\mathcal{B}_{\alpha\beta\gamma\delta}$ (coupled stiffness tensor) and $\mathcal{D}_{\alpha\beta\gamma\delta}$ (out-of-plane stiffness tensor)

$$\mathcal{A}_{\alpha\beta\gamma\delta} = \int\limits_{-h/2}^{h/2} C^*_{\alpha\beta\gamma\delta}dx_3,$$

$$\mathcal{B}_{\alpha\beta\gamma\delta} = \int\limits_{-h/2}^{h/2} C^*_{\alpha\beta\gamma\delta}x_3 dx_3,$$

$$\mathcal{D}_{\alpha\beta\gamma\delta} = \int\limits_{-h/2}^{h/2} C^*_{\alpha\beta\gamma\delta}x_3^2 dx_3$$

$C^*_{\alpha\beta\gamma\delta}$ is the reduced Hookean tensor which should be calculated similar to the classical laminate theory [4]. Like in the classical theory a constitutive equation for the transverse shear forces cannot be established. In this case these forces should be determined by the equilibrium equations.

4 Summary and Further Developments

The following main conclusion from the present studies can be made:

- A consistent von Kármán theory can be derived from three-dimensional nonlinear continuum mechanics.
- The reduction of the equations results in some constraints which are discussed, for example, in [29].
- Starting with the simplest kinematics (Kirchhoff) any extension can be expressed by power series.
- The physical interpretation of higher order terms is more and more difficult when the number of power series terms is increasing.

The next developments should be focussed on the consideration of inelastic material behavior.

References

1. L.A. Agalovyan. *Asymptotic Theory of Anisotropic Plates and Shells (in Russ.)*. Nauka, Moscow, 1997.
2. H. Altenbach, J. Altenbach, and K. Naumenko. *Ebene Flächentragwerke. Grundlagen der Modellierung und Berechnung von Scheiben und Platten.* Springer-Verlag, Berlin, 1998.

3. H. Altenbach. On the determination of transverse shear stiffnesses of orthotropic plates. *ZAMP*, 51:629–649, 2000.

4. H. Altenbach, J. Altenbach, and W. Kissing. *Mechanics of Composite Structural Elements*. Springer-Verlag, Berlin, 2004.

5. J. Altenbach, W. Kissing, and H. Altenbach. *Dünnwandige Stab- und Stabschalentragwerke*. Vieweg, Braunschweig/Wiesbaden, 1994.

6. S.A. Ambarcumjan. *Theory of Anisotropic Plates: Strength, Stability and Vibrations*. Hemisphere, New York, 2nd edition, 1991.

7. L. Bollé. Contribution au problème linéaire de flexin d'une plaque élastique. *Bull. Techn. Suisse Romande*, 73(21):281–285, 1947.

8. L. Bollé. Contribution au problème linéaire de flexin d'une plaque élastique. *Bull. Techn. Suisse Romande*, 73(22):293–298, 1947.

9. W.S. Burton and A.K. Noor. Assessment of computational models for sandwich panels and shells. *Comput. Meth. Appl. Mech. Eng.*, 124(1–2):125–151, 1995.

10. P.G. Ciarlet. *Plates and Junctions in Elastic Multi-Structures: An Asymptotic Analysis*. Volume 14 of Collection Recherches en Mathématiques Appliquées. Masson, Paris, 1990.

11. K. Girkmann. *Flächentragwerke*. Springer, Wien, 6th edition, 1986.

12. P.L. Gould. *Analysis of Shells and Plates*. Springer, New York et al., 1988.

13. A.E. Green, P.M. Naghdi, and W.L. Waniwright. A general theory of Cosserat surface. *Arch. Rat. Mech. Anal.*, 20:287–308, 1965.

14. E.I. Grigolyuk and A.F. Kogan. Present state of the theory of multilayered shells (in Russ.). *Prikl. Mekh.*, 8(6):3–17, 1972.

15. E.I. Grigolyuk and I.T. Seleznev. Nonclassical theories of vibration of beams, plates and shelles (in Russ.). In *Itogi nauki i tekhniki*, volume 5 of *Mekhanika tverdogo deformiruemogo tela*, VINITI, Moskva, 1973.

16. W. Günther. Analoge Systeme von Schalengleichungen. *Ingenieur–Archiv*, 30:160–188, 1961.

17. H. Hencky. Über die Berücksichtigung der Schubverzerrung in ebenen Platten. *Ingenieur–Archiv*, 16:72–76, 1947.

18. H. Irschik. On vibrations of layered beams and plates. *ZAMM*, 73:T34–T45, 1993.

19. Z. Kączkowski. *Płyty obliczenia statyczne*. Arkady, Warszawa, 1980.

20. T. von Kármán. Festigkeitsprobleme in Maschinenbau. In F. Klein and C. Müller, editors, *Encyclopädie der Matematisches Wissenschaften*, volume IV/4, pages 311–385. Teubner, Leipzig, 1910.

21. R. Kienzler. Erweiterung der klassischen Schalentheorie; der Einfluß von Dickenverzerrung und Querschnittsverwölbungen. *Ingenieur-Archiv*, 52:311–322, 1982.

22. R. Kienzler. On the consistent plate theories. *Arch. Appl. Mech.*, 72:229–247, 2002.

23. R. Kienzler, H. Altenbach, and I. Ott, editors. *Critical Review of the Theories of Plates and Shells, New Applications*, volume 16 of *Lect. Notes Appl. Comp. Mech.*, Springer, Berlin, 2004.

24. G.R. Kirchhoff. Über das Gleichgewicht und die Bewegung einer elastischen Scheibe. *Crelles J. für die reine und angewandte Mathematik*, 40:51–88, 1850.

25. A. Kromm. Verallgemeinerte Theorie der Plattenstatik. *Ingenieur-Archiv*, 21:266–286, 1953.

26. M. Levinson. An accurate, simple theory of the statics and dynamics of elastic plates. *Mech. Res. Commun.*, 7(6):343–350, 1980.

27. K.H. Lo, R.M. Christensen, and E.M. Wu. A high-order theory of plate deformation. Part I: Homogeneous plates. *Trans. ASME. J. Appl. Mech.*, 44(4):663–668, 1977.

28. K.H. Lo, R.M. Christensen, and E.M. Wu. A high-order theory of plate deformation. Part II: Laminated plates. *Trans. ASME. J. Appl. Mech.*, 44(4):669–676, 1977.

29. J. Meenen and H. Altenbach. A consistent deduction of von Kármán-type plate theories from threedimensional non-linear continuum mechanics. *Acta Mech.*, 147:1–17, 2001.

30. R.D. Mindlin. Influence of rotatory inertia and shear on flexural motions of isotropic, elastic plates. *Trans. ASME. J. Appl. Mech.*, 18:31–38, 1951.

31. P. Naghdi. The theory of plates and shells. In S. Flügge, editor, *Handbuch der Physik*, volume VIa/2, pages 425–640. Springer, Heidelberg, 1972.

32. A.K. Noor and W.S. Burton. Assessment of shear deformation theories for multilayered composite plates. *Appl. Mech. Rev.*, 42(1):1–13, Jan. 1989.

33. A.K. Noor and W.S. Burton. Stress and free vibration analysis of multilayered composite plates. *Compos. Struct.*, 11(3):183–204, 1989.

34. A.K. Noor and W.S. Burton. Assessment of computational models for multilayered anisotropic plates. *Compos. Struct.*, 14(3):233–265, 1990.

35. A.K. Noor and W.S. Burton. Assessment of computational models for multilayered composite shells. *Appl. Mech. Rev.*, 43(4):67–96, 1990.

36. A.K. Noor, W.S. Burton, and C.W. Bert. Computational models for sandwich panels and shells. *Appl. Mech. Rev.*, 49(3):155–199, 1996.

37. W.A. Palmow and H. Altenbach. Über eine Cosseratsche Theorie für elastische Platten. *Tech. Me.*, 3(3):5–9, 1982.

38. V. Panc. *Theories of Elastic Plates*. Noordhoff, Leyden, 1975.

39. W. Pietraszkiewicz and C. Szymczak, editors. *Shell Structures – Theory and Application*, London, Taylor & Francis/Balkema, 2005.

40. P. Podio-Guidugli. A new quasilinear model for plate buckling. *J. Elasticity*, 71:157–182, 2003.

41. G. Preußer. Eine systematische Herleitung verbesserter Plattentheorien. *Ingenieur-Archiv*, 54:51–61, 1984.

42. J.N. Reddy. A simple higher–order theory for laminated composite plates. *Trans. ASME. J. Appl. Mech.*, 51:745–752, 1984.

43. J.N. Reddy. A general non-linear third order theory of plates with transverse deformations. *J. Non-linear Mech.*, 25(6):667–686, 1990.

44. J.N. Reddy. *Mechanics of Laminated Composite Plates: Theory and Analysis*. CRC Press, Boca Raton, 1996.

45. E. Reissner. On the theory of bending of elastic plates. *J. Math. Phys.*, 23:184–194, 1944.

46. E. Reissner. The effect of transverse shear deformation on the bending of elastic plates. *J. Appl. Mech.*, 12(11):A69–A77, 1945.

47. E. Reissner. On bending of elastic plates. *Q. Appl. Math.*, 5:55–68, 1947.

48. E. Reissner. Reflection on the theory of elastic plates. *Appl. Mech. Rev.*, 38(11):1453–1464, 1985.

49. H. Rothert. Direkte Theorie von Linien- und Flächentragwerken bei viskoelastischen Werkstoffverhalten. Techn.-Wiss. Mitteilungen des Instituts für Konstruktiven Ingenieurbaus 73-2, Ruhr-Universität, Bochum, 1973.

50. M.B. Rubin. *Cosserat Theories: Shells, Rods and Points*, volume 79 of *Solid Mechanics and Its Applications*. Springer, Berlin, 2000.

51. K.P. Soldatos and P. Watson. A general theory for the accurate stress analysis of homogeneous and laminated composite beams. *Int. J. Solids Struct.*, 34(22):2857–2885, 1997.

52. S.P. Timoshenko and S. Woinowsky-Krieger. *Theory of Plates and Shells*. Mc-Graw Hill, New York, 1985.

53. M. Touratier. An efficient standard plate theory. *Int. J. Eng. Sci.*, 29(8):901–916, 1991. / by Tamaz S. Vashakmadze. - Dordrecht [u.a.] : 1999 Schriftenreihe: Mathematics and its applications; 476.

54. T.S Vashakmadze. *The Theory of Anisotropic Elastic Plates*. volume 476 of Mathematics and its Applications. Kluwer, Dordrecht, 1999.

55. I. Vekua. *Shell Theory: General Methods of Construction*. Pitman, Boston, 1985.

56. C. Woźniak. *Mechanik sprężystych płyt i powłok*, volume VIII of *Mechanika techniczna*. Wydawnictwo Naukowe PWN, Warszawa, 2001.

57. W. Wunderlich. Vergleich verschiedener Approximationen der Theorie dünner Schalen (mit numerischen Beispielen). Techn.-Wiss. Mitt. 73-1, Institut für Konstruktiven Ingenieurbau der Ruhr-Iniversität Bochum, 1973.

58. P.A. Zhilin. Mechanics of deformable directed surfaces. *Int. J. Solids Struct.*, 12:635–648, 1976.

59. P.A. Zhilin. Basic equations of non-classical theory of shells (in Russ.). In *Trudy LPI (Trans. Leningrad Polytechnical Institute) – Dinamika i prochnost mashin (Dynamics and strength of machines)*, Nr. 386, pages 29–46. Leningrad Polytechnical Institute, 1982.

60. P.A. Zhilin. *Applied Mechanics. Foundations of the Theory of Shells (in Russ.)*. St. Petersburg State Polytechnical University, 2007.

The Determination of Linear Frequencies of Bending Vibrations of Ferromagnetic Shell by Exact Space Treatment

Bagdoev A.G., Vardanyan A.V., and Vardanyan S.V.

Institute of Mechanics of Armenian National Academy of Sciences

Abstract The analytical and numerical solutions for the frequencies of free bending vibrations of ferromagnetic cylindrical shells in an axial magnetic field are obtained by the space method. The comparison with a solution under Kirchhoff hypothesis is carried out.

Keywords: 3D method, ferromagnetic shell, free vibrations frequencies

Linear bending vibrations of ferromagnetic cylindrical shells in an axial magnetic field are considered. The bending vibrations of magnetoelastic plates and shells by averaged treatment based on classical theory are considered in [1–4], and vibrations of ferromagnetic plates and shells are considered in [5]. By the new space treatment, at first developed for elastic plates in [6], the magnetoelastic vibrations of plates are considered in [7–9].

In the present paper by the space treatment the frequencies of free bending vibrations of ferromagnetic cylindrical shells are obtained analytically and numerically.

Let the infinite cylindrical shell is posed in axial initial magnetic field H_0 (see Fig. 1).

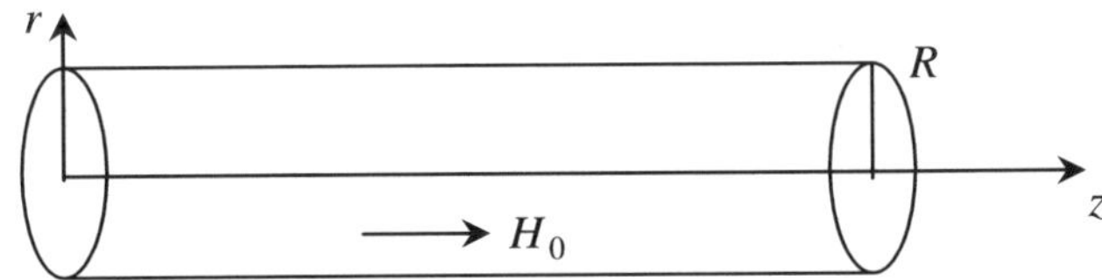

Fig. 1 Infinite cylindrical shell in an axial initial magnetic field

G. Jaiani, P. Podio-Guidugli (eds.), *IUTAM Symposium on Relations of Shell, Plate, Beam, and 3D Models,* © Springer Science+Business Media B.V. 2008

The equations of motion of ferromagnetic media for the case of axial symmetry are given as follows [10], [1]

$$a^2 \frac{\partial^2 u_r}{\partial r^2} + b^2 \frac{\partial^2 u_r}{\partial z^2} + (a^2 - b^2)\frac{\partial^2 u_z}{\partial r \partial z} + \frac{a^2}{r}\frac{\partial u_r}{\partial r} - \frac{a^2}{r^2}u_r$$

$$= \frac{\partial^2 u_r}{\partial t^2} - \frac{H_0 \mu_0}{\rho}\left(\frac{\partial h_r}{\partial z} - \frac{\partial h_z}{\partial r}\right) - H_0\mu_0\chi\frac{\partial h_r}{\partial z}, \tag{1}$$

$$b^2 \frac{\partial^2 u_z}{\partial r^2} + a^2 \frac{\partial^2 u_z}{\partial z^2} + (a^2 - b^2)\frac{\partial^2 u_z}{\partial r \partial z} + \frac{b^2}{r}\frac{\partial u_z}{\partial r} + \frac{a^2 - b^2}{r^2}\frac{\partial u_r}{\partial z}$$

$$= \frac{\partial^2 u_z}{\partial t^2} - H_0\mu_0\chi\frac{\partial h_z}{\partial z},$$

where $\chi = \mu - 1$, μ is the permeability of the magnetic field. Here b, $a-$ speed of elastic wave, magnetic field vector $\overline{H} = \overline{H}_0 + \overline{h}$, $\overline{H}_0 = (\overline{H}_0, 0)$ is the undisturbed field, $\overline{h} = (h_r, h_z)$ is the small disturbed magnetic field vector, (u_r, u_z) is the displacement vector ρ is density of shell.

In (1) the relations for the Lorent's force are taken into account:

$$\left(rot\,\overline{h}x\overline{H}_0\right)_r = \frac{\partial h_r}{\partial z} - \frac{\partial h_z}{\partial r},$$

$$\left(rot\,\overline{h}x\overline{H}_0\right)_z = 0. \tag{2}$$

Equations (1) can be rewritten as follows

$$a_1^2 = \frac{\mu_0 H_0^2}{\rho}, \quad \zeta = 1 - \frac{b^2}{a^2},$$

$$\frac{\partial^2 u_r}{\partial r^2} + \frac{b^2}{a^2}\frac{\partial^2 u_r}{\partial z^2} + \zeta\frac{\partial^2 u_z}{\partial r \partial z} + \frac{1}{r}\frac{\partial u_r}{\partial r} - \frac{1}{r^2}u_r$$

$$= \frac{1}{a^2}\frac{\partial^2 u_r}{\partial t^2} - \frac{a_1^2}{a^2}\left(\frac{\partial h_r}{\partial z} - \frac{\partial h_z}{\partial r}\right) - \frac{\chi}{H_0}\frac{a_1^2}{a^2}\frac{\partial h_r}{\partial z}, \tag{3}$$

$$\frac{b^2}{a^2}\frac{\partial^2 u_z}{\partial r^2} + \frac{\partial^2 u_z}{\partial z^2} + \zeta\frac{\partial^2 u_z}{\partial r \partial z} + \frac{b^2}{a^2 r}\frac{\partial u_z}{\partial r} + \frac{\zeta}{r}\frac{\partial u_r}{\partial z}$$

$$= \frac{1}{a^2}\frac{\partial^2 u_z}{\partial t^2} - \frac{\chi}{H_0}\frac{a_1^2}{a^2}\frac{\partial h_z}{\partial z},$$

The equation of electromagnetic induction is

$$\frac{\partial \overline{h}}{\partial t} = rot(\overline{v}x\overline{H}_0) + \nu_m \Delta\overline{h} \tag{4}$$

where $\overline{v} = \frac{\partial \overline{u}}{\partial t}$ is the vector of particles velocity, $\nu_m = \frac{c^2}{\mu_0\mu\sigma}$ is the magnetic viscosity, σ is the electroconductivity. On account $\Delta\overline{h} = \left(\Delta\overline{h}_r - \frac{h_r}{r^2}\right)\overline{e}_r + \Delta\overline{h}_z\overline{e}_z$, where $\overline{e}_r$, $\overline{e}_z$ are unit vectors along the axes r, z, $\Delta\varphi = \frac{1}{r}\frac{\partial}{\partial r}\left(r\frac{\partial \varphi}{\partial r}\right) + \frac{\partial^2 \varphi}{\partial z^2}$, the projection of (4) on the r, z axes yields the following equations

$$\frac{\partial h_r}{\partial t} = H_0 \frac{\partial^2 u_r}{\partial t \partial z} + \nu_m \left(\frac{\partial^2 h_r}{\partial r^2} + \frac{1}{r} \frac{\partial h_r}{\partial r} + \frac{\partial^2 h_r}{\partial z^2} - \frac{h_r}{r^2} \right)$$

$$\frac{\partial h_z}{\partial t} = -H_0 \left(\frac{\partial^2 u_r}{\partial t \partial r} + \frac{1}{r} \frac{\partial u_r}{\partial t} \right) + \nu_m \left(\frac{\partial^2 h_z}{\partial r^2} + \frac{1}{r} \frac{\partial h_z}{\partial r} + \frac{\partial^2 h_z}{\partial z^2} \right) \qquad (5)$$

Let us look for the solution of (3), (5) in form of a plane wave propagating along the direction of the z axis:

$$\xi_j = r\nu_j, \quad j = 1, 2, 3$$
$$u_r = A_j I_1(\xi_j) e^{-i\omega t + ikz} + A'_j K_1(\xi_j) e^{-i\omega t + ikz} + c.c.,$$
$$u_z = B_j I_0(\xi_j) e^{-i\omega t + ikz} + B'_j K_0(\xi_j) e^{-i\omega t + ikz} + c.c. \qquad (6)$$
$$h_z = C_j H_0 I_0(\xi_j) e^{-i\omega t + ikz} + C'_j H_0 K_0(\xi_j) e^{-i\omega t + ikz} + c.c.,$$
$$h_r = D_j H_0 I_1(\xi_j) e^{-i\omega t + ikz} + D'_j H_0 K_1(\xi_j) e^{-i\omega t + ikz} + c.c.,$$

where $I_{0,1}(\xi_j)$, $K_{0,1}(\xi_j)$ are Bessel functions of imaginary argument and j refers to a summation from 1 to 3.

On account of the relations

$$I'_0(\xi) = I_1(\xi), \quad K'_0(\xi) = -K_1(\xi),$$
$$\frac{dI_1(\xi)}{d\xi} + \frac{1}{\xi} I_1(\xi) = I_0(\xi), \quad \frac{dK_1(\xi)}{d\xi} + \frac{1}{\xi} K_1(\xi) = -K_0(\xi) \qquad (7)$$

one can obtained from (3), (5), (6):

$$A_j \left(\nu_j^2 - \frac{b^2}{a^2} k^2 + \frac{\omega^2}{a^2} \right) + \zeta i k \nu_j B_j = \frac{a_1^2}{a^2} (\nu_j C_j - ik D_j) - ik \frac{a_1^2}{a^2} D_j \chi,$$

$$\left(\frac{b^2}{a^2} \nu_j^2 - k^2 + \frac{\omega^2}{a^2} \right) B_j + \zeta i k \nu_j A_j = -ik \frac{a_1^2}{a^2} C_j \chi,$$

$$C_j = \frac{i\omega \nu_j}{\chi_j} A_j, \quad D_j = \frac{\omega k A_j}{\chi_j}, \quad \chi_j = -i\omega + \nu_m k^2 - \nu_m \nu_j^2, \qquad (8)$$

$$A'_j \left(\nu_j^2 - \frac{b^2}{a^2} k^2 + \frac{\omega^2}{a^2} \right) - \zeta i k \nu_j B'_j = \frac{a_1^2}{a^2} (-\nu_j C'_j - ik D'_j) - ik \frac{a_1^2}{a^2} D'_j \chi,$$

$$\left(\frac{b^2}{a^2} \nu_j^2 - k^2 + \frac{\omega^2}{a^2} \right) B'_j - \zeta i k \nu_j A'_j = -ik \frac{a_1^2}{a^2} C'_j \chi, \quad C'_j = -\frac{i\omega \nu_j}{\chi_j} A'_j, \quad D'_j = \frac{\omega k A'_j}{\chi_j},$$

where us summation is carried out over j. Thus the connections between A'_j, $-C'_j$, D'_j with $-B'_j$ are the same as the relations of A_j, C_j, D_j with respect to B_j and the final equation for $\bar{\nu} = \nu_j$ is the same for both cases:

$$\left(\bar{\nu}^2 - \frac{b^2}{a^2} k^2 + \frac{\omega^2}{a^2} + \frac{a_1^2}{a^2} \frac{\bar{\nu}^2 - k^2(1 + \chi)}{1 - \frac{k^2 - \bar{\nu}^2}{\theta}} \right) \left(\frac{b^2}{a^2} \bar{\nu}^2 - k^2 + \frac{\omega^2}{a^2} \right)$$

$$+ \zeta^2 k^2 \bar{\nu}^2 - \zeta \chi \frac{a_1^2}{a^2} \frac{k^2 \bar{\nu}^2}{1 - \frac{k^2 - \bar{\nu}^2}{\theta}} = 0, \quad \theta = \frac{i\omega}{\nu_m}, \quad \chi_j = -i\omega \left(1 - \frac{k^2 - \bar{\nu}^2}{\theta} \right). \qquad (9)$$

The obtained equation for $\bar{\nu}^2$ has three roots and coincides with the equation for plates in longitudinal field reported in [11]. For small $\frac{a_1^2}{a^2}$ and bounded ν_m one can obtain solutions of (9) of the following form [11]:

$$1 - \frac{k^2 - \nu_3^2}{\theta} = -\frac{a_1^2}{a^2}\left(1 + \frac{a^2}{b^2}\frac{\zeta + \chi}{\theta}k^2\right), \tag{10}$$

$$\nu_1^2 = k^2 - \frac{\omega^2}{a^2} - \frac{a_1^2}{a^2}\left(k^2 - \frac{\omega^2}{a^2}\right)\left(1 + \frac{\omega^2}{a^2\theta}\right), \tag{11}$$

$$\nu_2^2 = k^2 - \frac{\omega^2}{b^2} + \frac{k^2 a_1^2}{b^2}(1 + \chi)\left(1\frac{\omega^2}{b^2\theta}\right). \tag{12}$$

In order to obtain the dispersion relation for bending waves in the shell one must specify boundary conditions both on internal and external shell surfaces: $r = R - h$, $r = R + h$ [12], and for longitudinal magnetic field yield conditions for the normal and tangential stresses component [11]

$$\sigma_{rr} = 0, \quad \sigma_{rz} = 0. \tag{13}$$

Using Hooke's law, one obtains from (13)

$$a^2\frac{\partial u_r}{\partial r} + (a^2 - 2b^2)\left(\frac{u_r}{r} + \frac{\partial u_z}{\partial z}\right) = 0, \quad \frac{\partial u_r}{\partial z} + \frac{\partial u_z}{\partial r} = 0.$$

Substituting (6) into these conditions one obtains $\xi' = \xi - \frac{b^2}{a^2}$

$$A_j\nu_j I_1'(\xi_j^\pm) + \zeta A_j\nu_j\frac{1}{\xi_j^\pm}I_1(\xi_j^\pm) + A_j'\nu_j K_1'(\xi_j^\pm)$$

$$\zeta A_j'\nu_j\frac{1}{\xi_j^\pm}K_1(\xi_j^\pm) + \zeta B_j ik I_0(\xi_j^\pm) + \zeta B_j' ik K_0(\xi_j^\pm)) = 0, \tag{14}$$

$$ik A_j I_1(\xi_j^\pm) + ik A_j' K_1(\xi_j^\pm) + B_j I_1(\xi_j^\pm)\nu_j - B_j' K_1(\xi_j^\pm)\nu_j = 0,$$

where summation over j extends from 1 to 3, $\xi_j^\pm = (R \pm h)\nu_j$. The remaining conditions on the shell surfaces are [5, 11]

$$h_r = \mu\tilde{h}_r - \chi H_0\frac{\partial u_r}{\partial z}, \quad h_z = \tilde{h}_z, \quad \mu = \chi + 1, \tag{15}$$

where $\tilde{h}_r$, $\tilde{h}_z$ are the components of the disturbed magnetic field outside of shell in dielectric, where the Maxwell equations have form of the Laplace equation. For $r > R + h$ we have

$$\tilde{h}_z = \tilde{C}H_0 K_0(r\nu)e^{-i\omega t + ikz} + c.c.$$
$$\tilde{h}_r = \tilde{D}H_0 K_1(r\nu)e^{-i\omega t + ikz} + c.c. \tag{16}$$

For $r < R - h$

$$\tilde{h}_z = \tilde{\tilde{C}}H_0 I_0(r\nu)e^{-i\omega t + ikz} + c.c.$$
$$\tilde{h}_r = \tilde{\tilde{D}}H_0 I_1(r\nu)e^{-i\omega t + ikz} + c.c. \tag{17}$$

Substitution of (16) and (17) into

$$\frac{\partial \tilde{h}_r}{\partial r} + \frac{1}{r}\tilde{h}_r + \frac{\partial \tilde{h}_z}{\partial z} = 0$$

one obtained

$$\tilde{D} = i\tilde{C}, \quad \tilde{\tilde{D}} = -i\tilde{\tilde{C}}. \tag{18}$$

Then, the boundary conditions (15) after elimination of $\tilde{C}$, $\tilde{\tilde{C}}$ provide relations for $r = R + h$ and $r = R - h$

$$h_r = \mu i \frac{K_1\{(R+h)k\}}{K_0\{(R+h)k\}} h_z - H_0 \chi \frac{\partial u_r}{\partial z},$$

$$h_r = -\mu i \frac{I_1\{(R-h)k\}}{I_0\{(R-h)k\}} h_z - H_0 \chi \frac{\partial u_r}{\partial z},$$

or on account of (6), (8)

$$\frac{\omega k A_j}{\chi_j} I_1(\xi_j^+) + \frac{\omega k A_j'}{\chi_j} K_1(\xi_j^+)$$
$$= \mu i \frac{K_1\{(R+h)k\}}{K_0\{(R+h)k\}} \left\{ \frac{i\omega\nu_j}{\chi_j} A_j I_0(\xi_j^+) - \frac{i\omega\nu_j}{\chi_j} A_j' K_0(\xi_j^+) \right\}$$
$$- \chi ik \left\{ A_j I_1(\xi_j^+) + A_j' K_1(\xi_j^+) \right\}, \tag{19}$$
$$\frac{\omega k A_j}{\chi_j} I_1(\xi_j^-) + \frac{\omega k A_j'}{\chi_j} K_1(\xi_j^-)$$
$$= -\mu i \frac{I_1\{(R-h)k\}}{I_0\{(R-h)k\}} \left\{ \frac{i\omega\nu_j}{\chi_j} A_j I_0(\xi_j^-) - \frac{i\omega\nu_j}{\chi_j} A_j' K_0(\xi_j^-) \right\}$$
$$- \chi ik \left\{ A_j I_1(\xi_j^-) + A_j' K_1(\xi_j^-) \right\},$$

where the summation over j extends from 1 to 3. To (14) and (19) one must add

$$A_j = -\frac{\frac{b^2}{a^2}\nu_j^2 - k^2 + \frac{\omega^2}{a^2}}{\zeta ik\nu_j - \frac{\omega\nu_j k}{\chi_j}\frac{a_1^2}{a^2}\chi} B_j, \; A_j' = \frac{\frac{b^2}{a^2}\nu_j^2 - k^2 + \frac{\omega^2}{a^2}}{\zeta ik\nu_j - \frac{\omega\nu_j k}{\chi_j}\frac{a_1^2}{a^2}\chi} B_j', \tag{20}$$

where there is no summation over j. Here $\frac{B_j}{A_j}$, $\frac{B_j'}{A_j'}$, $j = 1, 2, 3$, by (20) and χ_j by (9). The systems (14) and (19) represent homogeneous equations with respect to $A_{1,2,3}$, $A_{1,2,3}'$. Setting the determinant equal to zero, yields

$$\begin{vmatrix} \Pi_1^+ & \Pi_2^+ & \Pi_3^+ & M_1^+ & M_2^+ & M_3^+ \\ \Pi_1^- & \Pi_2^- & \Pi_3^- & M_1^- & M_2^- & M_3^- \\ P_1^+ & P_2^+ & P_3^+ & \Omega_1^+ & \Omega_2^+ & \Omega_3^+ \\ P_1^- & P_2^- & P_3^- & \Omega_1^- & \Omega_2^- & \Omega_3^- \\ N_1^+ & N_2^+ & N_3^+ & \Lambda_1^+ & \Lambda_2^+ & \Lambda_3^+ \\ N_1^- & N_2^- & N_3^- & \Lambda_1^- & \Lambda_2^- & \Lambda_3^- \end{vmatrix} = 0 \tag{21}$$

54   A.G. Bagdoev et al.

with

$$\Pi_j^{\pm} = \nu I_1'(\xi_j^{\pm}) + \zeta \nu_j \frac{I_1(\xi_j^{\pm})}{\xi_j^{\pm}} + \zeta \frac{B_j}{A_j} ik I_0(\xi_j^{\pm}),$$

$$M_j^{\pm} = \nu K_1'(\xi_j^{\pm}) + \zeta \nu_j \frac{K_1(\xi_j^{\pm})}{\xi_j^{\pm}} + \zeta \frac{B_j'}{A_j'} ik I_0(\xi_j^{\pm}),$$

$$P_j^{\pm} = ik I_1(\xi_j^{\pm}) + \frac{B_j}{A_j} \nu_j I_1(\xi_j^{\pm}), \quad \Omega_j^{\pm} = ik K_1(\xi_j^{\pm}) - \frac{B_j'}{A_j'} \nu_j I_1(\xi_j^{\pm}),$$

$$N_j^{+} = \frac{\omega k}{\chi_j} I_1(\xi_j^{+}) + \mu \frac{\omega \nu_j}{\chi_j} \frac{K_1\{(R+h)k\}}{K_0\{(R+h)k\}} I_0(\xi_j^{+}) + \chi ik I_1(\xi_j^{+}),$$

$$N_j^{-} = \frac{\omega k}{\chi_j} I_1(\xi_j^{-}) + \mu \frac{\omega \nu_j}{\chi_j} I \frac{I_1\{(R-h)k\}}{I_0\{(R-h)k\}} I_0(\xi_j^{-}) + \chi ik I_1(\xi_j^{-}),$$

$$\Lambda_j^{+} = \frac{\omega k}{\chi_j} K_1(\xi_j^{+}) - \mu \frac{\omega \nu_j}{\chi_j} \frac{K_1\{(R+h)k\}}{K_0\{(R+h)k\}} K_0(\xi_j^{+}) + \chi ik K_1(\xi_j^{+}),$$

$$N_j^{-} = \frac{\omega k}{\chi_j} K_1(\xi_j^{-}) + \mu \frac{\omega \nu_j}{\chi_j} \frac{I_1\{(R-h)k\}}{I_0\{(R-h)k\}} K_0(\xi_j^{-}) + \chi ik K_1(\xi_j^{-}).$$

Numerical calculations of Equation (21) are carried out for the special case of magnetoelasticity, i.e. for $\chi = 0$.

In the solution of Equation (21) all χ_j can be divided by $-i\omega$, and because (10) the third and the sixth column of (21) can be multiplied by a_1^2 and besides, in the terms $a_1^2 N_3^{\pm}$ and $a_1^2 \Lambda_3^{\pm}$, it may be assumed that $\frac{a_1^2}{\chi_3} = -a^2 \frac{1}{\zeta \frac{a^2 k^2}{b^2 \theta} + 1}$. These manipulations are necessary to carry out the calculations in (21) also for the elastic case $a_1 = 0$, for which the frequency $\omega = \omega_{00}$ of the vibration of the cylindrical shell [12], is known:

$$\omega_{00} = h'b\sqrt{\frac{\zeta}{3}} \sqrt{k^4 + 12\frac{1-\nu_0^2}{R^2 h'^2}}, \quad h' = 2h, \tag{22}$$

where ν_0 is a Poisson's ratio.

The solution of the transcendent Equation (21) with the Equation (9) was obtained, where during iterations as zero approximation is taken solution (22) for small $\frac{a_1}{a}$. The calculations are made for magnetoelastic case $\chi = 0$, with values of typical parameters for aluminum $\frac{b^2}{a^2} = \frac{1}{3}$, $\zeta = \frac{2}{3}$, $a = 10^5$ cm/s., $\rho = 3\,\text{g/cm}^3$, $\nu_m = 1000\,\text{cm}^2/\text{s.}$, $h' = 0.1\,\text{cm}$, $R = 10^4$, 10^8 cm, $k = 0.1$, 0.2, 0.3, 0.4, $0.5\,\text{cm}^{-1}$. The results of the calculation for real part of ω, $Re\omega$, are listed in Tables 1, 2. The case $R = 10^8$ cm, represented in Table 2, is compared with results from calculations of the corresponding third-order determinant equation for plates [7–9]. The results coincide, which proves the accuracy the equations derived for the shell. We compare also Tables 1 and 3 with corresponding results from the averaged theory. The dispersion equation for the averaged treatment is given [12], and after some transformations can be written in the form, obtained in [9]:

$$\overline{\omega}^2 = \omega_{00}^2 + \frac{a_1^2 \overline{\omega} k^2}{i\nu_m \lambda_1^2}\left(1 - \frac{2}{kh'}\frac{\frac{i\overline{\omega}}{\nu_m}}{sh(\lambda_1 \frac{h'}{2}) + \frac{k}{\lambda_1}ch(\lambda_1 \frac{h'}{2})}\frac{sh(\lambda_1 \frac{h'}{2})}{\lambda_1^2}\right), \qquad (23)$$

where $\lambda_1 = \sqrt{k^2 - \frac{i\overline{\omega}}{\nu_m}}$. For values of parameters $|\lambda_1 \frac{h'}{2}| << 1$, (23) can be simplified as follows:

$$\overline{\omega}^2 = \omega_{00}^2 + \frac{a_1^2 \overline{\omega}}{i\nu_m}\frac{1}{1 - \frac{i\overline{\omega}}{\nu_m k}\frac{h'}{2}}. \qquad (24)$$

The calculations of $Re\overline{\omega}$ by (23) and (24) give almost the same results. They are listed in Tables 3, 4. The calculations are performed for $h' = 0.1\,\mathrm{cm}$, $R = 10^8$, 10^3, 10^2 cm. The results for $R = 10^3$, 10^8 cm for $Re\omega$, $Re\overline{\omega}$ are given in Tables 1, 2 based on (21) and in Tables 3, 4 for (24). The comparison of results from Tables 1, 2, obtained by exact treatment with results from Tables 3, 4 based on the Kirchhoff hypothesis show that, qualitatively, the character of the variation of the curves for $Re\omega(H_0)$ and $Re\overline{\omega}(H_0)$ corresponds, the last functions first decrease from $\omega = \omega_{00}$, $\overline{\omega} = \omega_{00}$, for $H_0 = 0$, and then increase, but the quantitative results of both tables are quite different. The results in Table 2 coincide with corresponding results for plates. In Table 2, for small $\frac{a_1}{a} = \frac{1}{10000}$ the values of $Re\omega$ for $k = 0.1\,\mathrm{cm}^{-1}$, $k = 0.2\,\mathrm{cm}^{-1}$ coincide with ω_{00}, but for $k = 0.2;\ 0.4;\ 0.5\mathrm{cm}^{-1}$ for small $\frac{a_1}{a}$ the obtained results decreased by a factor of 10 with respect to the ω_{00} values $Re\omega$. This discrepancy can be explained by the fact that for $H_0 = 0$ the Equation (21) for the magnetoelastic problem ($\chi = 0$), obtained from (19), which follows from (15), both sides of which are divided on H_0, is not equivalent to the equation resulting from the fourth-order determinant of elasticity when H_0 is zero identically. Thus by exact analytical and numerical methods values of $Re\omega$ are obtained compared with those obtained by the Kirchhoff hypothesis for magnetoelastic shells. It is shown that there is great difference of the obtained values from the purely elastic case. The Kirchhoff hypothesis is not applicable to the mentioned problems.

Table 1 Dependence of $Re\omega$ on k and $\frac{a_1}{a}$ by space treatment when $h' = 0.1$, $R = 10^3$

$\frac{a_1}{a}k$	0.1	0.2	0.3	0.4	0.5
$5/10^5$	95.25	142.071	34.912	56.7086	59.0876
$1/1000$	95.3561	141.834	261.411	445.1	686.297
$2/1000$	94.0374	140.728	261.772	445.556	687.275
$7/1000$	99.7618	146.206	291.762	399.149	498.628
$1/100$	141.459	276.627	425.007	563.777	686.181
$5/100$	706.671	1413.34	2119.97	2826.6	3533.28

Table 2 Dependence of $Re\omega$ on k and $\frac{a_1}{a}$ by space treatment when $h' = 0.1$, $R = 10^8$

$\frac{a_1}{a}k$	0.1	0.2	0.3	0.4	0.5
$5/10^5$	27.2162	108.861	244.912	435.349	680.126
$1/1000$	26.8179	108.876	245.064	435.587	680.444
$2.4/1000$	0.019999	105.698	244.374	435.952	681.456
$2.6/1000$	36.7695	104.288	243.927	435.861	681.567
$5/1000$	70.7102	141.42	213.726	423.299	676.911
$5/100$	706.665	1413.33	2119.999	2826.657	3533.328

Table 3 Dependence of $Re\omega$ on k and $\frac{a_1}{a}$ by Kirchhoff hupothesis when $h' = 0.1$, $R = 10^3$

$\frac{a_1}{a}k$	0.1	0.2	0.3	0.4	0.5
$5/10^5$	92.9628	140.546	260.579	444.445	686.196
$1/1000$	93.0612	140.633	260.748	444.694	686.519
$3/1000$	82.7625	134.544	258.591	444.678	687.835
$5/1000$	0	57.2232	232.336	433.044	683.306
$7/1000$	0	0	49.9928	378.297	656.017
$5/100$	7000.59	9798.89	11876.9	13570.8	15012.6

Table 4 Dependence of $Re\omega$ on k and $\frac{a_1}{a}$ by Kirchhoff hupothesis when $h' = 0.1$, $R = 10^8$

$\frac{a_1}{a}k$	0.1	0.2	0.3	0.4	0.5
$5/10^5$	27.2167	108.867	244.949	435.466	680.415
$1/1000$	26.8186	108.888	245.102	435.708	680.735
$2.4/1000$	0	105.71	244.425	436.09	681.775
$5/1000$	0	0	213.803	423.509	677.346
$7/1000$	0	0	0	366.929	649.631
$5/100$	7000.05	9798.52	11876.6	13570.5	15012.4

References

1. Ambartsumyan S.A., Bagdasaryan G.E., Belubekyan M.V. (1977) Magneto-elasticity of thin shells and plates. M. Nauka, 272p. (In Russian)
2. Ambartsumyan S.A., Bagdasaryan G.E. (1996) Electro conducting plates and shells in magnetic field. M. Phys.-Math. Literature, 286p. (In Russian)
3. Kaliski S. (1962) Magnetoelastic vibration of perfectly conducting plates and bars assuming the principle of plane sections. Proc. Vibr. Pol. Acad. Sci. V. 3, N4, pp225–234.

4. Bagdoev A.G., Movsisyan L.A. (1999) Modulation of thermomagnetoelastic waves in magnetic field. Izv. NAS Armenia, Mechanica, V.52, N1, pp25–30. (In Russian)
5. Sarkisyan V.S., Sarkisyan S.V., Dzilavyan S.A., Sarkisyan A.L. (1980) Investigation of electroconducting plates in magnetic field. Mezhvus. sbornik nauchnikh trudov, Mechanika, Yerevan State University, pp45–30. (In Russian)
6. Novatski V. (1975) Elasticity theory. M. Mir. 863p. (In Russian)
7. Bagdoev A.G., Sahakyan S.G. (2001) Stability of nonlinear modulation waves in magnctic ficld for spacc and averaged problems. Izv. RAS MTT., V.5, pp35–42.
8. Bagdoev A.G., Vantsyan A.A. (2002) Theoretical and experimental investigations of waves in plate in magnetic field for space and averaged problems. Int. J. Solids Struct. V.39. pp851–859.
9. Safaryan Yu. S. (2001) Investigations of vibrations of magnetoelastic plates vibrations in space and averaged treatment. Inform. Tech. Manage., V.2, pp. 17–49.
10. Kolski H. (1953) Stress waves in solids. Oxford. 192p.
11. Bagdoev A.G., Kevnakszyan L.S. (2004) Nonlinear modulation waves in ferromagnetic plates for arbitrary electroconductivity. Mathematics in Hagh school, Engineering State University of Armenia. N3. p.12–33. (In Russian)
12. Bagdasaryan G.E., Belubekyan M.V. (1967) Axialsymetric vibrations of cylindrical shell in magnetic field. Izv. AN Arm SSR, Mechanika, V.20, N5 pp21–27. (In Russian)

Stability of a Rectangular Plate Capable of Transverse Shear Deformations

Vagharshak M. Belubekyan

Yerevan State University, 1 Manookyan str, Yerevan 375049, Armenia,
vbelub@gmail.com

Abstract The stability of a rectangular plate, when its two opposite edges are hinged, and the third edge is under sliding contact conditions, are considered; for the fourth edge, two cases are studied: sliding contact conditions and "restricted" sliding contact conditions. In the first case, the buckling load is the same as the critical load for cylindrical buckling, calculated by means of Kirchhoff's theory. In the second case, the buckling load may be refined twice or more, with respect to the Kirchhoff's theory. The refinement term depends on geometry of the plate.

Keywords: plate bending, stability, buckling, transverse shear deformations

1 Introduction

In most stability problems for elastic plates, accounting for transverse shear deformations does not lead to a significant change in the critical value of the applied load. However, there are several problems, where accounting of transverse shear deformations leads to a change in the boundary conditions, or even to a need for additional boundary conditions. In such problems, the consequent refinement in the evaluation of the critical load may be significant. In the following, we consider the stability of a rectangular plate, when its two opposite edges are hinged, and the third edge is under sliding contact conditions; for the fourth edge, two cases are considered: sliding contact conditions and "restricted" sliding contact conditions. In the first case, the buckling load is the same as the critical load for cylindrical buckling, calculated by means of Kirchhoff's theory, up to the terms of second order relatively to the thickness. In the second case, the buckling load may be refined twice or more, with respect to the Kirchhoff's theory. The refinement term depends on geometry of the plate.

2 Governing Equations of Problem

Let the geometry of the plate in Cartesian coordinates to be $0 \leq x \leq a$, $0 \leq y \leq b$, $-h \leq z \leq h$. Edges $x = 0, a$ are under uniform compressive load.

$$p = 2h\sigma_0 \tag{1}$$

The stability equations in the frame of S.A. Ambartsunyan's refined bending theory are:

$$\frac{\partial \phi_1}{\partial x} + \frac{\partial \phi_2}{\partial y} = \frac{3}{2}\sigma_0 \frac{\partial^2 w}{\partial x^2} \tag{2}$$

$$D\frac{\partial}{\partial x}\Delta w - \frac{8h^3}{15}\left[\Delta\phi_1 + \theta\frac{\partial}{\partial x}\left(\frac{\partial\phi_1}{\partial x} + \frac{\partial\phi_2}{\partial y}\right)\right] + \frac{4h}{3}\phi_1$$
$$-\frac{2h^3}{3}\sigma_0\frac{\partial^2}{\partial x^2}\left(\frac{\partial w}{\partial x} - \frac{4}{5G}\phi_1\right) = 0$$

$$D\frac{\partial}{\partial y}\Delta w - \frac{8h^3}{15}\left[\Delta\phi_2 + \theta\frac{\partial}{\partial y}\left(\frac{\partial\phi_1}{\partial x} + \frac{\partial\phi_2}{\partial y}\right)\right] + \frac{4h}{3}\phi_2$$
$$-\frac{2h^3}{3}\sigma_0\frac{\partial^2}{\partial x^2}\left(\frac{\partial w}{\partial x} - \frac{4}{5G}\phi_2\right) = 0$$

And by E. Reissner's refined theory they are:

$$\Delta w - \frac{\partial\theta_1}{\partial x} - \frac{\partial\theta_2}{\partial y} = \frac{\sigma_0}{G}\frac{\partial^2 w}{\partial x^2} \tag{3}$$

$$D\left[\Delta\theta_1 + \theta\frac{\partial}{\partial x}\left(\frac{\partial\theta_1}{\partial x} + \frac{\partial\theta_2}{\partial y}\right)\right] + \frac{4Gh}{1-\nu}\left(\frac{\partial w}{\partial x} - \theta_1\right) = \frac{4h^3\sigma_0}{3(1-\nu)}\frac{\partial^2\theta_1}{\partial x^2}$$

$$D\left[\Delta\theta_2 + \theta\frac{\partial}{\partial y}\left(\frac{\partial\theta_1}{\partial x} + \frac{\partial\theta_2}{\partial y}\right)\right] + \frac{4Gh}{1-\nu}\left(\frac{\partial w}{\partial y} - \theta_2\right) = \frac{4h^3\sigma_0}{3(1-\nu)}\frac{\partial^2\theta_2}{\partial x^2}$$

Where ϕ_1, ϕ_2 are functions representing transverse shears, θ_1, θ_2 are functions representing rotation angles, and $\theta = (1+\nu)/(1-\nu)$. Applying the following transform, as in [2]

$$\phi_1 = \frac{\partial \Phi}{\partial x} + \frac{\partial \Psi}{\partial y}$$

$$\phi_2 = \frac{\partial \Phi}{\partial y} - \frac{\partial \Psi}{\partial x} \tag{4}$$

$$\theta_1 = \frac{\partial w}{\partial x} - \frac{2}{3G}\left(\frac{\partial \Phi}{\partial x} + \frac{\partial \Psi}{\partial y}\right)$$

$$\theta_2 = \frac{\partial w}{\partial y} - \frac{2}{3G}\left(\frac{\partial \Phi}{\partial y} - \frac{\partial \Psi}{\partial x}\right) \tag{5}$$

both systems of Equations (2), (3) yield to:

$$\Delta\Phi = \frac{3\sigma_0}{2}\frac{\partial^2 w}{\partial x^2} \tag{6}$$

$$-D\left(\Delta w - \frac{1-\nu}{2G}\sigma_0\frac{\partial^2 w}{\partial x^2}\right) + \frac{8\chi h^3}{3(1-\nu)}\left(\Delta\Phi - \frac{1-\nu}{2G}\sigma_0\frac{\partial^2 \Phi}{\partial x^2} - \frac{1-\nu}{2\chi h^2}\Phi\right) = 0$$

$$\Delta\Psi - \frac{\sigma_0}{G}\frac{\partial^2 \Psi}{\partial x^2} + \frac{1}{\chi h^2}\Psi = 0$$

where $\chi = 2/5$ by S.A. Ambartsumyan's theory, and $\chi = 1/3$ by E. Reissner's theory. Equation for the function Ψ is independent of the equations for w and Φ. However these functions are dependent on each other by means of boundary conditions. From the first two equations of system (6), eliminating Φ, equation for deflections w is obtained:

$$\Delta^2 w - \frac{1-\nu}{2G}\sigma_0\left(1 + \frac{6\chi}{1-\nu}\right)\frac{\partial^2}{\partial x^2}\Delta w$$

$$+ \frac{3\chi(1-\nu)}{2G^2}\sigma_0^2\frac{\partial^4 w}{\partial x^4} + \frac{2h\sigma_0}{D}\frac{\partial^2 w}{\partial x^2} = 0 \tag{7}$$

Similar equation is obtained for Φ, when w is eliminated from Equations (6). Equations (7) can be rewritten in more compact expression:

$$\left(\Delta - \frac{1-\nu}{2}\frac{\sigma_0}{G}\frac{\partial^2}{\partial x^2}\right)\left(\Delta w - \frac{3\chi\sigma_0}{G}\frac{\partial^2 w}{\partial x^2}\right) + \frac{2h\sigma_0}{D}\frac{\partial^2 w}{\partial x^2} = 0 \tag{8}$$

3 Sliding Contact Boundary Conditions

Consider rectangular plate with hinged edges $x = 0, a$, sliding contact conditions at edge $y = 0$, and different boundary conditions at $y = b$. According to [3, 4] the boundary conditions at $x = 0, a$ are:

$$w = 0 \tag{9}$$

$$\frac{\partial \Phi}{\partial y} - \frac{\partial \Psi}{\partial x} = 0$$

$$\frac{\partial^2 w}{\partial x^2} - \frac{2\chi}{G} \frac{\partial}{\partial x} \left(\frac{\partial \Phi}{\partial x} + \frac{\partial \Psi}{\partial y} \right) = 0$$

Sliding contact boundary conditions at $y = 0$ imply, that the edge of the plate can slide without friction in directions z and x, but cannot have displacements in y direction. Sliding contact boundary conditions at $y = 0$ are the following

$$\frac{\partial w}{\partial y} = 0 \tag{10}$$

$$\frac{\partial \Phi}{\partial y} - \frac{\partial \Psi}{\partial x} = 0$$

$$\frac{\partial}{\partial y} \left(\frac{\partial \Phi}{\partial x} + \frac{\partial \Psi}{\partial y} \right) = 0$$

Solution of the system (6) under conditions (9), (10) is searched in a form of series expansions:

$$w = \sum_{m=1}^{\infty} \left(a_m \cosh p_1 \mu_m y + b_m \cosh p_2 \mu_m y \right) \sin \mu_m x \tag{11}$$

$$\Psi = \Sigma_{m=1}^{\infty} c_m \sinh \mu_m p_3 y \cos \mu_m x$$

$$\Phi = -\frac{3\sigma_0}{2} \Sigma_{m=1}^{\infty} \left(\frac{a_m}{p_1^2 - 1} \cosh p_1 \mu_m y + \frac{b_m}{p_2^2 - 1} \cosh p_2 \mu_m y \right) \sin \mu_m x$$

where

$$p_{1,2} = \left\{ 1 - \frac{1-\nu}{4G} \sigma_0 \left(1 + \frac{6\chi}{1-\nu} \right) \right. \tag{12}$$

$$\left. \pm \sqrt{\eta_m^2 + \frac{1-\nu}{2G^2} \sigma_0^2 \left[\frac{1-\nu}{3} \left(1 + \frac{6\chi}{1-\nu} \right)^2 - 3\chi \right]} \right\}^{1/2}$$

$$\eta_m^2 = 2h\sigma_0/(D\mu_m^2), \quad \mu_m = m\pi/a \tag{13}$$

and a_m, b_m, c_m are undefined constants.

At first, consider the most simple case, when at $y = b$ also, the sliding contact conditions are applied, i.e. conditions (10) are valid. Substituting (11) into (10) at $y = b$ yield to the following linear system of equations, determining the constants a_m, b_m, c_m

$$p_1 a_m \sinh p_1 \mu_m b + p_2 b_m \sinh p_2 \mu_m b = 0$$

$$p_1 \left(p_1^2 - 1\right)^{-1} a_m \sinh p_1 \mu_m b + p_2 \left(p_2^2 - 1\right)^{-1} b_m \sinh p_2 \mu_m b = 0$$

$$\left(p_3^2 - 1\right) c_m \sinh p_3 \mu_m b = 0 \tag{14}$$

From the third equation of the system, since p_3 is a real number and $p_3 \neq 1$, follows that $c_m = 0$. Equating to zero the determinant of the system of two remaining equations, we obtain:

$$p_1 p_2 \left(p_1^2 - p_2^2\right) \sinh p_1 \mu_m b \sinh p_2 \mu_m b = 0 \tag{15}$$

The root $p_1^2 = p_2^2$ leads to a trivial solution $a_m = b_m = 0$. Since p_i are real numbers, from (15) follows

$$\sinh p_2 \mu_m b = 0, \quad or \quad p_2 \mu_m b = in\pi/a \tag{16}$$

Using expression (12), Equation (16) determining the critical value of parameter η_m can be written as:

$$\frac{2\chi}{3(1-\nu)}\mu_m^4 h^4 \eta^4 - \tag{17}$$

$$\left[1 + \left(1 + \left(\frac{na}{mb}\right)^2\right)\left(1 + \frac{6\chi}{1-\nu}\right)\frac{\mu_m^2 h^2}{3}\right]\eta_m^2 + \left(1 + \left(\frac{na}{mb}\right)^2\right)^2 = 0$$

In (17), to obtain a simpler expression, terms of order $\mu_m^4 h^4 \ll 1$ can be neglected

$$\eta_{mn}^2 \approx \left[1 + \left(1 + \left(\frac{na}{mb}\right)^2\right)\left(1 + \frac{6\chi}{1-\nu}\right)\frac{\mu_m^2 h^2}{3}\right]^{-1}\left(1 + \left(\frac{na}{mb}\right)^2\right)^2 \tag{18}$$

And minimal critical load is obtained

$$\eta_1^2 = \left[1 + \left(1 + \frac{6\chi}{1-\nu}\right)\frac{\mu_1^2 h^2}{3}\right]^{-1} \tag{19}$$

The same result is obtained also in [1] when $\chi = 2/5$. As in the frame of Kirchhoff's theory, the buckling shape is cylindrical. For this problem Kirchhoff's theory yields to correct results within its tolerance, that is for case when $\mu_m^2 h^2 \ll 1$.

4 Restricted Sliding Contact Boundary Conditions

Let the edge $y = b$ to be under conditions of "restricted" sliding contact
[2,5]. Restricted sliding contact boundary conditions at $y = b$ imply, that the
edge of the plate can slide without friction in direction z, but cannot have
displacements in directions x and y.

$$\frac{\partial w}{\partial y} = 0 \tag{20}$$

$$\frac{\partial \Phi}{\partial y} - \frac{\partial \Psi}{\partial x} = 0$$

$$\frac{\partial w}{\partial x} - \frac{2\chi}{G}\left(\frac{\partial \Phi}{\partial x} + \frac{\partial \Psi}{\partial y}\right) = 0$$

These conditions are intermediate between clumped and sliding contact
conditions, and they cannot be adequately represented in the frame of Kirchhoff's theory [5]. Substitution of (11) into boundary conditions (20), yields to
the following system of linear homogeneous algebraic equations for a_m, b_m, c_m:

$$a_m p_1 \sinh p_1 \mu_m b + b_m p_2 \sinh p_2 \mu_m b = 0 \tag{21}$$

$$\left[1 + \frac{3\chi\sigma_0}{G\left(p_1^2 - 1\right)}\right] a_m \cosh p_1 \mu_m b + \left[1 + \frac{3\chi\sigma_0}{G\left(p_2^2 - 1\right)}\right] b_m \cosh p_2 \mu_m b -$$
$$\frac{2\chi}{G} p_3 c_m \cosh p_3 \mu_m b = 0$$

$$\frac{3\sigma_0}{2}\left(\frac{p_1}{p_1^2 - 1} a_m \sinh p_1 \mu_m b + \frac{p_2}{p_2^2 - 1} b_m \sinh p_2 \mu_m b\right) - c_m \sinh p_3 \mu_m b = 0$$

Equating determinant of this system to zero, yields to equation for parameter η of critical load

$$\left[\left(p_2^2 - 1 + \frac{3\chi\sigma_0}{G}\right) p_2 \tanh p_2 \mu_m b - \left(p_1^2 - 1 + \frac{3\chi\sigma_0}{G}\right) p_1 \tanh p_1 \mu_m b\right] \times$$

$$\tanh p_3 \mu_m b + \frac{3\chi\sigma_0}{G} p_1 p_2 p_3 \left(p_1^2 - p_2^2\right) \tanh p_1 \mu_m b \tanh p_2 \mu_m b = 0 \tag{22}$$

Equation (22) within precision of order $\mu_m^2 h^2 \ll 1$ can be approximated
as follows:

$$\sqrt{\eta_m - 1}\tan\sqrt{\eta_m - 1}\mu_m b - \sqrt{\eta_m + 1}\tan\sqrt{\eta_m + 1}\mu_m b +$$

$$\frac{1\sqrt{\chi}\mu_m h}{1 - \nu}\eta_m^2\sqrt{\eta_m^2 - 1}\tan\sqrt{\eta_m - 1}\mu_m b\tanh\sqrt{\eta_m + 1}\mu_m b = 0 \qquad (23)$$

In Table 1 values of η

$$\eta_1 = \mu_1^{-1}\sqrt{2h\sigma_0/D} \qquad (24)$$

are shown versus parameters of relative thickness α and relative size β of the plate

$$\alpha = 2\sqrt{\chi}\mu_1 h\left(1 - \nu\right)_{-1}, \beta = \mu_1 b \qquad (25)$$

Minimal critical load, depending on β may correspond to values of m greater than 1.

Table 1 Parameter characterizing critical load

$\alpha \setminus \beta$	1	3	5
0	2.0314	1.1843	1.0763
0.05	1.8619	1.1779	1.0749
0.10	1.7639	1.1721	1.0734
0.15	1.6956	1.1667	1.0721
0.20	1.6437	1.1616	1.0707

5 Conclusion

The Kirchhoff's theory is not adequate to represent the boundary conditions of "restricted" sliding contact, because it leads to critical value of $\eta_1 = 1$. Table 1 shows, that refined boundary conditions lead to a significant growth of the critical load. With a growth of size β, the critical value of η tends to the value obtained by Kirchhoff's theory ($\eta = 1$).

References

1. Ambartsumyan S.A. (1987) Theory of anisotropic plates. Nauka, Moscow.
2. Belubekyan M.V. (2003) In collection: Problems of mechanics of thin deformable bodies. National Sc. Acad of Armenia 61–66.
3. Belubekyan V.M. (2004) MTT, Proc Rus. Sc Academy 2:126–131.
4. Belubekyan V.M., Belubekyan M.V. (1999) Proc. National Sc. Acad of Armenia, Mechanics 52:11–21.
5. Ivanova E.A. (1998) MTT, Proc Rus. Sc Academy, 2:163–174.

On a Problem of Thermal Stresses in the Theory of Cosserat Elastic Shells with Voids

Mircea Bîrsan

Faculty of Mathematics, "A.I. Cuza" University of Iaşi, Bvd. Carol I, no. 11, 700506, Iaşi, Romania, `bmircea@uaic.ro`

Abstract We consider a problem of thermal stresses in cylindrical Cosserat elastic shells made from a material with voids. The cylindrical shells have arbitrary cross-sections. The problem consists in finding the equilibrium of the shell under the action of a given temperature distribution. We assume that the temperature field is independent of the axial coordinate and we determine a closed-form solution expressed in terms of the displacement vector and porosity field.

Keywords: cosserat shell, cylindrical surface, temperature distribution, volume fraction field

1 Introduction

The theory of Cosserat shells is an interesting approach to the mechanics of elastic shell-like bodies, in which the thin three-dimensional body is modelled as a two-dimensional continuum (i.e. a surface) endowed with a deformable director assigned to every point. For a detailed analysis of the theory of Cosserat surfaces and its relation with other (hierarchical) shell theories, we refer to the classical monograph of Naghdi [1] and the modern book of Rubin [2]. According to [1], the Cosserat theory is also called the *direct approach* of shell theory, since its governing equations are deduced directly from the balance laws postulated for these two-dimensional continua (instead of deriving them starting from the three-dimensional theory). One advantage of this approach is that we can use methods analogous to those employed in the three-dimensional theory of elasticity to obtain corresponding results for Cosserat shells. Another feature of the Cosserat theory is that it can easily be extended to account for some important effects in the mechanical behavior of shells, such as thermal effects or porosity effects (see [3,4]). In our paper, we shall illustrate both of the advantages mentioned above.

In the context of linear theory for Cosserat elastic shells, the existence of solution can be proved on the basis of inequalities of Korn's type for Cosserat surfaces, using the method described by Ciarlet [5] in the classical shell theory (see [6]). Several general theorems (such as uniqueness, reciprocal and variational theorems) are obtained in [4, 7] via the same procedures as in the three-dimensional theory of elasticity. According to [8], the theory of Cosserat surfaces can also be used for the modelling of interphases in elastic media.

In this paper, we present an interesting application of the Cosserat theory for shells: we determine the static deformation of a porous cylindrical shell, due to a given temperature distribution in the body. To this aim, we employ the theory of thermoelastic Cosserat shells with voids established in [9]. Here, the porosity of the material is described by introducing the *volume fraction field* as a kinematical variable assigned to each material particle, in accordance to the Nunziato-Cowin theory for elastic media with voids [10]. The interpretation of the volume fraction field from the viewpoint of the theory of media with microstructure was given by Capriz and Podio-Guidugli [11].

We consider cylindrical shells made of isotropic and homogeneous materials. The cross-sections of the cylindrical surfaces are open curves of arbitrary shape. As is usual in the treatment of Saint-Venant's problem, we consider a relaxed formulation of the boundary conditions on the end edges of the cylindrical shell, in which the pointwise assignment of mechanical loads is replaced by prescribing the corresponding resultant force and resultant moment acting on these boundaries. In the classical theory of elasticity, the deformation of loaded (solid) cylinders has been intensively studied by many scientists (see e.g., [12–14]). We mention that the method to solve Saint-Venant's problem established in the context of three-dimensional elasticity by Ieşan [15] can also be applied for the corresponding problem in the theory of Cosserat shells (see [16, 17]).

In our work, we approach a thermal stresses problem for porous cylindrical shells. We assume that the mechanical loads are absent and thus, our goal is to determine the static deformation of the shell due to a given temperature field. We consider that the temperature distribution in the body is independent of the axial coordinate. The corresponding problem for solid cylinders has been solved in [18]. On the basis of some results presented in [17], we find a closed-form solution to our problem expressed in terms of the displacement vector and volume fraction field, which can be useful in practical situations.

2 Basic Equations and Formulation of the Problem

The linear theory of thermoelastic Cosserat shells with voids has been presented in [9]. In this paper, we confine our attention to cylindrical shells and we begin by recalling the basic field equations for this particular case.

Let $\mathcal{S}$ be the reference configuration of a cylindrical Cosserat surface and let (s, z) be the curvilinear material coordinate system on $\mathcal{S}$ such that z is the axial coordinate and s is the circumferential coordinate (i.e. s is the arc parameter along the cross-section curves of $\mathcal{S}$). The deformation of the porous thermoelastic Cosserat shell is defined by the functions

$$\mathbf{r} = \mathbf{r}(s, z, t), \quad \mathbf{d} = \mathbf{d}(s, z, t), \quad \nu = \nu(s, z, t), \quad \chi = \chi(s, z, t),$$
$$\theta = \theta(s, z, t), \quad \phi = \phi(s, z, t), \tag{1}$$

where $\mathbf{r}$ and $\mathbf{d}$ represent the position vector and the director attached to each point at time t, the scalars θ and ϕ denote the two temperature fields which describe the thermal properties of Cosserat shells (see [3]), while ν and χ are the volume fraction fields which account for the porosity of the shell-like body (see [4,9]). Let $\mathbf{R}$, $\mathbf{D}$, ν_0, χ_0, θ_0 and ϕ_0 designate, respectively, the reference values of the functions $\mathbf{r}$, $\mathbf{d}$, ν, χ, θ and ϕ (on $\mathcal{S}$). We refer the cylindrical surface $\mathcal{S}$ to a rectangular Cartesian coordinate frame $Ox_1 x_2 x_3$ such that Ox_3 is parallel to the generators and $\mathcal{S}$ is situated between the planes $x_3 = 0$ and $x_3 = \bar{z}$. Denote by $\mathbf{e}_i$ the unit vectors along the Ox_i axes ($i = 1, 2, 3$). Then, the parametric equation of $\mathcal{S}$ can be written as

$$\mathbf{R} = \mathbf{R}(s, z) = x_1(s)\mathbf{e}_1 + x_2(s)\mathbf{e}_2 + z\mathbf{e}_3 , \quad s \in [0, \bar{s}], \ z \in [0, \bar{z}],$$

where $x_\alpha(s)$ are known functions of class $C^3[0, \bar{s}]$. Let $\mathcal{C}_z$ be the cross-section curve of $\mathcal{S}$ lying in the plane $x_3 = z$. The cross-sections $\mathcal{C}_z$ are simple open curves of arbitrary shape ($0 \leq z \leq \bar{z}$). We designate by L_s the generator of $\mathcal{S}$ which points are characterized by the circumferential coordinate s. Clearly, the lateral edges are L_0 and $L_{\bar{s}}$, while the end edges of the cylindrical shells are $\mathcal{C}_0$ and $\mathcal{C}_{\bar{z}}$. Throughout the paper, the Greek indices range over the integers $\{1, 2\}$, while the Latin indices take the values $\{1, 2, 3\}$. The summation convention over the repeated indices is also employed.

In the linear theory, we introduce the infinitesimal displacement vector $\mathbf{u} = \mathbf{r} - \mathbf{R}$, the director displacement vector $\boldsymbol{\delta} = \mathbf{d} - \mathbf{D}$, the variations of volume fraction fields $\varphi = \nu - \nu_0$, $\psi = \chi - \chi_0$ and the changes in temperature fields $\tau = \theta - \theta_0$, $\sigma = \phi - \phi_0$. The displacement vectors can be decomposed as

$$\mathbf{u} = u_i \mathbf{e}_i = u_s \boldsymbol{\tau} + u_n \mathbf{n} + u_z \mathbf{e}_3 , \quad \boldsymbol{\delta} = \delta_i \mathbf{e}_i = \delta_s \boldsymbol{\tau} + \delta_n \mathbf{n} + \delta_z \mathbf{e}_3 ,$$

where $\boldsymbol{\tau}$ and $\mathbf{n}$ are the unit tangent vector and normal vector to $\mathcal{C}_z$. We have

$$\boldsymbol{\tau}(s) = x'_\alpha(s)\mathbf{e}_\alpha , \quad \mathbf{n}(s) = \epsilon_{\alpha\beta} x'_\beta(s)\mathbf{e}_\alpha , \quad r(s) = \left[\epsilon_{\alpha\beta} x'_\alpha(s) x''_\beta(s) \right]^{-1} .$$

Here, $\epsilon_{\alpha\beta}$ is the two-dimensional alternator ($\epsilon_{12} = -\epsilon_{21} = 1$, $\epsilon_{11} = \epsilon_{22} = 0$), $r(s)$ is the curvature radius of $\mathcal{C}_z$ and we use the notation $f' = df/ds$.

We consider the following thermal stresses problem: *determine the equilibrium of a cylindrical porous Cosserat shell, under the action of a given temperature field.* We assume that the mechanical loads are absent. As is usual in the treatment of Saint-Venant's problem, we consider a relaxed formulation of the problem in which the pointwise assignment of mechanical loads on the end edges of cylindrical shells is replaced by prescribing the corresponding resultant forces and resultant moments acting on these boundaries.

The linear strain measures for cylindrical shells are

$$e_{ss} = \tfrac{\partial}{\partial s} u_s + u_n r^{-1}, \quad e_{sz} = e_{zs} = \tfrac{1}{2}\left(\tfrac{\partial}{\partial z} u_s + \tfrac{\partial}{\partial s} u_z\right),$$

$$e_{zz} = \tfrac{\partial}{\partial z} u_z, \quad \gamma_s = \delta_s - u_s r^{-1} + \tfrac{\partial}{\partial s} u_n, \quad \gamma_z = \delta_z + \tfrac{\partial}{\partial z} u_n,$$

$$\gamma_n = \delta_n, \quad \rho_{ss} = \tfrac{\partial}{\partial s}\delta_s + r^{-1}\tfrac{\partial}{\partial s} u_s + u_n r^{-2}, \quad \rho_{zz} = \tfrac{\partial}{\partial z}\delta_z,$$

$$\rho_{sz} = \tfrac{\partial}{\partial z}\delta_s, \quad \rho_{zs} = \tfrac{\partial}{\partial s}\delta_z + r^{-1}\tfrac{\partial}{\partial z} u_s, \quad \rho_{ns} = \tfrac{\partial}{\partial s}\delta_n, \quad \rho_{nz} = \tfrac{\partial}{\partial z}\delta_n. \tag{2}$$

We designate by $\mathbf{N}$ the contact force vector, $\mathbf{M}$ the contact director couple and h, H the equilibrated stresses acting per unit length of the curves c included in $\mathcal{S}$. Then, for an arbitrary such curve c having the unit normal $\boldsymbol{\eta} = \eta_s \boldsymbol{\tau} + \eta_z \mathbf{e}_3$, the following relations of Cauchy type hold

$$\mathbf{N} = (N_{ss}\boldsymbol{\tau} + N_{sz}\mathbf{e}_3 + V_s\mathbf{n})\,\eta_s + (N_{zs}\boldsymbol{\tau} + N_{zz}\mathbf{e}_3 + V_z\mathbf{n})\,\eta_z,$$

$$\mathbf{M} = (M_{ss}\boldsymbol{\tau} + M_{sz}\mathbf{e}_3 + M_{sn}\mathbf{n})\,\eta_s + (M_{zs}\boldsymbol{\tau} + M_{zz}\mathbf{e}_3 + M_{zn}\mathbf{n})\,\eta_z,$$

$$h = h_s\eta_s + h_z\eta_z, \quad H = H_s\eta_s + H_z\eta_z.$$

The constitutive equations for isotropic and homogeneous materials are given by (see [9])

$$N_{ss} = (\alpha_1 + 2\alpha_2)\,e_{ss} + \alpha_1 e_{zz} + (\alpha_0\rho_{ss} + \alpha_5\rho_{zz})r^{-1} + \alpha_9\gamma_n + \beta_4\varphi + \beta_7\psi r^{-1}$$

$$+\bar\beta_1\tau + \bar\beta_4\sigma r^{-1}, \quad N_{zz} = \alpha_1 e_{ss} + (\alpha_1 + 2\alpha_2)\,e_{zz} + \alpha_9\gamma_n + \beta_4\varphi + \bar\beta_1\tau,$$

$$N_{sz} = 2\alpha_2 e_{sz}, \quad N_{zs} = N_{sz} + M_{sz}r^{-1}, \quad V_s = \alpha_3\gamma_s + \beta_9\tfrac{\partial}{\partial s}\psi,$$

$$V_z = \alpha_3\gamma_z + \beta_9\tfrac{\partial}{\partial z}\psi, \quad V_n = \alpha_9\,(e_{ss} + e_{zz}) + \alpha_4\gamma_n + \beta_5\varphi + \bar\beta_2\tau,$$

$$M_{ss} = \alpha_0\rho_{ss} + \alpha_5\rho_{zz} + \beta_7\psi + \bar\beta_4\sigma, \quad M_{zz} = \alpha_5\rho_{ss} + \alpha_0\rho_{zz} + \beta_7\psi + \bar\beta_4\sigma,$$

$$M_{sz} = \alpha_6\rho_{zs} + \alpha_7\rho_{sz}, \quad M_{zs} = \alpha_6\rho_{sz} + \alpha_7\rho_{zs}, \quad M_{sn} = \alpha_8\rho_{ns} + \beta_2\tfrac{\partial}{\partial s}\varphi,$$

$$M_{zn} = \alpha_8\rho_{nz} + \beta_2\tfrac{\partial}{\partial z}\varphi, \quad g = \beta_4\,(e_{ss} + e_{zz}) + \beta_5\gamma_n + \beta_3\varphi + \bar\beta_3\tau,$$

$$h_s = \beta_2\rho_{ns} + \beta_1\tfrac{\partial}{\partial s}\varphi, \quad h_z = \beta_2\rho_{nz} + \beta_1\tfrac{\partial}{\partial z}\varphi, \quad H_s = \beta_9\gamma_s + \beta_8\tfrac{\partial}{\partial s}\psi,$$

$$H_z = \beta_9\gamma_z + \beta_8\tfrac{\partial}{\partial z}\psi, \quad G = \beta_7(\rho_{ss} + \rho_{zz}) + \beta_6\psi + \bar\beta_5\sigma. \tag{3}$$

Here, $\alpha_1, ..., \alpha_9$, $\beta_1, ..., \beta_9$ and $\bar\beta_1, ..., \bar\beta_5$ are the constant constitutive coefficients of the Cosserat shell, and we denote by $\alpha_0 = \alpha_5 + \alpha_6 + \alpha_7$, $\beta_0 = \alpha_5\alpha_0^{-1}$. The fields g and G represent the internal equilibrated body forces.

The equilibrium equations in the absence of body loads can be written as

$$\tfrac{\partial}{\partial s}N_{ss} + \tfrac{\partial}{\partial z}N_{zs} + V_s r^{-1} = 0, \quad \tfrac{\partial}{\partial s}N_{sz} + \tfrac{\partial}{\partial z}N_{zz} = 0,$$

$$\tfrac{\partial}{\partial s}V_s + \tfrac{\partial}{\partial z}V_z - N_{ss}r^{-1} = 0, \quad \tfrac{\partial}{\partial s}M_{ss} + \tfrac{\partial}{\partial z}M_{zs} - V_s = 0,$$

$$\tfrac{\partial}{\partial s}M_{sz} + \tfrac{\partial}{\partial z}M_{zz} - V_z = 0, \quad \tfrac{\partial}{\partial s}M_{sn} + \tfrac{\partial}{\partial z}M_{zn} - V_n = 0,$$

$$\tfrac{\partial}{\partial s}h_s + \tfrac{\partial}{\partial z}h_z - g = 0, \quad \tfrac{\partial}{\partial s}H_s + \tfrac{\partial}{\partial z}H_z - G = 0. \tag{4}$$

Since the lateral edges are free of applied loads, we have the following boundary conditions

$$\mathbf{N} = \mathbf{0}, \quad \mathbf{M} = \mathbf{0}, \quad h = 0, \quad H = 0 \quad \text{on } L_0 \cup L_{\bar s}. \tag{5}$$

We define the vector-valued linear functionals $\mathcal{R}(\cdot)$ and $\mathcal{M}(\cdot)$ by

$$\mathcal{R}(v) = \int_{\mathcal{C}_0} \mathbf{N}(v)\,\mathrm{d}l, \quad \mathcal{M}(v) = \int_{\mathcal{C}_0} [\mathbf{R} \times \mathbf{N}(v) + \mathbf{D} \times \mathbf{M}(v)]\,\mathrm{d}l,$$

for any displacement and porosity field $v = \{\mathbf{u}, \boldsymbol{\delta}, \varphi, \psi\} \in C^1(\bar{\mathcal{S}})$. We mention that $\mathcal{R}(v)$ and $\mathcal{M}(v)$ represent the resultant force and the resultant moment about O of the contact forces and contact director couples acting on $\mathcal{C}_0$, corresponding to the field v. We consider the following boundary conditions on the end edges

$$\mathcal{R}(v) = \mathbf{0}, \qquad \mathcal{M}(v) = \mathbf{0}, \tag{6}$$

$$\int_{\mathcal{C}_0} h(v)\,\mathrm{d}l = \int_{\mathcal{C}_{\bar{z}}} h(v)\,\mathrm{d}l = 0, \quad \int_{\mathcal{C}_0} H(v)\,\mathrm{d}l = \int_{\mathcal{C}_{\bar{z}}} H(v)\,\mathrm{d}l = 0. \tag{7}$$

To resume, the problem of thermal stresses under consideration consists in finding the displacement and porosity field $v = \{\mathbf{u}, \boldsymbol{\delta}, \varphi, \psi\} \in C^2(\mathcal{S}) \cap C^1(\bar{\mathcal{S}})$ which satisfies the Equations (2)–(4) and the conditions (5)–(7), assuming that the temperature fields τ and σ are given. In this paper, we shall determine a solution of this problem in the case when the temperature distribution is independent of the axial coordinate z, i.e. we have

$$\tau = \tau(s), \qquad \sigma = \sigma(s). \tag{8}$$

In the next section, we present some auxiliary results.

3 Saint-Venant's Problem for Porous Shells

The solution of our problem is based on certain results concerning Saint-Venant's problem for porous shells, which we recall subsequently. Following the classical procedure (see [15]), we reduce our thermal stresses problem to an elastostatic problem for cylindrical porous shells, where the mechanical loads are expressed in terms of the given temperature fields. The static deformation of cylindrical porous Cosserat shells has been investigated in [17], within the isothermal theory. Suggested by the results of [17], we introduce the following displacement and porosity field with remarkable properties

$$u_\alpha = -\tfrac{1}{2}\,a_\alpha x_3^2 + W_\alpha[a_i](s), \quad u_3 = (a_\alpha x_\alpha + a_3)x_3, \quad \delta_\alpha = Z_\alpha[a_i](s),$$
$$\delta_3 = \epsilon_{\alpha\beta} a_\alpha x'_\beta x_3, \qquad \varphi = a_i \varphi_{(i)}(s), \qquad \psi = a_\alpha \psi_{(\alpha)}(s), \tag{9}$$

where a_i $(i = 1, 2, 3)$ are arbitrary constants and we have denoted by $W_\alpha[a_i](\cdot)$ and $Z_\alpha[a_i](\cdot)$ the functions

$$W_\alpha[a_i](s) = \beta_0 \epsilon_{\alpha\beta} \epsilon_{\gamma\delta} a_\gamma \int_0^s x'_\beta x_\delta \mathrm{d}s$$
$$+ a_i \int_0^s \left[x'_\alpha y_{(i)} + \epsilon_{\alpha\beta} x'_\beta \left(\int_0^s \left(y_{(i)} r^{-1} + \beta_7 \alpha_0^{-1} \psi_{(i)} \right) \mathrm{d}s - \beta_9 \alpha_3^{-1} \psi'_{(i)} \right) \right] \mathrm{d}s,$$
$$Z_\alpha[a_i](s) = \beta_0 \epsilon_{\beta\gamma} a_\gamma x_\beta x'_\alpha + a_i \left(\epsilon_{\alpha\beta} x'_\beta z_{(i)} - x'_\alpha \int_0^s \left(y_{(i)} r^{-1} + \beta_7 \alpha_0^{-1} \psi_{(i)} \right) \mathrm{d}s \right). \tag{10}$$

72   M. Bîrsan

We designate the displacement and porosity field (9) by $v[a_i]$, indicating thus its dependence on the constants a_i. The functions $\psi_{(\alpha)}(s)$, $s \in [0, \bar{s}]$, which appear in the expression of $v[a_i]$, are the unique solutions of the boundary-value problems

$$\left(\beta_8 - \beta_9^2 \alpha_3^{-1}\right) \psi''_{(\alpha)} - \left(\beta_6 - \beta_7^2 \alpha_0^{-1}\right) \psi_{(\alpha)} = \beta_7(1 - \beta_0)\epsilon_{\alpha\beta} x'_\beta \,,$$
$$\psi'_{(\alpha)}(0) = \psi'_{(\alpha)}(\bar{s}) = 0, \qquad \alpha = 1, 2.$$

Also, the functions $y_{(\alpha)}(s)$, $z_{(\alpha)}(s)$ and $\varphi_{(\alpha)}(s)$, $s \in [0, \bar{s}]$, are determined by the boundary-value problems

$$\alpha_8 z''_{(\gamma)} + \beta_2 \varphi''_{(\gamma)} - \alpha_4 z_{(\gamma)} - \beta_5 \varphi_{(\gamma)} - \alpha_9 y_{(\gamma)} = \alpha_9 x_\gamma \,,$$
$$\beta_2 z''_{(\gamma)} + \beta_1 \varphi''_{(\gamma)} - \beta_5 z_{(\gamma)} - \beta_3 \varphi_{(\gamma)} - \beta_4 y_{(\gamma)} = \beta_4 x_\gamma \,,$$
$$(\alpha_1 + 2\alpha_2)\, y_{(\gamma)} + \alpha_9 z_{(\gamma)} + \beta_4 \varphi_{(\gamma)} = -\alpha_1 x_\gamma \,,$$
$$z'_{(\gamma)}(0) = z'_{(\gamma)}(\bar{s}) = 0, \quad \varphi'_{(\gamma)}(0) = \varphi'_{(\gamma)}(\bar{s}) = 0, \quad \gamma = 1, 2.$$

On the other hand, $y_{(3)}$, $z_{(3)}$ and $\varphi_{(3)}$ denote the constants given by the algebraic system

$$(\alpha_1 + 2\alpha_2)\, y_{(3)} + \alpha_9 z_{(3)} + \beta_4 \varphi_{(3)} = -\alpha_1 \,,$$
$$\alpha_9 y_{(3)} + \alpha_4 z_{(3)} + \beta_5 \varphi_{(3)} = -\alpha_9 \,,$$
$$\beta_4 y_{(3)} + \beta_5 z_{(3)} + \beta_3 \varphi_{(3)} = -\beta_4 \,,$$

and we have used the notation $\psi_{(3)} = 0$, for simplicity.

The displacement and porosity field $v[a_i]$ defined by (9) admits the following properties:

(i) $\frac{\partial}{\partial z} v[a_i]$ is a rigid body displacement field of the Cosserat shell;
(ii) $v[a_i]$ satisfies the equations of equilibrium in the absence of body loads (4) and the boundary conditions on the lateral edges (5), in the isothermal theory;
(iii) $v[a_i]$ corresponds to the resultant force and resultant moment on the end edge $\mathcal{C}_0$ given by

$$\mathcal{R}(v[a_i]) = (I_{3i} a_i)\, \mathbf{e}_3, \qquad \mathcal{M}(v[a_i]) = (I_{\alpha i} a_i)\, \mathbf{e}_\alpha \,, \tag{11}$$

and satisfies the conditions (7).

The coefficients I_{ji} which appear in (11) are defined by

$$I_{\alpha\gamma} = \int_0^{\bar{s}} \left[2\alpha_2 \epsilon_{\alpha\beta} x_\beta (y_{(\gamma)} - x_\gamma) + x'_\alpha \left((\alpha_0 - \alpha_5 \beta_0)\epsilon_{\gamma\beta} x'_\beta + \beta_7(1 - \beta_0)\psi_{(\gamma)}\right)\right] \mathrm{d}s,$$

$$I_{\alpha 3} = 2\alpha_2(y_{(3)} - 1)\epsilon_{\alpha\beta} \int_0^{\bar{s}} x_\beta \mathrm{d}s, \qquad I_{3\alpha} = 2\alpha_2 \int_0^{\bar{s}} (y_{(\alpha)} - x_\alpha)\mathrm{d}s \,,$$

$$I_{33} = 2\alpha_2(y_{(3)} - 1)\bar{s} \,.$$

4 The Solution of the Thermal Stresses Problem

In what follows, we present a solution of the thermal stresses problem formulated in Section 2. We assume that the functions (8) are given. First, let us write our problem in the form of an elastostatic problem, where the mechanical loads are expressed in terms of the given temperature fields.

We observe that, if we separate the thermal terms (i.e., those involving τ and σ) from the elastic terms, then the constitutive Equations (3) can be written as

$$N_{ss} = N_{ss}^e + \bar{\beta}_1\tau + \bar{\beta}_4\sigma r^{-1}, \quad N_{zz} = N_{zz}^e + \bar{\beta}_1\tau, \quad N_{sz} = N_{sz}^e, \quad N_{zs} = N_{zs}^e,$$
$$V_s = V_s^e, \quad V_z = V_z^e, \quad V_n = V_n^e + \bar{\beta}_2\tau, \quad M_{ss} = M_{ss}^e + \bar{\beta}_4\sigma, \quad M_{sz} = M_{sz}^e,$$
$$M_{zz} = M_{zz}^e + \bar{\beta}_4\sigma, \quad M_{zs} = M_{zs}^e, \quad M_{sn} = M_{sn}^e \quad M_{zn} = M_{zn}^e, \quad h_s = h_s^e,$$
$$h_z = h_z^e, \quad g = g^e + \bar{\beta}_3\tau, \quad G = G^e + \bar{\beta}_5\sigma, \quad H_s = H_s^e, \quad H_z = H_z^e,$$

$$(12)$$

where the superscript e is used to indicate the *elastic* terms. In view of (3) and (12), the tensor components N_{ss}^e ,..., H_z^e have obvious expressions in terms of the strain measures (2) and the volume fraction fields φ and ψ.

By virtue of (4), (8) and (12), the equations of equilibrium for our problem have the form

$$\tfrac{\partial}{\partial s}N_{ss}^e + \tfrac{\partial}{\partial z}N_{zs}^e + \tfrac{1}{r}V_s^e = -\tfrac{\mathrm{d}}{\mathrm{d}s}(\bar{\beta}_1\tau + \tfrac{1}{r}\bar{\beta}_4\sigma), \quad \tfrac{\partial}{\partial s}N_{sz}^e + \tfrac{\partial}{\partial z}N_{zz}^e = 0,$$
$$\tfrac{\partial}{\partial s}V_s^e + \tfrac{\partial}{\partial z}V_z^e - \tfrac{1}{r}N_{ss}^e = \tfrac{1}{r}(\bar{\beta}_1\tau + \tfrac{1}{r}\bar{\beta}_4\sigma), \quad \tfrac{\partial}{\partial s}M_{ss}^e + \tfrac{\partial}{\partial z}M_{zs}^e - V_s^e = -\bar{\beta}_4\tfrac{\mathrm{d}}{\mathrm{d}s}\sigma,$$
$$\tfrac{\partial}{\partial s}M_{sz}^e + \tfrac{\partial}{\partial z}M_{zz}^e - V_z^e = 0, \quad \tfrac{\partial}{\partial s}M_{sn}^e + \tfrac{\partial}{\partial z}M_{zn}^e - V_n^e = \bar{\beta}_2\tau,$$
$$\tfrac{\partial}{\partial s}h_s^e + \tfrac{\partial}{\partial z}h_z^e - g^e = \bar{\beta}_3\tau, \quad \tfrac{\partial}{\partial s}H_s^e + \tfrac{\partial}{\partial z}H_z^e - G^e = \bar{\beta}_5\sigma.$$

$$(13)$$

The boundary conditions on the lateral edges (5) reduce to

$$N_{ss}^e = -(\bar{\beta}_1\tau + \tfrac{1}{r}\bar{\beta}_4\sigma), \quad N_{sz}^e = 0, \quad V_s^e = 0, \quad M_{ss}^e = -\bar{\beta}_4\sigma,$$
$$M_{sz}^e = 0, \quad M_{sn}^e = 0, \quad h_s^e = 0, \quad H_s^e = 0 \quad \text{for} \quad s = 0, \bar{s}.$$

$$(14)$$

On the other hand, the boundary conditions on the end edges (6), (7) can be written as

$$\int_{\mathcal{C}_0}\left(x_\alpha' N_{zs}^e + \epsilon_{\alpha\beta}x_\beta' V_z^e\right)\mathrm{d}l = 0, \quad \int_{\mathcal{C}_0}N_{zz}^e\,\mathrm{d}l = -\int_{\mathcal{C}_0}\bar{\beta}_1\tau\,\mathrm{d}l,$$
$$\int_{\mathcal{C}_0}(\epsilon_{\beta\alpha}x_\beta N_{zz}^e + x_\alpha' M_{zz}^e)\,\mathrm{d}l = -\int_{\mathcal{C}_0}(\bar{\beta}_4\sigma x_\alpha' + \bar{\beta}_1\tau\epsilon_{\beta\alpha}x_\beta)\mathrm{d}l,$$
$$\int_{\mathcal{C}_0}(\epsilon_{\alpha\beta}x_\alpha' x_\beta N_{zs}^e + x_\alpha x_\alpha' V_z^e - M_{zs}^e)\,\mathrm{d}l = 0,$$
$$\int_{\mathcal{C}_0}h_z^e\,\mathrm{d}l = \int_{\mathcal{C}_{\bar{z}}}h_z^e\,\mathrm{d}l = 0, \quad \int_{\mathcal{C}_0}H_z^e\,\mathrm{d}l = \int_{\mathcal{C}_{\bar{z}}}H_z^e\,\mathrm{d}l = 0.$$

$$(15)$$

Our problem consists in determining the displacement and porosity field $v = \{\mathbf{u}, \boldsymbol{\delta}, \varphi, \psi\}$ which satisfies the equilibrium Equations (13) and the conditions

(14), (15). Suggested by the corresponding results from the three-dimensional theory (see [15, 18]), we search for the solution v in the form

$$v = v[a_i] + w(s),\tag{16}$$

where $v[a_i]$ is the field defined in Sect. 3 and $w(s) = \{f_i(s)\mathbf{e}_i, g_i(s)\mathbf{e}_i, p(s), q(s)\}$ is an unknown field of class $C^2[0, \bar{s}]$. Let us determine the functions $f_i(s)$, $g_i(s)$, $p(s)$, $q(s)$ and the constants a_i ($i = 1, 2, 3$) such that the field v given by (16) represent a solution of the problem (13)–(15). Taking into account the property (ii) of $v[a_i]$ and the linearity of the theory in conjunction with (16), it follows that the field $w(s)$ must satisfy the Equations (13) and the boundary conditions (14).

If we write the Equation $(13)_2$ and the conditions $(14)_2$ for the field $w(s)$, we obtain

$$f_3''(s) = 0, \quad s \in [0, \bar{s}] \quad \text{and} \quad f_3'(0) = f_3'(\bar{s}) = 0.$$

Neglecting a rigid body displacement field of the Cosserat shell, we deduce that

$$f_3(s) = 0, \quad s \in [0, \bar{s}].\tag{17}$$

Analogously, from the Equation $(13)_5$ and the conditions $(14)_5$ we get

$$g_3(s) = 0, \quad s \in [0, \bar{s}].\tag{18}$$

Imposing that $w(s)$ verifies the Equations $(13)_{1,3,4,6,7,8}$ and the conditions $(14)_{1,3,4,6,7,8}$ we find (by a straightforward calculation) the following system for the determination of $f_\alpha(s)$, $g_\alpha(s)$, $p(s)$ and $q(s)$

$$\begin{aligned}
&(\alpha_1 + 2\alpha_2)(\mathbf{f}' \cdot \boldsymbol{\tau}) + \alpha_9(\mathbf{g} \cdot \mathbf{n}) + \beta_4 p = -\bar{\beta}_1 \tau, \\
&\alpha_3(\mathbf{f}' \cdot \mathbf{n}) + \alpha_3(\mathbf{g} \cdot \boldsymbol{\tau}) + \beta_9 q' = 0, \\
&\alpha_0 r^{-1}(\mathbf{f}' \cdot \boldsymbol{\tau}) + \alpha_0(\mathbf{g} \cdot \boldsymbol{\tau})' + \beta_7 q = -\bar{\beta}_4 \sigma, \\
&[\alpha_8(\mathbf{g} \cdot \mathbf{n})'' - \alpha_4(\mathbf{g} \cdot \mathbf{n})] + [\beta_2 p'' - \beta_5 p] - \alpha_9(\mathbf{f}' \cdot \boldsymbol{\tau}) = \bar{\beta}_2 \tau, \\
&[\beta_2(\mathbf{g} \cdot \mathbf{n})'' - \beta_5(\mathbf{g} \cdot \mathbf{n})] + [\beta_1 p'' - \beta_3 p] - \beta_4(\mathbf{f}' \cdot \boldsymbol{\tau}) = \bar{\beta}_3 \tau, \\
&\left(\beta_8 - \beta_9^2 \alpha_3^{-1}\right) q'' - \left(\beta_6 - \beta_7^2 \alpha_0^{-1}\right) q = \left(\bar{\beta}_5 - \bar{\beta}_4 \beta_7 \alpha_0^{-1}\right) \sigma, \quad s \in [0, \bar{s}],
\end{aligned}\tag{19}$$

together with the boundary conditions

$$(\mathbf{g} \cdot \mathbf{n})' = 0, \quad p' = 0, \quad q' = 0 \quad \text{for} \quad s = 0, \bar{s},\tag{20}$$

where we have denoted by $\mathbf{f} = f_\alpha \mathbf{e}_\alpha$ and $\mathbf{g} = g_\alpha \mathbf{e}_\alpha$.

In view of the restrictions on the constitutive coefficients deduced from the hypothesis of positive definiteness for the strain energy density of the porous Cosserat shell (see [17]), we notice that the differential Equation $(19)_6$ with the conditions $(20)_3$ admit an unique solution $q(s)$. Also, from the system of Equations $(19)_{1,4,5}$ and the conditions $(20)_{1,2}$ we can determine the functions $(\mathbf{f}' \cdot \boldsymbol{\tau})(s)$, $(\mathbf{g} \cdot \mathbf{n})(s)$ and $p(s)$, using the variation of constants method (see e.g., [19]). Next, from the Equation $(19)_3$ we obtain $(\mathbf{g} \cdot \boldsymbol{\tau})(s)$ and from the relation $(19)_2$ we get the function $(\mathbf{f}' \cdot \mathbf{n})(s)$. Hence, we can find the unknown functions $f_\alpha(s)$ and $g_\alpha(s)$. In what follows, we consider that $f_\alpha(s)$, $g_\alpha(s)$, $p(s)$ and $q(s)$ have been so determined and thus, the field $w(s)$ is known.

Let us find the constants a_i $(i = 1, 2, 3)$ by imposing that the field v expressed by (16) verifies the conditions on the end edges (15). In view of the constitutive equations, the property (iii) of $v[a_i]$ and the relations (17), (18), we deduce that the conditions $(15)_{1,4,5,6}$ are satisfied. On the other hand, by virtue of (11), the conditions $(15)_{2,3}$ reduce to the relations

$$
\begin{aligned}
I_{3i}\, a_i &= -2\alpha_2 \int_0^{\bar{s}} (\mathbf{f}' \cdot \boldsymbol{\tau})\mathrm{d}s, \\
I_{\alpha i}\, a_i &= -\int_0^{\bar{s}} \left[2\alpha_2(\mathbf{f}' \cdot \boldsymbol{\tau})\epsilon_{\alpha\beta}x_\beta + (1 - \beta_0)(\beta_7 q + \bar{\beta}_4\sigma)x'_\alpha \right] \mathrm{d}s.
\end{aligned}
\tag{21}
$$

The values of a_i are determined from the system of linear algebraic Equations (21).

In conclusion, on the basis of (16)–(18), we have obtained the following solution for our thermal stresses problem

$$
\begin{aligned}
u_\alpha &= -\tfrac{1}{2} a_\alpha x_3^2 + W_\alpha[a_i](s) + f_\alpha(s), & u_3 &= (a_\alpha x_\alpha + a_3)x_3\,, \\
\delta_\alpha &= Z_\alpha[a_i](s) + g_\alpha(s), & \delta_3 &= \epsilon_{\alpha\beta}a_\alpha x'_\beta x_3\,, \\
\varphi &= a_i\varphi_{(i)}(s) + p(s), & \psi &= a_\alpha\psi_{(\alpha)}(s) + q(s),
\end{aligned}
\tag{22}
$$

where W_α and Z_α are the functions defined by (10).

The solution (22) gives the displacement vector, the director displacement vector and the variations of volume fraction fields for the cylindrical Cosserat shell in equilibrium, subject to a temperature distribution independent of the axial coordinate. We observe that, in our case, the porosity fields φ and ψ do not depend on the axial coordinate.

Using the same technique, we can solve the thermal stresses problem also in the case of *closed* cylindrical shells, i.e. thin-walled tubes. The method can be extended for the treatment of more general problems, when the temperature distribution is a polynomial in the axial coordinate z, which coefficients depend only on the circumferential coordinate s (in this direction, see e.g. [20], Chap. 6).

Acknowledgements

The author acknowledges support from the Ministry of Education and Research through CEEX programm, Contract CERES No. 56/25.07.2006.

References

1. Naghdi P M (1972) The Theory of Shells and Plates. In: C. Truesdell (ed), Handbuch der Physik, Vol. VI a/2, pp. 425–640, Springer-Verlag, Berlin Heidelberg New York

2. Rubin M B (2000) Cosserat Theories: Shells, Rods, and Points. Kluwer Academic Publishers, Dordrecht

3. Green A E, Naghdi P M (1979) On thermal effects in the theory of shells. Proc. R. Soc. Lond. A365: 161–190

4. Bîrsan M (2006) On the theory of elastic shells made from a material with voids. Int. J. Solids Struct. 43: 3106–3123

5. Ciarlet P G (2000) Mathematical Elasticity, Vol. III: Theory of Shells. North-Holland, Amsterdam

6. Bîrsan M (2008) Inequalities of Korn's type and existence results in the theory of Cosserat elastic shells. J. Elasticity 90: 227–239

7. Bîrsan M (2006) Several results in the dynamic theory of thermoelastic Cosserat shells with voids. Mech. Res. Comm. 33: 157–176

8. Rubin M B, Benveniste Y (2004) A Cosserat shell model for interphases in elastic media. J. Mech. Phys. Solids 52: 1023–1052

9. Bîrsan M (2006) On a thermodynamic theory of porous Cosserat elastic shells. J. Therm. Stresses 29: 879–900

10. Nunziato J W, Cowin S C (1979) A nonlinear theory of elastic materials with voids. Arch. Ration. Mech. An. 72: 175–201

11. Capriz G, Podio-Guidugli P (1981) Materials with spherical structure. Arch. Ration. Mech. An. 75: 269–279

12. Vekua I N, Rukhadze A K (1933) Torsion problem for a circular cylinder reinforced by a longitudinal circular rod (in Russian). Izv. Akad. Nauk SSSR 3: 1297–1308

13. Lomakin V A (1976) Theory of Nonhomogeneous Elastic Bodies (in Russian). MGU, Moscow

14. Hatiashvili G M (1983) Almansi-Michell Problems for Homogeneous and Composed Bodies (in Russian). Izd. Metzniereba, Tbilisi

15. Ieşan D (1987) Saint-Venant's Problem. Lect. Notes Math., no. 1279, Springer-Verlag, Berlin

16. Bîrsan M (2004) The solution of Saint-Venant's problem in the theory of Cosserat shells. J. Elasticity 74: 185–214

17. Bîrsan M (2006) Extension, bending and torsion of cylindrical Cosserat shells made from a porous elastic material. In: CD-ROM Proceedings of the 3rd European Conference on Computational Mechanics (Portugal, Lisbon, 5–8 June 2006), 18pp., ISBN 978-1-4020-4994-1, Springer, Dordrecht

18. Ieşan D (2007) Thermal stresses in inhomogeneous porous elastic cylinders. J. Therm. Stresses 30: 145–164

19. Vrabie I I (2004) Differential Equations: An Introduction to Basic Concepts, Results and Applications. World Scientific, New Jersey

20. Ieşan D (2004) Thermoelastic Models of Continua. Kluwer Academic Publishers, Dordrecht Boston London

Vibration of an Elastic Plate Under the Action of an Incompressible Fluid

Natalia Chinchaladze

I. Vekua Institute of Applied Mathematics of Iv. Javakhishvili Tbilisi State University, 2, University St., 0186, Tbilisi, Georgia, natalic@viam.sci.tsu.ge

Abstract This paper deals with solid-fluid interaction problems where the solid part is an elastic plate considered in the $N = 0$ approximation of Vekua's hierarchical models [21], namely, with the cylindrical vibration of an elastic plate under the action of an incompressible fluid has been studied.

Keywords: hierarchical models, plates with variable thickness, prismatic shells, incompressible fluid, solid-fluid interaction problem

1 Introduction

I. Vekua [21] introduced linear hierarchical models for elastic prismatic shells which was based on the expansion of the three-dimensional displacement vector field and the strain and stress tensors into orthogonal Fourier-Legendre series with respect to the plate thickness variable. By taking into account only the first $N + 1$ terms of these expansions, he introduced the so-called N-th approximation. Each of these approximations ($N = 0, 1, ...$) can be considered as an independent mathematical model of plates. In particular, the approximation for $N = 1$, actually, corresponds to the classical Kirchhoff-Love plate models. In the 1960s, I. Vekua [22] developed the analogous mathematical model for thin shallow shells. All his results concerning plates and shells are collected in his monograph [23]. Works of I. Babuška, R. Gilbert, D. Gordeziani, V. Guliaev, I. Khoma, A. Khvoles, T. Meunargia, T. Vashakmadze, V. Zhgenti, G. Jaiani, C. Schwab, N. Chinchaladze, S. Kharibegashvili, D. Natroshvili, W.L. Wendland, and others (see e.g., [1,7,10, 14–17], and the references therein) are devoted to further analysis of I. Vekua's models (rigorous estimation of the modelling error, numerical solutions, etc.).

Sanchez-Palensia E., Wol'mir, A., Jäger, W., Mikelic, A., Bielak, J., MacCamy, R., Boström, A., Everstine, G., Au-Yang, M., and many others devoted attention to solid-fluid interaction problems (see e.g., [9,19,20], and the references therein). In [4–6] interaction problems are considered where the profile of an elastic part is a cusped one on some part of the plate projection boundary. The transmission conditions at the interface between the elastic cusped plate and the fluid, where the fluid induces cylindrical bending of the plate, were established in [4–6]. Within the Kirchhoff-Love theory, the cylindrical bending of such plates under the action of an incompressible ideal or viscous fluid has been considered; in particular, harmonic vibrations have been studied. In [3], the cylindrical bending of an elastic cusped plate with fixed edges under the action of an incompressible fluid has been investigated within the framework of the $N = 0$ approximation of Vekua's hierarchical models.

The aim of this paper is to study a solid-fluid interaction problem where continuity conditions of displacements and stresses are fulfilled and at the interface and the solid is an elastic plate in the $N = 0$ Vekua approximation.

2 Title Problem

Let the projection of the plate on the plane $x_3 = 0$ be

$$\Omega = \{(x_1, x_2, x_3) \in R^3 : -\infty < x_1 < \infty, \ 0 < x_2 < l, \ x_3 = 0\},$$

where l is a width of the plate; the thickness be

$$2h(x_2) := \overset{(+)}{h}(x_2) - \overset{(-)}{h}(x_2), \quad \overset{(+)}{h}(x_2) = -\overset{(-)}{h}(x_2) > 0,$$

$x_3 = \overset{(+)}{h}$ and $x_3 = \overset{(-)}{h}$ are s.c. plate faces which in the cross-section give the face curves.

In this paper we consider the cylindrical vibration of the plate of the infinite length along the axis x_1, i.e., the solid displacement along x_1:

$$u_1^s = 0 \tag{1}$$

in the middle plane, under the action of a plane flow of the fluid parallel to the plane Ox_2x_3. Hence, the fluid velocity component $v_1^f \equiv 0$ (the upper indices s and f mean the solid and the fluid parts respectively). So, we actually have the plane deformation parallel to the plane $x_1 = 0$.

In the zero approximation of I. Vekua's hierarchical models the governing equations for the elastic plate by the cylindrical deformation has the following form (see, [11–13]):

$$(\lambda^s + 2\mu^s)(2h(x_2)u_{20,2}^s(x_2,t))_{,2} + X_{20}^s(x_2,t) = 2h(x_2)\,\rho^s\,u_{20,tt}^s(x_2,t), \tag{2}$$

$$\mu^s\,(2h(x_2)\,u_{30,2}^s(x_2,t))_{,2} + X_{30}^s(x_2,t) = 2h(x_2)\,\rho^s\,u_{30,tt}^s(x_2,t), \tag{3}$$

$$x_2 \in\,]0, l[, \ \ t > 0,$$

where $u_{j0}^s(x_2)$ $(j = 2, 3)$ are the components of the displacement vector, λ^s and μ^s are Lamé constants, ρ^s is a density,

$$X_{i0}^s(x_2,t) := Q_{\underset{n}{(+)}_i}^s(x_2,t)\sqrt{1+(\overset{(+)}{h},_2)^2} + Q_{\underset{n}{(-)}_i}^s(x_2,t)\sqrt{1+(\overset{(-)}{h},_2)^2}$$
$$+ \Phi_{i0}^s(x_2,t), \quad i=2,3,$$

here

$$\Phi_{i0}^s(x_2,t) := \int\limits_{\underset{h}{(-)}}^{\overset{(+)}{h}} \Phi_i^s(x_2,x_3,t)dx_3, \quad i=2,3,$$

are the zero moments of the component of volume forces $\Phi_i^s \; i=2,3,$

$$Q_{\underset{n}{(\pm)}_j}^s(x_2,t) := \sum_{i=2}^{3} \sigma_{ij}^s(x_2,\overset{(\pm)}{h},t)\cos(\overset{(\pm)}{n},x_i) = \sigma_{2j}^s(x_2,\overset{(\pm)}{h},t)\cos(\overset{(+)}{n},x_2)$$
$$+ \sigma_{3j}^s(x_2,\overset{(\pm)}{h},t)\cos(\overset{(\pm)}{n},x_3), \quad j=2,3,$$

$\sigma_{ij}^s \; (i=2,3, \; j=2,3)$ are stresses.

In the case under consideration Hooke's law has the form

$$\sigma_{130}^s(x_2,t) = \mu^s h\frac{\partial u_{30}^s(x_2,t)}{\partial x_1} \equiv 0,$$

$$\sigma_{230}^s(x_2,t) = \mu^s h\frac{\partial u_{30}^s(x_2,t)}{\partial x_2}, \tag{4}$$

where $\sigma_{\alpha 30}^s(x_2,t)$ is the so called zero moment of the stress $\sigma_{\alpha 3}^s(x_2,x_3,t)$ which in the zero approximation is equal to $\frac{\sigma_{\alpha 30}^s(x_2,t)}{2h(x_2)}$ $(\alpha=1,2)$.

Let in the fluid part

$$p(x_2,x_3,t) \to p_\infty(t) \quad \text{when} \quad x_2^2+x_3^2 \to +\infty, \tag{5}$$

and

$$v_j^f(x_2,x_3,t) = v_{j\infty}^f(t) \quad \text{when} \quad x_2^2+x_3^2 \to +\infty, \tag{6}$$

where $p(x_2,x_3,t)$ is the pressure, $v^f := (v_2^f,v_3^f)$ is a velocity vector of the fluid in the plane $x_1=0$, p_∞ and $v_{j\infty}^f$ are the prescribed at infinity values of the pressure and velocity vector component correspondingly.

In what follows we consider the incompressible Newton fluid, i.e.,

$$\text{div } v^f(x_2,x_3,t) = 0, \; t \geq 0, \tag{7}$$

and (see e.g., [8], p. 5)

$$\sigma_{jk}^f = -p\delta_{jk} + \mu^f\left(\frac{\partial v_j^f}{\partial x_k} + \frac{\partial v_k^f}{\partial x_j}\right), \quad j,k=2,3, \tag{8}$$

in the domain occupied by the fluid, where σ^f_{jk} is the stress tensor, μ^f is the coefficient of viscosity, δ_{jk} is the Kronecker delta.

Let I be the segment $[0, l]$ of the axis $0x_2$ and Ω^f be the plane $0x_2x_3$ except I.

Since the plate thickness is sufficiently small, we can make the following assumptions:

(i) the fluid occupies Ω^f;

(ii) the plate is identified with its middle plane, i.e., its section by the plane $x_1 = 0$ coincides with I (the plate profile geometry depending on the thickness variation is reflected in the coefficients of Equations (2), (3));

(iii) because of smallness of $\overset{(\pm)}{h}(x_2)$, the transmission conditions

$$\sum_{i=2}^{3} \sigma^f_{ij}(x_2, \overset{(+)}{h}, t)\cos(\overset{(+)}{n}, x_i) = Q^s_{\overset{(+)}{n}_j},$$

$$\sum_{i=2}^{3} \sigma^f_{ij}(x_2, \overset{(-)}{h}, t)\cos(\overset{(-)}{n}, x_i) = Q^s_{\overset{(-)}{n}_j}, \quad j = 2, 3,$$

$$v^f_j(x_2, \overset{(\pm)}{h}, t) = u^s_{j,t}(x_2, \overset{(\pm)}{h}, t), \quad j = 2, 3,$$

on the face curves $\Big[$since the normals of I are $(0, 0, 1)$ and $(0, 0, -1)$, we can take $\sigma^f_{33}(x_2, 0\pm, t) \approx \sigma^f_{33}(x_2, \overset{(\pm)}{h}, t),\ u^s_{3,t}(x_2, \overset{(\pm)}{h}, t) \approx w^s_{30,t}(x_2, t)\Big]$ can be written in the plate middle plane, more precisely, on I:

$$\sigma^f_{32}(x_2, 0+, t) - \sigma^f_{32}(x_2, 0-, t) = X^s_{20} - \Phi^s_{20} = Q^s_{\overset{(+)}{n}_2} + Q^s_{\overset{(-)}{n}_2},$$

$$\sigma^f_{33}(x_2, 0+, t) - \sigma^f_{33}(x_2, 0-, t) = X^s_{30} - \Phi^s_{30} = Q^s_{\overset{(+)}{n}_3} + Q^s_{\overset{(-)}{n}_3}, \tag{9}$$

$$v^f_j(x_2, 0\pm, t) = v^f_j(x_2, 0\pm, t) = u^s_{j0,t}(x_2, t), \quad x_2 \in]0, l[,\ t > 0,\ j = 2, 3. \tag{10}$$

(9), in view of (7), (8) and (10), can be rewritten as

$$\frac{\partial v_2(x_2, 0+, t)}{\partial x_3} - \frac{\partial v_2(x_2, 0-, t)}{\partial x_3} = X^s_{20}(x_2, t) - \Phi^s_{20}(x_2, t), \tag{11}$$

$$-p(x_2, 0+, t) - 2\mu^f \frac{\partial v_2(x_2, 0+, t)}{\partial x_2}$$

$$-\left(-p(x_2, 0-, t) - 2\mu^f \frac{\partial v_2(x_2, 0-, t)}{\partial x_2}\right) = -p(x_2, 0+, t) + p(x_2, 0-, t)$$

$$= X^s_{30}(x_2, t) - \Phi^s_{30}(x_2, t), \quad x_2 \in]0, l[,\ t > 0. \tag{12}$$

We add to the transmission conditions (10–12) the following conditions at points $x_2 = 0$ and $x_2 = l$, respectively,

$$[2h(0)]^{-1} \int_{-h(0)}^{h(0)} v_j^f(l,x_3,t)\,dx_3 = u_{j0,t}^s(0,t), \quad j = 2,3, \tag{13}$$

$$[2h(l)]^{-1} \int_{-h(l)}^{h(l)} v_j^f(l,x_3,t)\,dx_3 = u_{j0,t}^s(l,t), \quad j = 2,3.. \tag{14}$$

Let the motion of the fluid be sufficiently slow, i.e., v_j and $v_{j,k}$ $(j,k = 2,3)$ be so small that linearized Navier-Stokes, i.e., Stokes equations can be applied:

$$\frac{\partial v_j^f(x_2,x_3,t)}{\partial t} = -\frac{1}{\rho^f}\frac{\partial p(x_2,x_3,t)}{\partial x_j} + \nu\Delta v_j^f(x_2,x_3,t), \quad j = 2,3, \tag{15}$$

where $\nu = \mu^f/\rho^f$, $\Delta := \frac{\partial^2}{\partial x_2^2} + \frac{\partial^2}{\partial x_3^2}$, $v_j^f \in C_t^1(]0,+\infty[) \cap C_t([0,+\infty[) \cap C^2(\Omega^f)$. As usual from (7), and (15), there follows

$$\Delta p(x_2,x_3,t) = 0. \tag{16}$$

Let us now consider the case of the harmonic vibration with the oscillation frequency ω, i.e.,

$$p(x_2,x_3,t) = \overset{0}{p}(x_2,x_3)\begin{cases}\cos\omega t,\\ \sin\omega t,\end{cases}$$

$$u_j^f(x_2,x_3,t) = \overset{0}{u_j^f}(x_2,x_3)\begin{cases}\cos\omega t,\\ \sin\omega t,\end{cases} \quad j = 2,3,$$

$$u_{j0}^s(x_2,t) = \overset{0}{u_{j0}^s}(x_2)\begin{cases}\cos\omega t,\\ \sin\omega t,\end{cases} \quad j = 2,3, \quad X_{30}^s(x_2,t) := \overset{0}{X_{30}^s}(x_2)\begin{cases}\cos\omega t,\\ \sin\omega t,\end{cases}$$

$$p_\infty = \overset{0}{p_\infty}\begin{cases}\cos\omega t,\\ \sin\omega t,\end{cases} \quad \Phi_{j0}^s(x_2,t) = \overset{0}{\Phi_{j0}^s}(x_2)\begin{cases}\cos\omega t,\\ \sin\omega t,\end{cases} \quad j = 2,3,$$

where $\overset{0}{\Phi_{20}^s}(x_2) \in C([0,l])$, $\overset{0}{\Phi_{30,2}^s}(x_2) \in C^{0,\gamma}([0,l])$, $0 < \gamma < 1$, by $u_j^f(x_2,x_3,t)$, $j = 2,3$, we denote components of the displacement vector in the fluid part, ω is an arbitrary constant.

Then

$$v_j^f(x_2,x_3,t) = \omega\,\overset{0}{u_j^f}(x_2,x_3)\begin{cases}-\sin\omega t,\\ \cos\omega t,\end{cases} \quad j = 2,3.$$

In this case from Equations (15), (7), (2), and (3) we obtain the following system

$$\omega^2\overset{0}{u_j^f}(x_2,x_3) = \frac{1}{\rho^f}\frac{\partial\overset{0}{p}(x_2,x_3)}{\partial x_j}, \quad j = 2,3, \tag{17}$$

$$\Delta\overset{0}{u_j^f}(x_2,x_3) = 0, \quad j = 2,3, \tag{18}$$

$$\overset{0}{u_{j,j}^f}(x_2,x_3) = 0, \quad j = 2,3, \quad (x_2,x_3) \in \Omega^f, \tag{19}$$

$$(\lambda^s + 2\mu^s)\,(2h(x_2)\,\overset{0}{u^s_{20,2}}(x_2))_{,2} + \overset{0}{X^s_{20}} = -2h(x_2)\omega^2\rho^s\overset{0}{u^s_{20}}(x_2), \qquad (20)$$

$$\mu^s\,(2h(x_2)\,\overset{0}{u^s_{30,2}}(x_2))_{,2} + \overset{0}{X^s_{30}} = -2h(x_2)\omega^2\rho^s\overset{0}{u^s_{30}}(x_2), \quad x_2 \in]0, l[. \qquad (21)$$

From the transmission conditions (10–14) we get

$$\overset{0}{u^f_j}(x_2, 0) = \overset{0}{u^s_{j0}}(x_2), \quad x_2 \in]0, l[, \quad j = 2, 3, \qquad (22)$$

$$\overset{0}{u^s_{j0}}(0) = [2h(0)]^{-1} \int_{-h(0)}^{h(0)} \overset{0}{u^f_j}(0, x_3)\,dx_3, \quad j = 2, 3, \qquad (23)$$

$$\overset{0}{u^s_{j0}}(l) = [2h(l)]^{-1} \int_{-h(l)}^{h(l)} \overset{0}{u^f_j}(l, x_3)\,dx_3, \quad j = 2, 3, \qquad (24)$$

$$\overset{0}{u^f_{2}}_{,3}(x_2, 0+) - \overset{0}{u^f_{2}}_{,3}(x_2, 0-) = 0, \qquad (25)$$

$$\overset{0}{X^s_{20}}(x_2) - \overset{0}{\Phi^s_{20}}(x_2) = 0, \qquad (26)$$

$$\overset{0}{X^s_{30}}(x_2) - \overset{0}{\Phi^s_{30}}(x_2) = -\overset{0}{p}(x_2, 0+) + \overset{0}{p}(x_2, 0-). \qquad (27)$$

From conditions (5), (6) at infinity we have

$$\overset{0}{p}(x_2, x_3) \to \overset{0}{p_\infty}, \quad \overset{0}{u^f_j}(x_2, x_3) \to 0, \ j = 2, 3, \ when \ (x_2^2 + x_3^2) \to +\infty. \quad (28)$$

Evidently, by virtue of (17),

$$\overset{0}{u^f_{2,3}}(x_2, 0+) - \overset{0}{u^f_{2,3}}(x_2, 0-) = \frac{1}{\rho^f\,\omega^2}\,(\overset{0}{p}_{,32}(x_2, 0+) - \overset{0}{p}_{,32}(x_2, 0-))$$

$$= \overset{0}{u^f_{3,2}}(x_2, 0+) - \overset{0}{u^f_{3,2}}(x_2, 0-) = 0,$$

i.e., (25) automatically fulfilled.

So, we arrive at the following Problem:

Find

$$\overset{0}{p}(x_2, x_3) \in C^2(\Omega^f),$$

$$\overset{0}{u^f_j}(x_2, x_3) \in C^2(\Omega^f) \cap C(\Omega^f \cup]0, l[), \quad j = 2, 3,$$

$$\overset{0}{u^s_{j0}}(x_2) \in C^2(]0,l[) \cap C([0,l]), \quad j = 2,3,$$

$$\overset{0}{X^s_{20}}(x_2) \in C([0,l]),$$

and

$$\overset{0}{X^s_{30,2}}(x_2) \in C^{0,\gamma}([0,l]), \quad 0 < \gamma < 1,$$

satisfying the system (17–21), and conditions (22–24), (26–28).

Taking into account (27) and (28), it is easy to see that the solution of Equation (16), has the following form [18]

$$\overset{0}{p}(x_2,x_3) = -\frac{x_3}{2\pi} \int_0^l \frac{(\overset{0}{X^s_{30}}(\xi_2) - \overset{0}{\Phi^s_{30}}(\xi_2))d\xi_2}{(\xi_2 - x_2)^2 + x_3^2} + \overset{0}{p_\infty}. \tag{29}$$

From (17), in view of (29), there follows

$$\overset{0}{u^f_2}(x_2,x_3) = \frac{x_3}{\pi\omega^2\rho^f} \int_0^l \frac{\left\{\overset{0}{X^s_{30}}(\xi_2) - \overset{0}{\Phi^s_{30}}(\xi_2)\right\}(\xi_2 - x_2)d\xi_2}{[(\xi_2 - x_2)^2 + x_3^2]^2}, \tag{30}$$

$$\overset{0}{u^f_3} = \frac{1}{2\pi\omega^2\rho^f} \int_0^l \frac{\left\{\overset{0}{X^s_{30}}(\xi_2) - \overset{0}{\Phi^s_{30}}(\xi_2)\right\}[x_3^2 - (\xi_2 - x_2)^2]\,d\xi_2}{[(\xi_2 - x_2)^2 + x_3^2]^2}. \tag{31}$$

So, $\overset{0}{u^f_2}(x_2,x_3)$ and $\overset{0}{u^f_3}(x_2,x_3)$, by virtue of (30) and (31), are expressed by means of $\overset{0}{X^s_{30}}$.

Substituting (30), (31) in (23) and (24), by virtue of (17), (19), we get

$$\overset{0}{u^s_{20}}(0) = \overset{0}{u^s_{20}}(l) = 0, \tag{32}$$

$$\overset{0}{u^s_{30}}(0) = -\frac{1}{2\pi} \int_0^l \frac{\overset{0}{X^s_{30}}(\xi_2) - \overset{0}{\Phi^s_{30}}(\xi_2)}{\xi_2^2 + h^2(0)}d\xi_2, \tag{33}$$

$$\overset{0}{u^s_{30}}(l) = -\frac{1}{2\pi} \int_0^l \frac{\overset{0}{X^s_{30}}(\xi_2) - \overset{0}{\Phi^s_{30}}(\xi_2)}{(\xi_2 - l)^2 + h^2(l)}d\xi_2. \tag{34}$$

Let us now consider the limit of (31) when $x_3 \to 0$ and $x_2 \in [0,l]$. Evidently,

$$\lim_{x_3 \to 0} \int_0^l \frac{\left[\overset{0}{X}{}^s_{30}(\xi_2) - \overset{0}{\Phi}{}^s_{30}(\xi_2) \right] \left[x_3^2 - (\xi_2 - x_2)^2 \right] d\xi_2}{\left[(\xi_2 - x_2)^2 + x_3^2 \right]^2}$$

$$= \lim_{x_3 \to 0} \left\{ \left[\overset{0}{X}{}^s_{30}(l) - \overset{0}{\Phi}{}^s_{30}(l) \right] \frac{l - x_2}{(l - x_2)^2 + x_3^2} \right.$$

$$+ \left[\overset{0}{X}{}^s_{30}(0) - \overset{0}{\Phi}{}^s_{30}(0) \right] \frac{x_2}{(x_2^2 + x_3^2)} - \left. \int_0^l \frac{\left[\overset{0}{X}{}^s_{30}(\xi_2) - \overset{0}{\Phi}{}^s_{30}(\xi_2) \right]_{,\xi_2} (\xi_2 - x_2) d\xi_2}{(\xi_2 - x_2)^2 + x_3^2} \right\}$$

$$= \frac{\overset{0}{X}{}^s_{30}(l) - \overset{0}{\Phi}{}^s_{30}(l)}{l - x_2} + \frac{\overset{0}{X}{}^s_{30}(0) - \overset{0}{\Phi}{}^s_{30}(0)}{x_2}$$

$$+ \lim_{x_3 \to 0} \left\{ - \int_0^l \frac{\left[\overset{0}{X}{}^s_{30,\xi_2}(\xi_2) - \overset{0}{\Phi}{}^s_{30,\xi_2}(\xi_2) - \overset{0}{X}{}^s_{30,2}(x_2) + \overset{0}{\Phi}{}^s_{30,2}(x_2) \right] (\xi_2 - x_2) d\xi_2}{(\xi_2 - x_2)^2 + x_3^2} \right.$$

$$\left. - \frac{\overset{0}{X}{}^s_{30,2}(x_2) - \overset{0}{\Phi}{}^s_{30,2}(x_2)}{2} \int_0^l \left\{ ln[(\xi_2 - x_2)^2 + x_3^2] \right\}_{,\xi_2} d\xi_2 \right\}$$

$$= \frac{\overset{0}{X}{}^s_{30}(l) - \overset{0}{\Phi}{}^s_{30}(l)}{l - x_2} + \frac{\overset{0}{X}{}^s_{30}(0) - \overset{0}{\Phi}{}^s_{30}(0)}{x_2} - \left[\overset{0}{X}{}^s_{30,2}(x_2) - \overset{0}{\Phi}{}^s_{30,2}(x_2) \right] ln\frac{l - x_2}{x_2}$$

$$- \int_0^l \frac{\overset{0}{X}{}^s_{30,\xi_2}(\xi_2) - \overset{0}{\Phi}{}^s_{30,\xi_2}(\xi_2) - \overset{0}{X}{}^s_{30,2}(x_2) + \overset{0}{\Phi}{}^s_{30,2}(x_2)}{\xi_2 - x_2} d\xi_2,$$

because of $\overset{0}{X}{}^s_{30,2}(x_2)$, $\overset{0}{\Phi}{}^s_{30,2}(x_2) \in C^{0,\gamma}[0,l]$.

On the other hand, if we define the following supersingular integral as an integral in the Hadamard sense [2], we have

$$\int\limits_0^l \frac{\overset{0}{X^s_{30}}(\xi_2) - \overset{0}{\varPhi^s_{30}}(\xi_2)}{(\xi_2 - x_2)^2} d\xi_2 = \lim_{\varepsilon \to 0} \left(\int\limits_0^{x_2-\varepsilon} \frac{\overset{0}{X^s_{30}}(\xi_2) - \overset{0}{\varPhi^s_{30}}(\xi_2)}{(\xi_2 - x_2)^2} d\xi_2 \right.$$

$$+ \int\limits_{x_2+\varepsilon}^l \frac{\overset{0}{X^s_{30}}(\xi_2) - \overset{0}{\varPhi^s_{30}}(\xi_2)}{(\xi_2 - x_2)^2} d\xi_2$$

$$\left. + \frac{2(\overset{0}{X^s_{30}}(x_2) - \overset{0}{\varPhi^s_{30}}(x_2))}{\varepsilon} \right) = \frac{\overset{0}{X^s_{30}}(l) - \overset{0}{\varPhi^s_{30}}(l)}{l - x_2} + \frac{\overset{0}{X^s_{30}}(0) - \overset{0}{\varPhi^s_{30}}(0)}{x_2}$$

$$- \left[\overset{0}{X^s_{30,2}}(x_2) - \overset{0}{\varPhi^s_{30,2}}(x_2) \right] ln\frac{l - x_2}{x_2}$$

$$- \int\limits_0^l \frac{[\overset{0}{X^s_{30,\xi_2}}(\xi_2) - \overset{0}{\varPhi^s_{30,\xi_2}}(\xi_2)] - [\overset{0}{X^s_{30,2}}(x_2) - \overset{0}{\varPhi^s_{30,2}}(x_2)]}{\xi_2 - x_2} d\xi_2.$$

Therefore, for $\overset{0}{u_3^f}(x_2, 0)$ we get the following expression

$$\overset{0}{u_3^f}(x_2, 0) = \frac{1}{2\pi\omega^2\rho^f} \int\limits_0^l \frac{\overset{0}{X^s_{30}}(\xi_2) - \overset{0}{\varPhi^s_{30}}(\xi_2)}{(\xi_2 - x_2)^2} d\xi_2, \quad x_2 \in]0, l[. \tag{35}$$

Integrating twice (21), we obtain

$$\overset{0}{u^s_{30}}(x_2) + \frac{\omega^s\rho^s}{\mu^s} \int\limits_{x_2}^l \frac{1}{h(\eta)} \int\limits_\eta^l h\overset{0}{u^s_{30}}(\xi_2)d\xi_2 d\eta$$

$$= -\frac{1}{2\mu^s} \int\limits_{x_2}^l \frac{1}{h(\eta)} \int\limits_\eta^l \overset{0}{X^s_{30}}(\xi_2)d\xi_2 d\eta + \int\limits_{x_2}^l \frac{c_1}{h(\xi_2)} d\xi_2 + c_2. \tag{36}$$

Substituting (36) into boundary conditions (33) and (34) and solving the obtained system of algebraic equations for c_1, c_2, we get

$$c_1 = \Bigg[-\frac{1}{2\pi\rho^f\omega^2} \int_0^l \frac{\overset{0}{X^s_{30}}(\xi_2) - \overset{0}{\Phi^s_{30}}(\xi_2)}{\xi^2 + h^2(0)} d\xi_2$$

$$+\frac{1}{2\pi\rho^f\omega^2} \int_0^l \frac{\overset{0}{X^s_{30}}(\xi_2) - \overset{0}{\Phi^s_{30}}(\xi_2)}{(\xi_2 - l)^2 + h^2(l)} d\xi_2$$

$$+\frac{\omega^2\rho^s}{\mu^s} \int_0^l h^{-1}(\eta) \int_\eta^l h(\xi_2)\overset{0}{u^s_{30}}(\xi_2)d\xi_2\, d\eta$$

$$+\frac{1}{2\mu^s} \int_0^l h^{-1}(\eta) \int_\eta^l \overset{0}{X^s_{30}}(\xi_2)d\xi_2\, d\eta\Bigg] \Big[\int_0^l \frac{d\xi_2}{h(\xi_2)}\Big]^{-1},$$

$$c_2 = -\frac{1}{2\pi\,\omega^2\rho^f} \int_0^l \frac{\overset{0}{X^s_{30}}(\xi_2) - \overset{0}{\Phi^s_{30}}(\xi_2)}{(\xi_2 - l)^2 + h^2(l)}\, d\xi_2.$$

Substituting obtained expressions of c_1 and c_2 into (36) we get a second kind Fredholm type integral equation with respect to $\overset{0}{u^s_{30}}(x_2)$:

$$\overset{0}{u^s_{30}}(x_2) + \omega^2 \int_0^l K(x_2, \xi_2)h(\xi_2)\overset{0}{u^s_{30}}(\xi_2)d\xi_2 = F_1(x_2) + F_2(x_2), \qquad (37)$$

where

$$K(x_2, \xi_2) := \begin{cases} -\dfrac{\rho^s}{\mu^s} H_1(\xi_2) \displaystyle\int_0^{x_2} h^{-1}(\eta)d\eta, & x_2 \le \xi_2 \le l, \\[2ex] -\dfrac{\rho^s}{\mu^s} H_1(x_2) \displaystyle\int_0^{\xi_2} h^{-1}(\eta)d\eta, & 0 \le \xi_2 \le x_2, \end{cases}$$

$$F_1(x_2) := -\frac{H_1(x_2)}{2\pi\rho^f\omega^2} \int_0^l \frac{\overset{0}{X^s_{30}}(\xi_2)}{\xi_2^2 + h^2(0)} d\xi_2 - \frac{1}{2\mu^s}\Big[\int_{x_2}^l h^{-1}(\eta) \int_0^\eta \overset{0}{X^s_{30}}(\xi_2)\, d\xi_2 d\eta$$

$$-H_1(x_2) \int_0^l h^{-1}(\eta) \int_\eta^l \overset{0}{X^s_{30}}(\xi_2)\, d\xi_2 d\eta\Big] + \frac{H_1(x_2) - 1}{2\pi\rho^f\omega^2} \int_0^l \frac{\overset{0}{X^s_{30}}(\xi_2)}{(\xi_2 - l)^2 + h^2(l)}d\xi_2,$$

$$F_2(x_2) := \frac{1 - H_1(x_2)}{2\pi\rho^f\omega^2} \int_0^l \frac{\overset{0}{\Phi^s_{30}}(\xi_2)}{(\xi_2 - l)^2 + h^2(l)}d\xi_2 - \frac{H_1(x_2)}{2\pi\rho^f\omega^2} \int_0^l \frac{\overset{0}{\Phi^s_{30}}(\xi_2)}{\xi_2^2 + h^2(0)}d\xi_2,$$

with

$$H_1(x_2) := \frac{\int\limits_{x_2}^{l} h^{-1}(\eta)d\eta}{\int\limits_{0}^{l} h^{-1}(\eta)d\eta}.$$

It can be shown that

1. $K(x_2, \xi_2)$ is a symmetric, positive defined kernel;
2. Equation (37) can be reduced to the integral equation with a symmetric kernel, whose solution can be written as follows

$$\overset{0}{u}{}^{s}_{30}(x_2) = F_1(x_2) + F_2(x_2) + \omega^2 \int\limits_{0}^{l} \Gamma(x_2, \xi_2, \omega^2)[F_1(\xi_2) + F_2(\xi_2)]d\xi_2, \quad (38)$$

$$= \int\limits_{0}^{l} K_1(x_2, \xi_2)\overset{0}{X}{}^{s}_{30}(\xi_2)d\xi_2 + f(x_2),$$

where $\Gamma(x_2, \xi_2, \omega^2)$ is a resolvent of integral Equation (37),

$$f(x_2) := F_2(x_2) + \omega^2 \int\limits_{0}^{l} \Gamma(x_2, \xi_2, \omega^2)\, F_2(\xi_2)\, d\xi_2,$$

$$K_1(x_2, \xi_2) = \frac{1}{2\mu^s} \left(H_1(x_2) \int\limits_{0}^{\xi_2} h^{-1}(\eta)d\eta - \int\limits_{\xi_2}^{l} h^{-1}(\eta)d\eta \right)$$

$$+ \frac{1}{2\pi\rho^f \omega^2} \left[\frac{H_1(x_2) - 1}{(\xi_2 - l)^2 + h^2(l)} - \frac{H_1(x_2)}{\xi_2^2 + h^2(0)} \right] + H_2(x_2, \xi_2)$$

$$+ \omega^2 \int\limits_{0}^{l} \Gamma(x_2, \xi, \omega^2) \left\{ \frac{1}{2\mu^s} \left[H_1(\xi) \int\limits_{0}^{\xi_2} h^{-1}(\eta)d\eta - \int\limits_{\xi_2}^{l} h^{-1}(\eta)\, d\eta \right] \right.$$

$$\left. + \frac{1}{2\pi\rho^f \omega^2} \left[\frac{H_1(\xi) - 1}{(\xi_2 - l)^2 + h^2(l)} - \frac{H_1(\xi)}{\xi_2^2 + h^2(0)} \right] + H_2(\xi, \xi_2) \right\} d\xi,$$

with

$$H_2(x_2, \xi_2) := \begin{cases} \dfrac{1}{2\mu^s} \int\limits_{\xi_2}^{x_2} h^{-1}(\eta)d\eta, & 0 \leq \xi_2 \leq x_2, \\[2ex] 0, & x_2 \leq \xi_2 \leq l. \end{cases}$$

So, we have found $\overset{0}{u}{}^{s}_{30}$ by means of $\overset{0}{X}{}^{s}_{30}$ and $\overset{0}{\Phi}{}^{s}_{30}$. For $\overset{0}{X}{}^{s}_{30}$, by virtue of (22) (for $j = 3$), taking into account (38) and (35), we obtain the following integral equation

$$\frac{1}{2\pi\omega^2\rho^f}\int_0^l \frac{\overset{0}{X}_{30}^s(\xi_2) - \overset{0}{\Phi}_{30}^s(\xi_2)}{(\xi_2 - x_2)^2}d\xi_2 - \int_0^l K_1(x_2,\xi_2)\overset{0}{X}_{30}^s(\xi_2)d\xi_2 = f(x_2). \quad (39)$$

Now, we find an approximate solution of (39) $\overset{0}{X}_{30,2}^s(x_2) \in C^{0,\gamma}([0,l])$, $0 < \gamma < 1$, using the method given in [2] (where the segment $[-1,1]$ should be replaced by $[0,l]$).

Let us divide the interval $[0,l]$ into N parts as follows

$$y_k' := \frac{lk}{N}, \quad k = \overline{0,N}, \quad y_k := \frac{lk}{N} + \frac{l}{2N}, \quad k = \overline{0,N-1}.$$

The expression

$$\overset{0}{X}_{30N} := (\overset{0}{X}_{30}(y_0), ..., \overset{0}{X}_{30}(y_{N-1})),$$

will be called an approximate solution of (39).

For $\overset{0}{X}_{30}(y_i)$ we get the following system of linear equations (see [2])

$$a_{\underline{i}\,i}\overset{0}{X}_{30}(y_i) - \sum_{\substack{j=0 \\ j\neq i-1,i,i+1}}^{N-1} \overset{0}{X}_{30}(y_j)\left[\frac{1}{y_{j+i}' - y_i} - \frac{1}{y_j' - y_i}\right]$$

$$- \frac{2\pi\omega^2\rho^f l}{N}\sum_{j=0}^{N-1} K_1(y_i,y_j)\overset{0}{X}_{30}(y_j) = f(y_i), \quad i = \overline{0,N-1}, \qquad (40)$$

where

$$a_{\underline{i}\,i} := -\frac{4N}{l}\int_{\Delta_{\underline{i}\,i}} \frac{d\xi_2}{(\xi_2 - y_i)^2}, \quad \Delta_{\underline{i}\,i} := [0,l] \cap \left[y_i' - \frac{n}{N}, y_{i+1}' + \frac{n}{N}\right], \quad n := \sqrt{N},$$

the bar under repeated indices means that we do not sum with respect to these indices.

Because of $\overset{0}{X}_{30,2}^s \in C^{0,\gamma}([0,l])$ there exists $A = \mathrm{const} > 0$ such that

$$|\overset{0}{X}_{30,2}^s(x_2^1) - \overset{0}{X}_{30,2}^s(x_2^2)| \leq A|x_2^1 - x_2^2|^\gamma \quad \text{for any} \ \ x_2^1, x_2^2 \in [0,l].$$

Following [2], we get

$$|\overset{0}{X^*}_{30} - \overset{0}{X^*}_{30N}| \leq 2^{-\gamma}A\left(\frac{l}{n}\right)^\gamma,$$

where $\overset{0}{X^*}_{30}$ and $\overset{0}{X^*}_{30N}$ are the solutions of the Equations (39) and (40), respectively.

Using the expression of $\overset{0}{X}_{30N}$, from (29) we get the approximate representation for $\overset{0}{p}$ by trapezoid rule,

$$\overset{0}{p}(x_2,x_3) = -\frac{x_3 l}{2\pi N}\sum_{j=0}^{N-1}\frac{\overset{0}{X}_{30}(y_j) - \overset{0}{\varPhi}_{30}^{s}(y_j)}{(y_j - x_2)^2 + x_3^2} + p_\infty^0.$$

Further, substituting $\overset{0}{X}_{30N}$ in (30), (31), and (37) we obtain the expressions for $\overset{0}{u}_2^f(x_2,x_3)$, $\overset{0}{u}_3^f(x_2,x_3)$, and $\overset{0}{u}_{30}^s(x_2)$.

$\overset{0}{u}_{20}^s$ can be easily found from the Equation (20) (where in view of (26) $\overset{0}{X}_{20}^s$ should be replaced by $\overset{0}{\varPhi}_{20}^s$), which will be solved under conditions (32).

Remark. *A solid-fluid interaction problem where the solid body is an elastic cusped plate whose thickness is given by the relation*

$$h(x_2) = h_0 x_2^\alpha, \quad h_0, \alpha = const > 0$$

can be dealt with in an analogous manner.

Acknowledgements

The research described in this paper was made possible in part by Award No. GEP1-3339-TB-06 of the Georgian Research and Development Foundation (GRDF) and the U.S. Civilian Research & Development Foundation for the Independent States of the Former Soviet Union (CRDF), and in part by the Georgian National Science Foundation (Project GNSF/ST06/3-035).

I would like to express my gratitude to Prof. George Jaiani for many useful discussions.

References

1. Babuška, I., Li, L. (1991) Hierarchic modelling of plates. Computers and Structures, 40, 419–430
2. Boikov, I.V., Dobrynin, N.F., Domnin, L. (1998) Approximation Methods for Calculating of Hadamard Integral Equations. Penza, (Russian)
3. Chinhaladze, N., Gilbert, R. (2006) Vibration of an elastic plate under action of an incompressible fluid in case of N=0 approximation of I. Vekua's hierarchical models. Applicable Analysis, 85, 9, 1177–1187
4. Chinhaladze, N., Jaiani, G. (2001) On a cusped elastic solid-fluid interaction problem. Applied Mathematics and Informatics, 6, 2, 25–64
5. Chinhaladze, N. (2002) Bending of an isotropic cusped elastic plates under uction of an incompressible fluid. Reports of Seminar of I. Vekua Institute of Applied Mathematics of Tbilisi State University, 28, 52–60
6. Chinhaladze, N. (2002) On a cusped elastic solid-icompressible fluid interaction problem. Harmonic Vibration. Mechanics and Mechanical Engineering, Technical University of Lodz, 6, 5–29

7. Chinhaladze, N. (2002) Vibration of the plate with two cusped edges. Proceedings of I. Vekua Institute of Applied Mathematics of Tbilisi State University, 52, 30–48

8. Dautray, R., Lions, J.L. (1990) Mathematical Analysis and Numerical Methods for Science and Technology. Springel-Verlag, Berlin, Heidelber, New-York, London, Paris, Tokyo, Hohg Kong, vol.1

9. Gilbert, R.P., Scotti, T., Wirgin, A., Youngzhi, S. Xu (1998) The unidentified object problem in a shallow ocean. Journal of the Acoustic Society of America, 103, 1320–1327

10. Gordeziani, D.G. (1974) To the exactness of one variant of the theory of thin shells. Soviet Mathematical Doklady, 215, 4, 751–754

11. Jaiani, G.V. (1980) On a physical interpretation of Fichera's function. Academia Nazionale dei Lincei, Rendiconti della Scienza Fisica, Matematica e Naturale, Serie VIII, Vol. LXVIII, fasc.5, 426–435

12. Jaiani, G.V. (1982) Solution of some Problems for a Degenerate Elliptic Equation of Higher Order and their Applications to Prismatic Shells, Tbilisi University Press (in Russian with Georgian and English summaries)

13. Jaiani, G.V. (1999) Initial and Boundary Value Problems for Singular Differential Equations and Applications to the Theory of Cusped Bars and Plates. Complex Methods for Partial Differential Equations. (ISAAC Serials, Vol. 6) Eds.: H. Begehr, O. Celebi, W. Tutschke. Kluwer, Dordrecht, 113–149

14. Jaiani, G.V. (2002) Relation of hierarchical models of cusped elastic plates and beams to the three-dimensional models. Reports of the Seminar of I. Vekua Institute of Applied Mathematics, 28, 40–51

15. Jaiani, G.V., Kharibegashvili, S.S., Natroshvili, D.G., Wendland, W.L. (2003) Hierarchical models for elastic cusped plates and beams. Lecture Notes of TICMI, 4 (electronic version: http://www.viam.sci.tsu.ge/others/TICMI)

16. Jaiani, G.V., Kharibegashvili, S.S., Natroshvili, D.G., Wendland, W.L. (2004) Two-dimensional hierarchical models for prismatic shells with thickness vanishing at the boundary. Journal of Elasticity, 77, 2, 95–122

17. Meunargia, T.V. (1998) On nonlinear and nonshallow shells. Bulletin of TICMI, 2, 46–49

18. Muskhelishvili N.I. (1946) Singular Integral Equations. P. Noordhoff N.V. – Groningen-Holland

19. Natroshvili, D., Kharibegashvili, S., Tediashvili, Z. (2000) Direct and inverse fluid-structure interaction problems. Rendiconti di Matematica, Serie VII, 20, Roma, 57–92

20. Sanchez-Palensia E. (1984) Non-Homogeneous Media and Vibration Theory. Moscow, Mir (Russian)

21. Vekua, I.N. (1955) Shell Theory: General Methods of Construction. Pitman Advanced Publishing Program, Boston-London-Melbourne

22. Vekua, I.N. (1965) The theory of thin shallow shells of variable thickness. Proceedings of A. Razmadze Institute of Mathematics of Georgian Academy of Sciences, 30, 5–103 (Russian)

23. Vekua, I.N. (1985) Shell Theory: General Methods of Construction. Pitman Advanced Publishing Program, Boston-London-Melbourne

Some Remarks on Anisotropic Singular Perturbation Problems

Michel Chipot

University of Zurich, Institute of Mathematics, Winterthurerstrasse 190, 8057 Zurich, e-mail: `m.m.chipot@math.unizh.ch`

Abstract The goal of this note is to study anisotropic singular elliptic perturbation problems. We will investigate in particular the asymptotic behaviour of the solution in the case of an anisotropic elastic membrane and in the case of an anisotropic plate.

Keywords: singular perturbations, anisotropic, elliptic problems, membrane, plate

1 Introduction

The goal of this note is to analyse diffusion or plate displacements for anisotropic materials. That is to say one assumes – in case of a diffusion problem – slow diffusion in one direction and try to see what are the consequences at the limit. Let us make this more precise.

For any $a > 0$ we denote by Ω_a the rectangle $\Omega_a = (-a, a) \times (-1, 1)$. If we denote by $x = (x_1, x_2)$ the points in $\mathbb{R}^2$ let

$$A(x) = \begin{pmatrix} a_{11}(x) & a_{12}(x) \\ a_{21}(x) & a_{22}(x) \end{pmatrix} \tag{1}$$

be a two-by-two matrix such that

$$a_{ij} \in L^\infty(\Omega_1), \quad |a_{ij}(x)| \leq \Lambda \qquad \text{a.e. } x \in \Omega_1, \quad \forall i, j = 1, 2, \tag{2}$$

$$\lambda |\xi|^2 \leq (A(x)\xi \cdot \xi) \qquad \text{a.e. } x \in \Omega_1, \quad \forall \xi \in \mathbb{R}^2, \tag{3}$$

for some positive constants λ, Λ. In the above inequalities $(\cdot)$ is the usual scalar product in $\mathbb{R}^2$, $A(x)\xi$ denotes the vector obtained by applying $A(x)$ to ξ. For $\varepsilon > 0$ we set

$$A_\varepsilon(x) = \begin{pmatrix} \varepsilon^2 a_{11}(x) & \varepsilon a_{12}(x) \\ \varepsilon a_{21}(x) & a_{22}(x) \end{pmatrix}. \tag{4}$$

G. Jaiani, P. Podio-Guidugli (eds.), *IUTAM Symposium on Relations of Shell, Plate, Beam, and 3D Models*, © Springer Science+Business Media B.V. 2008

Then for $f \in H^{-1}(\Omega_1)$ we would like to consider the following model problem

$$\begin{cases} \int_{\Omega_1} A_\varepsilon(x) \nabla u_\varepsilon \cdot \nabla v(x)\, dx = \langle f, v \rangle & \forall\, v \in H_0^1(\Omega_1), \\ u_\varepsilon \in H_0^1(\Omega_1), \end{cases} \tag{5}$$

more precisely we would like to study the asymptotic behaviour of u_ε when $\varepsilon \to 0$. If we choose for instance

$$A = \mathrm{Id} \tag{6}$$

where Id is the identity matrix, then u_ε is the weak solution to the problem

$$\begin{cases} -\varepsilon^2 \partial_{x_1}^2 u_\varepsilon - \partial_{x_2}^2 u_\varepsilon = f & \text{in } \Omega_1, \\ u_\varepsilon = 0 & \text{on } \partial\Omega_1, \end{cases} \tag{7}$$

where $\partial\Omega_1$ denotes the boundary of Ω_1. From the point of view of the applications u_ε could be a density of a population subjected to move in Ω_1 with a diffusion velocity very small in the direction of x_1 or the displacement of an elastic anisotropic membrane when some force f is applied (see [7]). Suppose to simplify that $f = f(x_2)$ – i.e. f is a function independent of x_1 – then a natural candidate for the limit of u_ε is u_0 the solution to

$$\begin{cases} -\partial_{x_2}^2 u_0 = f & \text{in } (-1,1), \\ u_0(-1) = u_0(1) = 0. \end{cases} \tag{8}$$

This is what we will establish in a more general context. Note that this kind of problem was not addressed – except for a short example – in [13]. The paper is divided as follows. In the next section we give a simple convergence result. Then we show how to obtain some rate of convergence for the solution. We choose to do it in the case of an anisotropic plate explaining how the problem is related to problems set in cylinders becoming infinite in some directions. For this kind of issues we refer the reader to [1, 3, 5, 6, 12, 14].

2 A Convergence Result

For the sake of simplification we suppose here

$$f \in L^2(\Omega_1). \tag{9}$$

We denote by ω the open set $(-1,1)$. Since for a.e. $x_1 \in \omega$, $f(x_1, \cdot) \in L^2(\omega)$ for a.e. x_1 in ω there exists a unique $u_0 = u_0(x_1, \cdot)$ weak solution to

$$\begin{cases} -\partial_{x_2}(a_{22}(x_1, x_2)\partial_{x_2} u_0) = f & \text{in } \omega, \\ u_0(x_1, \cdot) = 0 & \text{on } \partial\omega, \end{cases} \tag{10}$$

where $\partial\omega = \{-1, 1\}$. Note that by (3), $a_{22} \geq \lambda > 0$ and the existence of u_0 follows by the Lax–Milgram theorem. From (3) one also derives that

$$\lambda(\varepsilon^2 \xi_1^2 + \xi_2^2) \leq (A_\varepsilon(x)\xi \cdot \xi) \quad \text{a.e. } x \in \Omega_1, \quad \forall \xi \in \mathbb{R}^2. \tag{11}$$

Then the existence of u_ε solution to (5) is just also a consequence of the Lax–Milgram theorem. We have in addition:

Theorem 1. *Under the above assumptions, when $\varepsilon \to 0$,*

$$u_\varepsilon \rightharpoonup u_0, \qquad \partial_{x_2} u_\varepsilon \rightharpoonup \partial_{x_2} u_0 \quad in \ L^2(\Omega_1) \tag{12}$$

where u_0 is the solution to (10).

Proof. Let us choose $v = u_\varepsilon$ in (5). Recalling (9), (11) we get

$$\lambda \int_{\Omega_1} \varepsilon^2 (\partial_{x_1} u_\varepsilon)^2 + (\partial_{x_2} u_\varepsilon)^2 \, dx \leq |f|_2 |u_\varepsilon|_2 \tag{13}$$

where $|\cdot|_2$ denotes the usual $L^2(\Omega_1)$-norm. Due to the fact that Ω_1 is bounded in the x_2 direction we have a Poincaré inequality which is here (see [3])

$$|u_\varepsilon|_2 \leq \sqrt{2}|\partial_{x_2} u_\varepsilon|_2 \tag{14}$$

and from (13) we derive

$$\varepsilon^2 |\partial_{x_1} u_\varepsilon|_2^2 + |\partial_{x_2} u_\varepsilon|_2^2 \leq \sqrt{2}\frac{|f|_2}{\lambda}|\partial_{x_2} u_\varepsilon|_2. \tag{15}$$

From this it is easy to get

$$|\partial_{x_2} u_\varepsilon|_2 \leq \sqrt{2}\frac{|f|_2}{\lambda}, \qquad \varepsilon|\partial_{x_1} u_\varepsilon|_2 \leq \sqrt{2}\frac{|f|_2}{\lambda}. \tag{16}$$

Combining this with (14) we also obtain

$$|u_\varepsilon|_2 \leq 2\frac{|f|_2}{\lambda}. \tag{17}$$

Thus – up to a subsequence – there exists $v_0 \in L^2(\Omega_1)$ such that

$$u_\varepsilon \rightharpoonup v_0, \qquad \partial_{x_2} u_\varepsilon \rightharpoonup \partial_{x_2} v_0 \quad in \ L^2(\Omega_1). \tag{18}$$

In addition we have

$$\epsilon\partial_{x_1} u_\varepsilon \rightharpoonup 0 \quad in \ L^2(\Omega_1). \tag{19}$$

If we denote by $L^2(\omega; X)$ the L^2-space of functions from ω into the Banach space X – see [2, 10] – we derive from (16), (17)

$$|u_\varepsilon|_{L^2(\omega; L^2(\omega))}, \ |u_\varepsilon|_{L^2(\omega; H_0^1(\omega))} \leq 2\frac{|f|_2}{\lambda}, \tag{20}$$

(we selected $|\partial_{x_2} v|_2$ as the norm in $H_0^1(\omega)$). Thus up to a subsequence for some $\bar{v}_0$

$$u_\varepsilon \rightharpoonup \bar{v}_0 \quad \text{in} \quad L^2(\omega; H_0^1(\omega)) \quad \text{and} \quad L^2(\omega; L^2(\omega)). \tag{21}$$

Since $L^2(\omega; L^2(\omega)) \simeq L^2(\Omega_1)$, by (18) we have $\bar{v}_0 = v_0$ and in particular

$$v_0 \in L^2(\omega; H_0^1(\omega)). \tag{22}$$

Next, going back to (5) we have for any $v \in H_0^1(\Omega_1)$

$$\varepsilon^2 \int_{\Omega_1} a_{11} \partial_{x_1} u_\varepsilon \partial_{x_1} v \, dx + \varepsilon \int_{\Omega_1} a_{12} \partial_{x_2} u_\varepsilon \partial_{x_1} v \, dx + \varepsilon \int_{\Omega_1} a_{21} \partial_{x_1} u_\varepsilon \partial_{x_2} v \, dx$$

$$+ \int_{\Omega_1} a_{22} \partial_{x_2} u_\varepsilon \partial_{x_2} v \, dx = \int_{\Omega_1} f v \, dx. \tag{23}$$

Thus, passing to the limit we get that $v_0 \in L^2(\omega; H_0^1(\omega))$ satisfies

$$\int_{\Omega_1} a_{22} \partial_{x_2} v_0 \partial_{x_2} v \, dx = \int_{\Omega_1} f v \, dx \quad \forall v \in H_0^1(\Omega_1). \tag{24}$$

Taking $v = \varphi w$ where $\varphi \in \mathcal{D}(\omega)$, $w \in H_0^1(\omega)$ we obtain

$$\int_\omega \varphi(x_1) \int_\omega a_{22}(x_1, x_2) \partial_{x_2} v_0 \partial_{x_2} w \, dx_2 \, dx_1$$
$$= \int_\omega \varphi(x_1) \int_\omega f(x_1, x_2) w \, dx_2 \, dx_1. \tag{25}$$

It follows that for almost every $x_1 \in \omega$ we have

$$\begin{cases} \int_\omega a_{22}(x_1, x_2) \partial_{x_2} v_0 \partial_{x_2} w \, dx_2 = \int_\omega f(x_1, x_2) w \, dx_2 \quad \forall w \in H_0^1(\omega), \\ v_0(x_1, \cdot) \in H_0^1(\omega), \end{cases} \tag{26}$$

and thus, for a.e. $x_1 \in \omega$, $v_0(x_1, \cdot) = u_0$ and v_0 is uniquely determined. It implies that the whole sequence u_ε converges toward v_0 and this completes the proof. $\qquad \square$

Remark 1. One can allow the dependence of A in ε – i.e.

$$A(x) = \begin{pmatrix} a_{11}^\varepsilon(x) & a_{12}^\varepsilon(x) \\ a_{21}^\varepsilon(x) & a_{22}^\varepsilon(x) \end{pmatrix}$$

provided that (2), (3) hold independently of ε. One can also allow $A = A^\varepsilon(x, u)$ under suitable conditions.

Remark 2. The theory that we explained here in a two dimensional square domain can be extended to a domain Ω of $\mathbb{R}^n$. We refer the reader to [4] for details. It is also possible to show strong convergence.

3 Convergence Estimates

In this section we would like to estimate the rate of convergence of u_ε solution to (5) towards u_0 the solution to (10), that is to say we would like to estimate, for some norm, $||u_\epsilon - u_0||$ in terms of ϵ. To show the generality of our technique we do that for the problem of a thin clamped plate. This plate is for instance occupying a domain $\Omega_\ell \times (-\varepsilon, \varepsilon)$ where ε is very small and $\Omega_\ell = (-\ell, \ell) \times (-1, 1)$. (We changed the notation of the index of Ω, ℓ being supposed to go to $+\infty$.) Then the displacement u_ℓ of this thin plate under the action of a density of forces f is given by the weak solution to

$$\begin{cases} \Delta^2 u_\ell = f & \text{in } \Omega_\ell, \\ u_\ell = \dfrac{\partial u_\ell}{\partial \nu} = 0 & \text{on } \partial\Omega_\ell. \end{cases} \tag{27}$$

We refer the reader to [8, 9] for a derivation of this system via a particular scaling, ν denotes the outward unit normal to $\partial\Omega_\ell$. Suppose now that

$$f = f(x_2) \in L^2(\omega) \tag{28}$$

where $\omega = (-1, 1)$ i.e. the applied force is the same on each x_1-section of the plate. It is expected then, when ℓ goes to infinity, that the displacement becomes independent of the section of the plate. To make this precise let us introduce some notation. Denote by $H^2(\Omega_\ell)$ the space defined as

$$H^2(\Omega_\ell) = \{\, v \in L^2(\Omega_\ell) \mid \partial^\alpha v \in L^2(\Omega_\ell) \,\forall\, \alpha,\, |\alpha| \leq 2 \,\}. \tag{29}$$

$(\alpha = (\alpha_1, \alpha_2) \in \mathbb{N}^2, |\alpha| = \alpha_1 + \alpha_2, \partial^\alpha = \partial_{x_1}^{\alpha_1} \partial_{x_2}^{\alpha_2}$, the derivative is taken in the distributional sense – see [10, 11]). Moreover if $\mathcal{D}(\Omega_\ell)$ denotes the space of C^∞-functions with compact support in Ω_ℓ, denote by $H_0^2(\Omega_\ell)$ the closure of $\mathcal{D}(\Omega_\ell)$ in $H^2(\Omega_\ell)$ equipped with the norm

$$||v||_{2,2} = \left\{ \sum_{|\alpha| \leq 2} |\partial^\alpha v|_2^2 \right\}^{\frac{1}{2}} = ||v||_{H^2(\Omega_\ell)}. \tag{30}$$

Then the weak formulation of (27) is simply

$$\begin{cases} u_\ell \in H_0^2(\Omega_\ell), \\ \displaystyle\int_{\Omega_\ell} \Delta u_\ell \Delta v \, dx = \int_{\Omega_\ell} f v \, dx \quad \forall\, v \in H_0^2(\Omega_\ell). \end{cases} \tag{31}$$

It is easy to show that on $H_0^2(\Omega_\ell)$ the norms

$$|\Delta u|_{2,\Omega_\ell}, \qquad ||u||_{H^2(\Omega_\ell)} \tag{32}$$

are equivalent ($|\quad|_{2,\Omega_\ell}$ denotes the $L^2(\Omega_\ell)$-norm). Thus, the existence of a solution to (31) follows simply from the Lax–Milgram theorem. In fact, we will be in need of a more precise equivalence result for the norms (32), namely

Proposition 1. *There exist constants $c, C > 0$ independent of ℓ such that*

$$c||u||_{H^2(\Omega_\ell)} \leq |\Delta u|_{2,\Omega_\ell} \leq C||u||_{H^2(\Omega_\ell)} \quad \forall\, u \in H_0^2(\Omega_\ell). \tag{33}$$

Proof. By density of $\mathcal{D}(\Omega_\ell)$ in $H_0^2(\Omega_\ell)$ it is enough to show (33) for $u \in \mathcal{D}(\Omega_\ell)$. Now, the inequality of the right-hand side is easy to show and we prove only the one of the left-hand side. For that, we notice first that by integration by parts we have

$$\int_{\Omega_\ell} \partial_{x_1}^2 u\, \partial_{x_2}^2 u\, dx = \int_{\Omega_\ell} (\partial_{x_1 x_2}^2 u)^2\, dx \quad \forall\, u \in \mathcal{D}(\Omega_\ell). \tag{34}$$

Then it follows that

$$|\Delta u|_{2,\Omega_\ell}^2 = \int_{\Omega_\ell} (\Delta u)^2\, dx = \sum_{|\alpha|=2} |\partial^\alpha u|_{2,\Omega_\ell}^2 \quad \forall\, u \in \mathcal{D}(\Omega_\ell). \tag{35}$$

Applying the Poincaré inequality in ω for the functions

$$u(x_1, \cdot), \partial_{x_1} u(x_1, \cdot) \in \mathcal{D}(\omega)$$

we obtain easily (see (14))

$$|u(x_1, \cdot)|_{2,\omega}^2 \leq 2|\partial_{x_2} u(x_1, \cdot)|_{2,\omega}^2,$$
$$|\partial_{x_i}(x_1, \cdot)|_{2,\omega}^2 \leq 2|\partial_{x_2 x_i}^2 u(x_1, \cdot)|_{2,\omega}^2.$$

Integrating these inequalities with respect to x_1 on $(-\ell, \ell)$ we derive

$$|u|_{2,\Omega_\ell}^2 + ||\nabla u||_{2,\Omega_\ell}^2 \leq C \sum_{|\alpha|=2} |\partial^\alpha u|_{2,\Omega_\ell}^2 \quad \forall\, u \in \mathcal{D}(\Omega_\ell)$$

for some constant C independent of ℓ. The inequality (33) follows then from (35). This completes the proof of the proposition. $\qquad\square$

A natural candidate for the limit of u_ℓ is of course u_0 the solution to

$$\begin{cases} u_0 \in H_0^2(\omega), \\[2mm] \displaystyle\int_\omega \partial_{x_2}^2 u_0 \partial_{x_2}^2 v\, dx = \int_\omega fv\, dx_2 \quad \forall\, v \in H_0^2(\omega). \end{cases} \tag{36}$$

Note that the existence and uniqueness of u_0 follows again simply from the Lax–Milgram theorem. Moreover, we have:

Theorem 2. *Under the above assumptions for any $\ell_0 > 0$, $r > 0$ there exists a constant C independent of $\ell \geq 1$ such that*

$$||u_\ell - u_0||_{H^2(\Omega_{\ell_0})} \leq \frac{C}{\ell^r}, \tag{37}$$

i.e. $u_\ell \to u_0$ locally with an arbitrary speed of convergence in term of $\frac{1}{\ell}$.

Proof. We consider ρ a smooth function such that

$$0 \leq \rho \leq 1, \ \rho = 1 \text{ on } (-1+\delta, 1-\delta), \ \rho = 0 \text{ outside } (-1,1), \ |\rho'|, |\rho''| \leq C \quad (38)$$

(C might depend on δ). It is clear that for any $\ell_1 \leq \ell$ we have

$$(u_\ell - u_0)\rho^2\left(\frac{x_1}{\ell_1}\right) \in H_0^2(\Omega_\ell).$$

Using this function in (31) and (36) – after eventually approximating it first by a function in $\mathcal{D}(\Omega_\ell)$ we get

$$\int_{\Omega_\ell} \Delta(u_\ell - u_0)\Delta\{(u_\ell - u_0)\rho^2\}\,dx = 0$$

where for simplicity we set $\rho = \rho\left(\frac{x_1}{\ell_1}\right)$.

Using the equality

$$\Delta\{(u_\ell - u_0)\rho^2\} = \Delta\{(u_\ell - u_0)\rho\}\rho + 2\partial_{x_1}\{(u_\ell - u_0)\rho\}\partial_{x_1}\rho + (u_\ell - u_0)\rho\Delta\rho$$

we obtain – noting that ρ vanishes outside of Ω_{ℓ_1} –

$$\int_{\Omega_{\ell_1}} \rho\Delta(u_\ell - u_0)\Delta\{(u_\ell - u_0)\rho\}\,dx$$

$$= -\int_{\Omega_{\ell_1}} \Delta(u_\ell - u_0)[2\partial_{x_1}\{(u_\ell - u_0)\rho\}\partial_{x_1}\rho + (u_\ell - u_0)\rho\Delta\rho]\,dx.$$

Then since

$$\rho\Delta(u_\ell - u_0) = \Delta\{(u_\ell - u_0)\rho\} - 2\partial_{x_1}(u_\ell - u_0)\partial_{x_1}\rho - (u_\ell - u_0)\Delta\rho$$

we obtain

$$\int_{\Omega_{\ell_1}} \Delta\{(u_\ell - u_0)\rho\}^2\,dx$$

$$= -\int_{\Omega_{\ell_1}} \Delta(u_\ell - u_0)[2\partial_{x_1}\{(u_\ell - u_0)\rho\}\partial_{x_1}\rho + (u_\ell - u_0)\rho\Delta\rho]\,dx$$

$$+ \int_{\Omega_{\ell_1}} \Delta\{(u_\ell - u_0)\rho\}[2\partial_{x_1}(u_\ell - u_0)\partial_{x_1}\rho + (u_\ell - u_0)\Delta\rho]\,dx.$$

From (38) we clearly have for $\ell_1 \geq 1$

$$|\partial_{x_1}\rho| = |\partial_{x_1}\left\{\rho\left(\frac{x_1}{\ell_1}\right)\right\}| \leq \frac{C}{\ell_1}, \quad |\Delta\rho| \leq \frac{C}{\ell_1^2} \leq \frac{C}{\ell_1}$$

and we easily obtain

$$|\Delta\{(u_\ell - u_0)\rho\}|^2_{2,\Omega_{\ell_1}} \leq \frac{C}{\ell_1}\|(u_\ell - u_0)\rho\|_{H^2(\Omega_{\ell_1})}\|u_\ell - u_0\|_{H^2(\Omega_{\ell_1})}$$

for some constant C independent of ℓ_1. Using (33) and the fact that $\rho = 1$ on $(-1+\delta, 1-\delta)$ we obtain:

$$\|u_\ell - u_0\|_{H^2(\Omega_{\ell_1(1-\delta)})} \leq \frac{C}{\ell_1}\|u_\ell - u_0\|_{H^2(\Omega_{\ell_1})}.$$

Iterating this process we get

$$\|u_\ell - u_0\|_{H^2(\Omega_{\ell(1-\delta)^k})} \leq \frac{C}{\ell^k}\|u_\ell - u_0\|_{H^2(\Omega_\ell)}. \tag{39}$$

Now, taking $v = u_\ell$ in (31) and $v = u_0$ in (36) we easily obtain

$$\|u_\ell\|^2_{H^2(\Omega_\ell)}, \ \|u_0\|^2_{H^2(\Omega_\ell)} \leq C\ell|f|^2_{2,\omega}. \tag{40}$$

Choosing in (39) k such that $k - 1 > r$, δ such that $(1-\delta)^k \geq \frac{1}{2}$, ℓ_0 such that $\ell_0 \leq \frac{\ell}{2}$ we arrive to

$$\|u_\ell - u_0\|_{H^2(\Omega_{\ell_0})} \leq \|u_\ell - u_0\|_{H^2(\Omega_{\frac{\ell}{2}})} \leq \frac{C}{\ell^r} \tag{41}$$

which completes the proof of the theorem. $\qquad\qquad\qquad\Box$

We recast now our results in the framework of an anisotropic plate. For that we set

$$\hat{u}_\ell(x_1, x_2) = u_\ell(\ell x_1, x_2). \tag{42}$$

It is clear that $\hat{u}_\ell$ is defined on Ω_1 and we have

$$\partial^2_{x_1}\hat{u}_\ell(x_1, x_2) = \ell^2\partial^2_{x_1}u_\ell(\ell x_1, x_2), \quad \partial^2_{x_2}\hat{u}_\ell(x_1, x_2) = \partial^2_{x_2}u_\ell(\ell x_1, x_2). \tag{43}$$

Thus, changing x_1 in ℓx_1 in (31) we obtain that $\hat{u}_\ell$ satisfies

$$\begin{cases} \hat{u}_\ell \in H^2_0(\Omega_1), \\ \displaystyle\int_{\Omega_1}\left(\frac{1}{\ell^2}\partial^2_{x_1} + \partial^2_{x_2}\right)\hat{u}_\ell(x)\Delta v(\ell x_1, x_2)\, dx \\ \displaystyle\qquad = \int_{\Omega_1} f(x_2)v(\ell x_1, x_2)\, dx \quad \forall v \in H^2_0(\Omega_\ell). \end{cases} \tag{44}$$

For $w \in H^2_0(\Omega_1)$, setting

$$v(x_1, x_2) = w\left(\frac{x_1}{\ell}, x_2\right)$$

the equation in (44) becomes

$$\int_{\Omega_1}\left(\frac{1}{\ell^2}\partial^2_{x_1} + \partial^2_{x_2}\right)\hat{u}_\ell\left(\frac{1}{\ell^2}\partial^2_{x_1} + \partial^2_{x_2}\right)w\, dx = \int_{\Omega_1} fw\, dx \quad \forall w \in H^2_0(\Omega_1). \tag{45}$$

From (39)–(41) we derive that

$$|\Delta(u_\ell - u_0)|_{2,\Omega_{\ell(1-\delta)^k}} \leq \frac{C}{\ell^{k-1}}. \tag{46}$$

By a change of variable we obtain

$$\left|\left(\frac{1}{\ell^2}\partial^2_{x_1} + \partial^2_{x_2}\right)(\hat{u}_\ell - u_0)\right|_{2,\Omega_{(1-\delta)^k}} \leq \frac{C}{\ell^{k-2}}. \tag{47}$$

We set then

$$\Delta_\varepsilon = \varepsilon^2 \partial^2_{x_1} + \partial^2_{x_2} \tag{48}$$

and consider u_ε the solution to

$$\begin{cases} u_\varepsilon \in H^2_0(\Omega_1), \\ \displaystyle\int_{\Omega_1} \Delta_\varepsilon u_\varepsilon \Delta_\varepsilon v\, dx = \int_{\Omega_1} fv\, dx \quad \forall v \in H^2_0(\Omega_1). \end{cases} \tag{49}$$

We have

Theorem 3. *For any $a \in (0,1)$, $r > 0$ there exists a constant C independent of ε such that*

$$\|u_\varepsilon - u_0\|_{H^2(\Omega_a)} \leq C\varepsilon^r. \tag{50}$$

Proof. From (44), (45) it is clear that

$$u_\varepsilon = \hat{u}_{\frac{1}{\varepsilon}}.$$

Then the result is a simple consequence of (47) choosing first k large enough and then δ small enough. $\qquad\square$

Remark 3. With a similar technique we would obtain for the solution to (5) or (7)

$$\|u_\varepsilon - u_0\|_{H^1(\Omega_a)} \leq C\varepsilon^r.$$

We refer the reader to [4] for details.

4 Conclusions

In this note we have shown how it is possible to predict the asymptotic behaviour of anisotropic singular perturbation problems. We have indeed clearly singled out the limit problem. In addition, we have shown – in the case of an anisotropic plate – how to evaluate the speed of convergence of the solution of the singular perturbation problem towards its limit. This is done through a scaling which transform our fixed domain into a cylinder becoming unbounded in one direction. The interplay between the problem set in a fixed domain and the one scaled is essential to understand them both.

Acknowledgment

This research has been supported by the Swiss National Science Foundation under the contracts #20-111543/1 and #20-117614/1. We thank this institution for its support.

References

1. B. BRIGHI, S. GUESMIA: On elliptic boundary value problems of order $2m$ in cylindrical domain of large size, Advances in Mathematical Sciences and Applications, (2008).

2. M. CHIPOT: *Elements of Nonlinear Analysis*, Birkhäuser, 2000.

3. M. CHIPOT: *ℓ goes to Plus Infinity*, Birkhäuser, 2002.

4. M. CHIPOT: On some anisotropic singular perturbation problems, Asymptotic Analysis 55, (2007), pp. 125–144.

5. M. CHIPOT, A. ROUGIREL: On the asymptotic behavior of the solution of parabolic problems in domains of large size in some directions, DCDS Series B, 1 (2001), pp. 319–338.

6. M. CHIPOT, A. ROUGIREL: On the asymptotic behavior of the solution of elliptic problems in cylindrical domains becoming unbounded, Communications in Contemporary Math. 4, 1 (2002), pp. 15–24.

7. P. G. CIARLET: *Mathematical Elasticity, Vol 1: Three Dimensional Elasticity*, North-Holland, 1988.

8. P. G. CIARLET: *Mathematical Elasticity, Vol 2: Theory of Plates*, North-Holland, 1997.

9. P. G. CIARLET, P. DESTUYNDER: A justification of the two-dimensional plate model, J. Mécanique, 18, (1979), pp. 315–344.

10. R. DAUTRAY, J. L. LIONS: *Mathematical Analysis and Numerical Methods for Science and Technology*, Springer-Verlag, 1988.

11. D. GILBARG, N. S. TRUDINGER: *Elliptic Partial Differential Equations of Second Order*, Springer Verlag, 1983.

12. S. GUESMIA: *Etude du comportement asymptotique de certaines équations aux dérivées partielles dans des domaines cylindriques*. Thèse Université de Haute Alsace, December 2006.

13. J. L. LIONS: *Perturbations singulières dans les problèmes aux limites et en contrôle optimal*, Lecture Notes in Mathematics # 323, Springer-Verlag, 1973.

14. Y. XIE: *On Asymptotic Problems in Cylinders and Other Mathematical Issues*. Thesis University of Zürich, May 2006.

On the Variational Derivation
of the Kinematics for Thin-Walled
Closed Section Beams

Lorenzo Freddi[1], Antonino Morassi[2], and Roberto Paroni[3]

[1] Dipartimento di Matematica e Informatica, via delle Scienze 206, 33100 Udine,
   Italy, lorenzo.freddi@dimi.uniud.it
[2] Dipartimento di Georisorse e Territorio, via Cotonificio 114, 33100 Udine, Italy,
   antonino.morassi@uniud.it
[3] Dipartimento di Architettura e Pianificazione, Università degli Studi di Sassari,
   Palazzo del Pou Salit, Piazza Duomo, 07041 Alghero, Italy, paroni@uniss.it

Abstract The kinematics of thin-walled closed cross section beams is studied by
comparing the behavior of a closed section with an open section which differs from
the former by a "cut" on one side.

Keywords: Young measures, nonlinear elasticity, Gamma-convergence,
dimension reduction

1 Introduction

The definition of Γ-convergence given by De Giorgi [6], see also [4] and [5], has
found significant applications in dimension reduction problems in mechanics.
Strings, beams, membranes, plates and shells have found rigorous justifica-
tions, [1, 3, 9, 10]. Anzellotti et al. [2], see also Percivale [14], starting from
the three-dimensional linear theory of elasticity deduced, by Γ-convergence,
the De Saint-Venant beam theory. The three dimensional body considered in
these works is a cylinder with diameter much smaller than the length. In engineering applications, to minimize the weight of the structure, quite often are
used beams with cross section having "walls" of thickness much smaller than
the diameter of the cross section: the so called thin-walled beams. This kind of
beams have been studied by Rodriguez and Viano [15, 16] and, starting from
De Saint-Venant problem, by Morassi [11–13]. Only recently the fully three-
dimensional problem has been studied by means of Γ-convergence. In [7] the
authors of this note have considered a cantilever beam of finite length with a
rectangular cross section of sides proportional to ε and ε^2, with $0 < \varepsilon < 1$,
so to model the "thin-wall". Like in the previous papers, also in [7] the Γ-
limit, as the parameter ε goes to zero, was found to be the De Saint-Venant

beam model. More complex open sections have been studied in [8], where it is shown that the Γ-limit is either the De Saint-Venant theory or Vlassov theory according to the shape of the cross section. Closed thin-walled cross sections, which have not been considered in this last paper, are the main concern of the present note. We do not address the Γ-convergence problem here, but we study what can be considered as a preliminary step: the compactness of the displacements or, in mechanical terms, the kinematics of the model. To outline the differences between closed and open section we consider two sections: the first closed, and the second, which simply differs from the first by "a cut", open. We then outline the main steps needed to derive the kinematical description of the beam. In doing so we omit proofs by heavily relying on the similarity of the problem considered in this note with the one considered in [8]. Briefly, the section is decomposed in four rectangles and it is shown that each rectangle undergoes to a Bernoulli-Navier type of displacement. Thus the motion of each rectangle is described by four kinematical fields, hence, since the section comprises four rectangles, the motion of the beam, at this stage, is fully determined by means of sixteen fields. Relations between these fields are obtained by studying the kinematics on the regions where the rectangles overlap. Here is the main difference between the two sections considered: in the open section the rectangles overlap in three regions, while in the closed section they overlap in four regions (the fourth region gives rise to a compatibility condition that we call the supplementary junction condition). From the study of these junction conditions we deduce that in the open cross section thin-walled beam the displacement in the plane of the section is a rigid motion, while in the longitudinal direction it is the sum of a Bernoulli-Navier displacement and a quantity proportional to the derivative of the angle of rotation of the section. In other words, the open closed section undergoes to displacements of the type considered in the Vlassov theory [17]. The results found for the open cross section still hold for the closed section once that also the supplementary junction condition is satisfied. We show that this further condition imposes that the angle of rotation of the cross section to be equal to sero. We interpret this result not as a statement that the section does not rotate about the axis of the beam, but instead as a sign that, at the scale that we are looking at, the rotations are too small to be captured. In other words we believe that the sequence that generates the rotation field in the limit problem should be rescaled differently from the one considered in the open section case. This is mechanically evident: it is much harder to twist closed sections than open sections.

2 The 3-Dimensional Problem

The aim of this paper is to discuss the differences between open and closed section thin walled beams. Accordingly we consider two cylindrical three-dimensional bodies with cross sections as in Fig. 1. The open section that we consider differs from the closed section just by "a cut". Given the similarity of the two sections we shall describe in some detail only one of them.

Let us denote by $\Omega_\varepsilon \subset \mathbb{R}^3$ the reference configuration of the thin walled beam with closed section. We can write $\Omega_\varepsilon := \omega_\varepsilon \times (0, \ell)$, and $\omega_\varepsilon := \cup_{i=1}^4 \omega_\varepsilon^{(i)}$, where

$$\omega_\varepsilon^{(1)} := (\varepsilon q_1 - \varepsilon b/2, \varepsilon q_1 + \varepsilon b/2) \times (\varepsilon q_2 - \varepsilon h/2, \varepsilon q_2 - \varepsilon h/2 + \varepsilon^2 s),$$

$$\omega_\varepsilon^{(2)} := (\varepsilon q_1 + \varepsilon b/2 - \varepsilon^2 s, \varepsilon q_1 + \varepsilon b/2) \times (\varepsilon q_2 - \varepsilon h/2, \varepsilon q_2 + \varepsilon h/2),$$

$$\omega_\varepsilon^{(3)} := (\varepsilon q_1 - \varepsilon b/2, \varepsilon q_1 + \varepsilon b/2) \times (\varepsilon q_2 + \varepsilon h/2 - \varepsilon^2 s, \varepsilon q_2 + \varepsilon h/2),$$

$$\omega_\varepsilon^{(4)} := (\varepsilon q_1 - \varepsilon b/2, \varepsilon q_1 - \varepsilon b/2 + \varepsilon^2 s) \times (\varepsilon q_2 - \varepsilon h/2, \varepsilon q_2 + \varepsilon h/2),$$

are four non-empty rectangles.

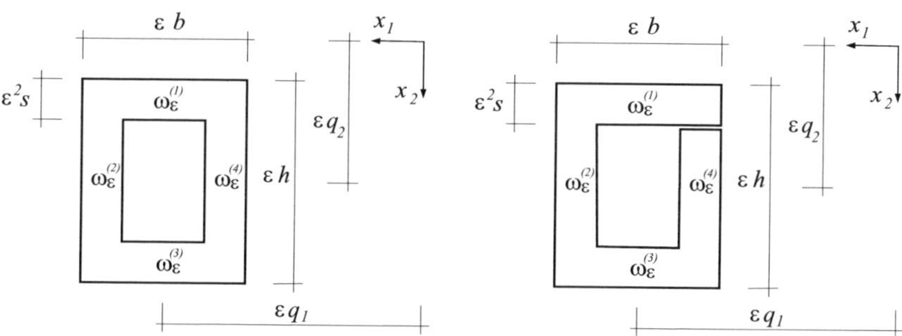

Fig. 1 The closed and the open sections

For later convenience we also set

$$\Omega_\varepsilon^{(i)} := \omega_\varepsilon^{(i)} \times (0, \ell), \quad i = 1, 2, 3, 4,$$

and we note that $\Omega_\varepsilon = \cup_{i=1}^4 \Omega_\varepsilon^{(i)}$ and that they are not pairwise disjoint.

We assume the beam clamped on one of its ends: we thus consider the spaces

$$H_\#^1(\Omega_\varepsilon; \mathbb{R}^3) := \left\{ \mathbf{w} \in H^1(\Omega_\varepsilon; \mathbb{R}^3) : \mathbf{w} = \mathbf{0} \text{ on } \omega_\varepsilon \times \{0\} \right\},$$

and $H_\#^1(\Omega_\varepsilon^{(i)}; \mathbb{R}^3)$ defined in a similar way. We further denote by

$$\mathbf{Ew}(\mathbf{x}) := \mathrm{sym}(D\mathbf{w}(\mathbf{x})) := \frac{D\mathbf{w}(\mathbf{x}) + D\mathbf{w}^T(\mathbf{x})}{2},$$

$$\mathbf{Ww}(\mathbf{x}) := \mathrm{skw}(D\mathbf{w}(\mathbf{x})) := \frac{D\mathbf{w}(\mathbf{x}) - D\mathbf{w}^T(\mathbf{x})}{2},$$

(1)

the strain of $\mathbf{w} : \Omega_\varepsilon \to \mathbb{R}^3$ and the skew symmetric part of the jacobian $D\mathbf{w}$.

Hereafter we denote by $\mathbf{u}^\varepsilon \in H^1_\#(\Omega_\varepsilon; \mathbb{R}^3)$ the solution of an equilibrium problem (not stated for brevity) posed on Ω_ε.

To discuss the convergence of the displacements $\mathbf{u}^\varepsilon$ it is convenient to work on domains which do not depend on ε. We denote by $\omega^{(i)} := \omega_1^{(i)}$ and $\Omega^{(i)} := \Omega_1^{(i)}$, and we let

$$p_\varepsilon^{(i)} : \Omega^{(i)} \to \Omega_\varepsilon^{(i)}, \quad i = 1, 2, 3, 4,$$

be defined by

$$p_\varepsilon^{(1)}(y_1, y_2, y_3) = \left(\varepsilon y_1, \varepsilon^2(y_2 - q_2 + \frac{h}{2}) + \varepsilon q_2 - \varepsilon\frac{h}{2}, y_3\right),$$

$$p_\varepsilon^{(2)}(y_1, y_2, y_3) = \left(\varepsilon^2(y_1 - q_1 - \frac{b}{2}) + \varepsilon q_1 + \varepsilon\frac{b}{2}, \varepsilon y_2, y_3\right),$$

$$p_\varepsilon^{(3)}(y_1, y_2, y_3) = \left(\varepsilon y_1, \varepsilon^2(y_2 - q_2 - \frac{h}{2}) + \varepsilon q_2 + \varepsilon\frac{h}{2}, y_3\right),$$

$$p_\varepsilon^{(4)}(y_1, y_2, y_3) = \left(\varepsilon^2(y_1 - q_1 + \frac{b}{2}) + \varepsilon q_1 - \varepsilon\frac{b}{2}, \varepsilon y_2, y_3\right).$$

For each $\varepsilon > 0$, from the solutions $\mathbf{u}^\varepsilon \in H^1_\#(\Omega_\varepsilon; \mathbb{R}^3)$ we define four functions $\mathbf{u}_\varepsilon^{(i)} \in H^1_\#(\Omega^{(i)}; \mathbb{R}^3)$ by

$$\mathbf{u}_\varepsilon^{(i)} := \mathbf{u}^\varepsilon \circ p_\varepsilon^{(i)}, \quad i = 1, 2, 3, 4.$$

Of course in the regions where the domains overlap we have

$$\mathbf{u}_\varepsilon^{(1)} \circ p_\varepsilon^{(1)}{}^{-1} = \mathbf{u}_\varepsilon^{(2)} \circ p_\varepsilon^{(2)}{}^{-1} \quad \text{in } \Omega_\varepsilon^{(1)} \cap \Omega_\varepsilon^{(2)},$$

$$\mathbf{u}_\varepsilon^{(3)} \circ p_\varepsilon^{(3)}{}^{-1} = \mathbf{u}_\varepsilon^{(2)} \circ p_\varepsilon^{(2)}{}^{-1} \quad \text{in } \Omega_\varepsilon^{(3)} \cap \Omega_\varepsilon^{(2)}, \tag{2}$$

$$\mathbf{u}_\varepsilon^{(3)} \circ p_\varepsilon^{(3)}{}^{-1} = \mathbf{u}_\varepsilon^{(4)} \circ p_\varepsilon^{(4)}{}^{-1} \quad \text{in } \Omega_\varepsilon^{(3)} \cap \Omega_\varepsilon^{(4)}.$$

The above equations hold for the closed section but also for the open section. For the former section we have a further junction condition, hereafter called the *supplementary junction condition*:

$$\mathbf{u}_\varepsilon^{(1)} \circ p_\varepsilon^{(1)}{}^{-1} = \mathbf{u}_\varepsilon^{(4)} \circ p_\varepsilon^{(4)}{}^{-1} \quad \text{in } \Omega_\varepsilon^{(1)} \cap \Omega_\varepsilon^{(4)}. \tag{3}$$

Setting

$$\begin{aligned}
\Omega^{(1)} \cap \Omega^{(2)} &= (q_1 + b/2 - s, q_1 + b/2) \times (q_2 - h/2, q_2 - h/2 + s) \times (0, \ell), \\
\Omega^{(3)} \cap \Omega^{(2)} &= (q_1 + b/2 - s, q_1 + b/2) \times (q_2 + h/2 - s, q_2 + h/2) \times (0, \ell), \\
\Omega^{(1)} \cap \Omega^{(4)} &= (q_1 - b/2, q_1 - b/2 + s) \times (q_2 - h/2, q_2 - h/2 + s) \times (0, \ell), \\
\Omega^{(3)} \cap \Omega^{(4)} &= (q_1 - b/2, q_1 - b/2 + s) \times (q_2 + h/2 - s, q_2 + h/2) \times (0, \ell).
\end{aligned}$$

we can rewrite conditions (2) as

$$\mathbf{u}^{(1)}\left(\varepsilon(z_1 - q_1 - \tfrac{b}{2}) + q_1 + \tfrac{b}{2}, z_2, z_3\right) = \mathbf{u}^{(2)}\left(z_1, \varepsilon(z_2 - q_2 + \tfrac{h}{2}) + q_2 - \tfrac{h}{2}, z_3\right),$$

$$\mathbf{u}^{(3)}\left(\varepsilon(z_1 - q_1 - \tfrac{b}{2}) + q_1 + \tfrac{b}{2}, z_2, z_3\right) = \mathbf{u}^{(2)}\left(z_1, \varepsilon(z_2 - q_2 - \tfrac{h}{2}) + q_2 + \tfrac{h}{2}, z_3\right),$$

$$\mathbf{u}^{(3)}\left(\varepsilon(z_1 - q_1 + \tfrac{b}{2}) + q_1 - \tfrac{b}{2}, z_2, z_3\right) = \mathbf{u}^{(4)}\left(z_1, \varepsilon(z_2 - q_2 - \tfrac{h}{2}) + q_2 + \tfrac{h}{2}, z_3\right),$$

$$(4)$$

which hold for $z \in \Omega^{(1)} \cap \Omega^{(2)}$, $z \in \Omega^{(3)} \cap \Omega^{(2)}$, $z \in \Omega^{(3)} \cap \Omega^{(4)}$, respectively. The supplementary junction condition (3), which we recall holds for the closed section but not for the open, can be rewritten as

$$\mathbf{u}^{(1)}\left(\varepsilon(z_1 - q_1 + \frac{b}{2}) + q_1 - \frac{b}{2}, z_2, z_3\right) = \mathbf{u}^{(4)}\left(z_1, \varepsilon(z_2 - q_2 + \frac{h}{2}) + q_2 - \frac{h}{2}, z_3\right),$$

where $z \in \Omega^{(1)} \cap \Omega^{(4)}$.

Let us consider the following 3×3 matrix valued differential operators

$$\mathbf{H}_\varepsilon^{(i)}\mathbf{w} := \left(\frac{D_1\mathbf{w}}{\varepsilon^{\alpha(i)}}, \frac{D_2\mathbf{w}}{\varepsilon^{\alpha(i+1)}}, D_3\mathbf{w}\right)$$

where $D_i\mathbf{u}$ denotes the column vector of the partial derivatives of $\mathbf{w}$ with respect to y_i, and $\alpha(i)$ is the parity of i, that is $\alpha(i) = 1$ if i is odd and $\alpha(i) = 2$ if i is even. The above definition is motivated by the following trivial result

$$\mathbf{H}_\varepsilon^{(i)}\mathbf{u}_\varepsilon^{(i)} = D\mathbf{u}^\varepsilon \circ p_\varepsilon^{(i)}, \quad i = 1, 2, 3, 4.$$

We also set

$$\mathbf{E}_\varepsilon^{(i)}\mathbf{w} := \mathrm{sym}(\mathbf{H}_\varepsilon^{(i)}\mathbf{w}), \quad \mathbf{W}_\varepsilon^{(i)}\mathbf{w} := \mathrm{skw}(\mathbf{H}_\varepsilon^{(i)}\mathbf{w}). \tag{5}$$

Under appropriate assumptions on the external loads, see [8], we may assume that the sequence $(\mathbf{u}_\varepsilon^{(1)}, \mathbf{u}_\varepsilon^{(2)}, \mathbf{u}_\varepsilon^{(3)}, \mathbf{u}_\varepsilon^{(4)}) \in \times_{i=1}^4 H^1_\#(\Omega^{(i)}; \mathbb{R}^3)$ satisfies

$$\sum_{i=1}^4 \|\mathbf{E}_\varepsilon^{(i)}\mathbf{u}_\varepsilon^{(i)}\|_{L^2(\Omega^{(i)}; \mathbb{R}^{3\times 3})} \leq C\varepsilon^2,$$

for some constant C and every $0 < \varepsilon \leq 1$. Then, by using an appropriate Korn inequality, see [8], we deduce that for any sequence of positive numbers ε_n converging to 0 there exist a subsequence (not relabeled) and 4-tuples of functions $(\mathbf{v}^{(1)}, \mathbf{v}^{(2)}, \mathbf{v}^{(3)}, \mathbf{v}^{(4)}) \in \times_{i=1}^4 H^1_\#(\Omega^{(i)}; \mathbb{R}^3)$ and $(\vartheta^{(1)}, \vartheta^{(2)}, \vartheta^{(3)}, \vartheta^{(4)}) \in \times_{i=1}^4 L^2(\Omega^{(i)})$ such that (as $n \to \infty$)

$$\mathbf{u}_{\varepsilon_n}^{(i)} \rightharpoonup \mathbf{v}^{(i)}, \text{ in } H^1(\Omega^{(i)}; \mathbb{R}^3), \tag{6}$$

$$(\mathbf{W}_{\varepsilon_n}^{(i)}\mathbf{u}_{\varepsilon_n}^{(i)})_{12} \rightharpoonup -\vartheta^{(i)} \text{ in } L^2(\Omega^{(i)}; \mathbb{R}^3). \tag{7}$$

Moreover it can be shown that the displacements $\mathbf{v}^{(i)}$ are of Bernoulli-Navier type, that is

$$v_\alpha^{(i)} = \xi_\alpha^{(i)}(y_3), \ \alpha = 1,2, \quad v_3^{(i)} = \xi_3^{(i)}(y_3) - y_\alpha \xi_\alpha^{(i)'}(y_3), \tag{8}$$

for some functions

$$\xi_\alpha^{(i)} \in H_\#^2(0,\ell) := \{\xi \in H^2(0,\ell) : \xi(0) = \xi'(0) = 0\}$$

and

$$\xi_3^{(i)} \in H_\#^1(0,\ell).$$

To describe the kinematics of the beam we have, at the moment, four fields for each rectangular component of the cross section, thus a total of sixteen kinematical fields.

Lemma 1. *For both the open and closed sections, with the notation above and for almost every $y_3 \in (0,\ell)$, we have*

1. $\vartheta^{(1)} = \vartheta^{(2)} = \vartheta^{(3)} = \vartheta^{(4)} =: \vartheta$;

2. $\xi_2^{(i)}(y_3) = 0, \qquad i = 1,3$;

3. $\xi_1^{(i)}(y_3) = 0, \qquad i = 2,4$;

4. $\xi_3^{(1)}(y_3) - (q_1 + b/2)\xi_1^{(1)'}(y_3) = \xi_3^{(2)}(y_3) - (q_2 - h/2)\xi_2^{(2)'}(y_3)$;

5. $\xi_3^{(3)}(y_3) - (q_1 + b/2)\xi_1^{(3)'}(y_3) = \xi_3^{(2)}(y_3) - (q_2 + h/2)\xi_2^{(2)'}(y_3)$;

6. $\xi_3^{(3)}(y_3) - (q_1 - b/2)\xi_1^{(3)'}(y_3) = \xi_3^{(4)}(y_3) - (q_2 + h/2)\xi_2^{(4)'}(y_3)$;

7. $\xi_1^{(1)}(y_3) - \xi_1^{(3)}(y_3) = h\vartheta(y_3)$;

8. $\xi_2^{(2)}(y_3) - \xi_2^{(4)}(y_3) = b\vartheta(y_3)$.

Moreover $\vartheta \in H_\#^2(0,\ell)$.

For brevity we omit the proof, see [8], we simply mention that they can be obtained by appropriately taking the limit of (4). The above "junction conditions" considerably reduce the number of independent kinematical variables. The first result of the Lemma states that the rotations of the four rectangles comprising the cross section are the same, the second and the third instead show that some of the kinematical variables have no importance. The fourth, fifth and sixth simply say that the longitudinal displacement on three of the four overlapping regions in the closed section (the fourth is taken into account by the supplementary junction condition) and on all overlapping regions in the open section are the same. Finally the last two conditions put into relation the displacements on opposite rectangles with the rotation of the section.

Lemma 2. *For the closed section thin walled beam we also have that*

1. $\xi_3^{(1)}(y_3) - (q_1 - b/2)\xi_1^{(1)'}(y_3) = \xi_3^{(4)}(y_3) - (q_2 - h/2)\xi_2^{(4)'}(y_3)$,

for almost every $y_3 \in (0,\ell)$.

This last Lemma, which holds only for the closed section, is deduced from the supplementary junction condition.

3 Kinematics of the Open Cross Section Thin-Walled Beam

In this short section we show that, by taking into account Lemma 1, the sixteen kinematical variables reduce to only four. We define ϑ as in Lemma 1, and we set

$$\eta_1 := \xi_1^{(1)} + (q_2 - c_2 - \frac{h}{2})\vartheta, \quad \eta_2 := \xi_2^{(2)} - (q_1 - c_1 + \frac{b}{2})\vartheta,$$

$$\eta_3 := \xi_3^{(1)} + (q_2 - \frac{h}{2})\xi_2^{(2)'} - K\vartheta',$$

where $(c_1, c_2) \in \mathbb{R}^2$ and $K \in \mathbb{R}$. We then find

$$\xi_1^{(1)} = \eta_1 - (q_2 - c_2 - \frac{h}{2})\vartheta, \; \xi_2^{(1)} = 0,$$

$$\xi_1^{(2)} = 0, \qquad\qquad\qquad \xi_2^{(2)} = \eta_2 + (q_1 - c_1 + \tfrac{b}{2})\vartheta,$$

$$\xi_1^{(3)} = \eta_1 - (q_2 - c_2 + \frac{h}{2})\vartheta, \; \xi_2^{(3)} = 0,$$

$$\xi_1^{(4)} = 0, \qquad\qquad\qquad \xi_2^{(4)} = \eta_2 + (q_1 - c_1 - \tfrac{b}{2})\vartheta,$$

and

$$\xi_3^{(1)} = \eta_3 - (q_2 - \frac{h}{2})\eta_2' - (q_2 - \frac{h}{2})(q_1 - c_1 + \frac{b}{2})\vartheta' + K\vartheta',$$

$$\xi_3^{(2)} = \eta_3 - (q_1 + \frac{b}{2})\eta_1' + (q_1 + \frac{b}{2})(q_2 - c_2 - \frac{h}{2})\vartheta' + K\vartheta',$$

$$\xi_3^{(3)} = \eta_3 - (q_2 + \frac{h}{2})\eta_2' - [(q_1 + \frac{h}{2})(q_1 - c_1 + \frac{b}{2}) + h(q_1 + \frac{b}{2}) - K]\vartheta',$$

$$\xi_3^{(4)} = \eta_3 - (q_1 - \frac{b}{2})\eta_1' - [h(q_1 + \frac{b}{2}) + b(q_2 + \frac{h}{2}) - (q_1 - \frac{b}{2})(q_2 - c_2 + \frac{h}{2}) - K]\vartheta'.$$

Substituting these quantities in (8) we find the following displacements

$$v_1^{(1)} = \eta_1 - (q_2 - c_2 - \frac{h}{2})\vartheta, \qquad\qquad v_2^{(1)} = 0,$$

$$v_3^{(1)} = \eta_3 - y_1\eta_1' - (q_2 - \frac{h}{2})\eta_2' + \psi^{(1)}(y_1)\vartheta',$$

$$v_1^{(2)} = 0, \qquad\qquad v_2^{(2)} = \eta_2 + (q_1 - c_1 + \tfrac{b}{2})\vartheta,$$

$$v_3^{(2)} = \eta_3 - (q_1 + \frac{b}{2})\eta_1' - y_2\eta_2' + \psi^{(2)}(y_2)\vartheta',$$

$$v_1^{(3)} = \eta_1 - (q_2 - c_2 + \frac{h}{2})\vartheta, \qquad\qquad v_2^{(3)} = 0,$$

$$v_3^{(3)} = \eta_3 - y_1\eta_1' - (q_2 + \frac{h}{2})\eta_2' + \psi^{(3)}(y_1)\vartheta',$$

$$v_1^{(4)} = 0, \qquad\qquad v_2^{(4)} = \eta_2 + (q_1 - c_1 - \tfrac{b}{2})\vartheta,$$

$$v_3^{(4)} = \eta_3 - (q_1 - \tfrac{b}{2})\eta_1' - y_2\eta_2' + \psi^{(4)}(y_2)\vartheta'.$$

where the so-called "sector coordinates"

$$\psi^{(1)}(y_1) := y_1(q_2 - c_2 - \tfrac{h}{2}) - (q_2 - \tfrac{h}{2})(q_1 - c_1 + \tfrac{b}{2}) + K,$$

$$\psi^{(2)}(y_2) := -y_2(q_1 - c_1 + \tfrac{b}{2}) + (q_1 + \tfrac{b}{2})(q_2 - c_2 - \tfrac{h}{2}) + K,$$

$$\psi^{(3)}(y_1) := y_1(q_2 - c_2 + \tfrac{h}{2}) - (q_1 + \tfrac{h}{2})(q_1 - c_1 + \tfrac{b}{2}) - h(q_1 + \tfrac{b}{2}) + K,$$

$$\psi^{(4)}(y_2) := -y_2(q_1 - c_1 - \tfrac{b}{2}) - h(q_1 + \tfrac{b}{2}) - b(q_2 + \tfrac{h}{2})$$
$$+ (q_1 - \tfrac{b}{2})(q_2 - c_2 + \tfrac{h}{2}) + K,$$

are defined up to an additive constant K.

We notice that in the plane of the section, plane 1–2, the motion is described by a translation (η_1, η_2) and a rotation ϑ about the point of coordinates (c_1, c_2). In direction 3 instead the motion is a Bernoulli-Navier displacement plus a quantity proportional to ϑ'.

4 On the Kinematics of the Closed Cross Section Thin Walled Beam

In the case of the closed section not only Lemma 1 holds but we also have the supplementary junction condition which is taken into account in Lemma 2. Thus the results of the previous section still hold. We now study the consequences of Lemma 2.

Subtracting Equation (5) from Equation (4) of Lemma 1 we find

$$\xi_3^{(1)} - \xi_3^{(3)} - (q_1 + b/2)(\xi_1^{(1)'} - \xi_1^{(3)'}) = h\xi_2^{(2)'}, \tag{9}$$

and subtracting Equation (6) of Lemma 1 from Equation (1) of Lemma 2 we deduce

$$\xi_3^{(1)} - \xi_3^{(3)} - (q_1 + b/2)(\xi_1^{(1)'} - \xi_1^{(3)'}) = h\xi_2^{(4)'}. \tag{10}$$

Hence, taking the difference of (9) and (10) we obtain

$$b(\xi_1^{(1)'} - \xi_1^{(3)'}) = h(\xi_2^{(4)'} - \xi_2^{(2)'}),$$

which leads, after taking into account Equations (7) and (8) of Lemma 1 to

$$bh\vartheta' = -bh\vartheta'.$$

Thus $\vartheta' = 0$, and since $\vartheta(0) = 0$ we have $\vartheta = 0$. Hence, the results of the previous section hold with $\vartheta = 0$.

We interpret the result $\vartheta = 0$ not as a statement that the section does not rotate about the axis of the beam, but instead as a sign that for a closed section thin-walled beam one should not look, as for the open sections, at the sequence $(\mathbf{W}_{\varepsilon_n}^{(i)}\mathbf{u}_{\varepsilon_n}^{(i)})_{12}$ to deduce the twist of the section. This sequence delivers a trivial result, $\vartheta = 0$, and to deduce the twist one should more likely consider the sequence generated by $(\mathbf{W}_{\varepsilon_n}^{(i)}\mathbf{u}_{\varepsilon_n}^{(i)})_{12}$ divided by some power of ε_n. To find the right power it is necessary to deduce an *ad hoc* Korn inequality for closed thin walled sections beams. This will be the aim of a future work.

References

1. E. Acerbi, G. Buttazzo, and D. Percivale, A variational definition of the strain energy for an elastic string, *J. Elasticity*, **25**, (1991), 137–148 .
2. G. Anzellotti, S. Baldo, and D. Percivale, Dimension reduction in variational problems, asymptotic development in Γ-convergence and thin structures in elasticity, *Asymptot. Anal.* **9**(1) (1994), 61–100.
3. F. Bourquin, P.G. Ciarlet, G. Geymonat, and A. Raoult, Gamma-convergence et analyse asymptotique des plaques minces, *C.R. Acad. Sci. Paris, t.* **315**, Série I, (1992), 1017–1024.
4. A. Braides, *Γ-convergence for beginners*, Oxford Lecture Series in Mathematics and its Applications, 22. Oxford University Press, Oxford, 2002.
5. G. Dal Maso, *An introduction to Γ-convergence*, Birkhäuser, Boston, 1993.
6. E. De Giorgi and T. Franzoni, Su un tipo di convergenza variazionale, *Rend. Sem. Mat. Brescia* **3** (1979), 63–101.
7. L. Freddi, A. Morassi, and R. Paroni, Thin-walled beams: the case of the rectangular cross-section, *J. Elasticity* **76** (2004), 45–66.
8. L. Freddi, A. Morassi and R. Paroni, Thin-walled beams: a derivation of Vlassov theory via Γ–convergence, *J. Elasticity* **86** (2007), 263–296.
9. H. Le Dret and A. Raoult, The nonlinear membrane model as variational limit of nonlinear three-dimensional elasticity, *J. Math. Pures Appl.*, **74**, (1995), 549–578.
10. M.G. Mora and S. Müller, A nonlinear model for inextensible rods as a low energy Γ-limit of three-dimensional nonlinear elasticity, *Ann. I. H. Poincaré*, **21**, (2004), 271–293.
11. A. Morassi, Torsion of thin tubes: a justification of some classical results, *J. Elasticity* **39** (1995), 213–227.
12. A. Morassi, Torsion of thin tubes with multicell cross-section, *Meccanica* **34** (1999), 115–132.
13. A. Morassi, An asymptotic analysis of the flexure problem for thin tubes, *Math. Mech. Solids* **4** (1999), 357–390.
14. D. Percivale, Thin elastic beams: the variational approach to St. Venant's problem, *Asymptot. Anal.* **20** (1999), 39–59.

15. J.M. Rodríguez and J.M. Viaño, Asymptotic derivation of a general linear model for thin-walled elastic rods, *Comput. Methods Appl. Mech. Engrg.* **147** (1997), 287–321.
16. J.M. Rodríguez and J.M. Viaño, Asymptotic analysis of Poisson's equation in a thin domain and its application to thin-walled elastic beams and tubes, *Math. Methods Appl. Sci.* **21** (1998), 187–226.
17. B.Z. Vlassov, *Pièces Longues en Voiles Minces*, Éditions Eyrolles, Paris, 1962.

Variational Dimension Reduction in Nonlinear Elasticity: A Young Measure Approach

Lorenzo Freddi[1] and Roberto Paroni[2]

[1] Dipartimento di Matematica e Informatica, via delle Scienze 206, 33100 Udine, Italy, `lorenzo.freddi@dimi.uniud.it`
[2] Dipartimento di Architettura e Pianificazione, Università degli Studi di Sassari, Palazzo del Pou Salit, Piazza Duomo, 07041 Alghero, Italy, `paroni@uniss.it`

Abstract Starting form 3D elasticity, we deduce the variational limit of the string and of the membrane on the space of one and two-dimensional gradient Young measures, respectively. The physical requirement that the energy becomes infinite when the volume locally vanishes is taken into account in the string model. The rate at which the energy density blows up characterizes the effective domain of the limit energy. The limit problem uniquely determines the energy density of the thin structure.

Keywords: thin-walled cross-section beams, linear elasticity, Gamma-convergence, dimension reduction

1 Introduction

A first variational derivation of the energy of a string starting from the 3D nonlinear elasticity is due to Acerbi, Buttazzo and Percivale [1]. Following the same leading ideas, Le Dret and Raoult [17] derived the energy of a thin film. The integrands involved in the bulk energy by them derived are quasi-convex. More precisely they are the quasi-convex envelope QW_0 of a function W_0 (denoted f_0 for the string), which is obtained from the 3D free energy density by solving a suitable minimization problem.

It is well known that quasi-convex integrands with appropriate growth conditions lead to the existence of minimizers of the total energy. If we are dealing with martensitic materials, we can not conclude that QW_0 is the energy density, since the infimum of the total free energy is not attained, in general. Hence the question: what is the energy density to be considered for the thin structure?

It must be noticed that from the results of Acerbi, Buttazzo and Percivale, or of Le Dret and Raoult as well, we can not deduce that W_0 is the free energy, since there are an infinite number of functions Z such that $QZ = QW_0$. It

G. Jaiani, P. Podio-Guidugli (eds.), *IUTAM Symposium on Relations of Shell, Plate, Beam, and 3D Models*, © Springer Science+Business Media B.V. 2008

is also well known, in phase transition theory, that it is the free energy that determines the microstructure and not its quasi-convex envelope. Thus the problem at hand is of noticeable importance in applications.

To explain how we derive the energy density of the thin structure, we must be slightly more specific. The asymptotic methodology initiated in [1] is the following: a sequence of bodies given in a cylindrical configuration of diameter or thickness ε is considered. For each of these bodies the total energy is known. Under quite general assumptions on the energy density there are different topologies which ensure compactness to the family of minimizers (or quasi-minimizers) of these energies. Once chosen one of these topologies, the thin structure model is obtained by passing to the limit as $\varepsilon \to 0$ in an appropriate variational sense (Γ-convergence). Roughly speaking, this variational limit ensures the convergence, in such topology, of minimizers of the energy at level ε to the minimizers of the thin structure problem. The obtained limit problem depends on the chosen topology. As said before, typically the infimum of the total free energy of a martensitic material is not attained. The minimizing sequences shall, in general, develop fine scale oscillations, which, according to the interpretation due to Ball and James [3, 4], model the microstructure experimentally observed in specimens of phase transforming materials. Thus, in phase transforming problems the "main properties" of the minimizing sequences are to be determined. This suggests that when we pass to the limit as $\varepsilon \to 0$ we should try to use a topology which, loosely speaking, ensures the convergence of the "main properties" of the minimizing sequences at level ε to the "main properties" of the minimizing sequences of the thin structure problem. We achieve this by embedding the 3D problems into a space of Young measures, see L.C. Young [22], which is one of the most successful tools used to characterize the oscillatory behaviour of sequences of functions. In this way we derive a limit functional which has a feature missing in all the other previously obtained variational limits: it uniquely determines the energy density of the thin structure.

This methodology introduces several difficulties which are completely missing in the work of Acerbi, Buttazzo and Percivale. Like them, we perform the computation of the Γ-limit under the requirement that the energy becomes infinite when the volume locally vanishes, that is $\lim_{\det F \to 0^+} f(F) = +\infty$. The same requirement is also met in a recent paper of Ben Belgacem [7]. In contrast with [1, 7], we need to specify the rate at which the energy blows up when the volume decreases. In fact, in the cited papers, the obtained Γ-limit involves the convexification of the energy density, which completely disregards the behaviour of the energy near vanishing-volume deformations. On the other hand, in the Young measure setting, where no convexification appears, the growth near such small deformations is as much important as the growth for large deformations. This reflects on the fact that the domain of the limit functional strongly depends on the prescribed growths.

The martensitic thin film model is obtained under the usual growth conditions of order p. This problem has been studied previously by Bhattacharya and James [8] who have considered a body characterized not only by a free energy but also by an interfacial energy which, mathematically speaking, behaves as a viscosity term and hence the limit problem does not contain any quasi-convex envelope. About 1 year later, Shu [21] has shown that letting the interfacial energy go to zero the variational limit coincides with that obtained by Le Dret and Raoult. Bělík and Luskin [5,6] have observed that with the energy considered by Bhattacharya and James [8] the deformations with finite energy cannot have sharp interfaces between compatible variants. For this reason they set the problem within the framework of functions with bounded Hessian and consider an interfacial energy proportional to the square of the total variation of the deformation gradient. Other interesting results on the Young measure theory of thin films have also been obtained, from a slightly different point of view, by Bocea and Fonseca [9].

The paper is written in a quite concise manner and all technical details, for which we refer to Freddi and Paroni [12,13], have been left out.

2 The Γ-Convergence Tool

The energies of the thin structures are obtained by taking a limit in the variational sense of Γ-convergence. Roughly speaking, the Γ-limit ensures convergence of minimizers of the energy at level ε to minimizers of the thin structure problem. In fact, we use a variant of De Giorgi's Γ-convergence, which has been introduced by Anzellotti, Baldo and Percivale in [2] and allows to treat families of functionals defined on a space which may be different from the domain of the limit. Let us recall here just the definition, referring for a precise formulation of the variational properties to [2] and to the books of Braides [10] and Dal Maso [11]. Let X be a set, let (Y, τ) be a topological space and let $q : X \to Y$. Given a sequence $F_n : X \to \overline{\mathbb{R}}$ and a point $y \in Y$, let us denote by

$$\begin{aligned}
\Gamma(q, \tau Y) \liminf_{n \to \infty} F_n(y) &:= \inf\{\liminf_{n \to \infty} F_n(x_n) \ : \ q(x_n) \xrightarrow{\tau} y\}, \\
\Gamma(q, \tau Y) \limsup_{n \to \infty} F_n(y) &:= \inf\{\limsup_{n \to \infty} F_n(x_n) \ : \ q(x_n) \xrightarrow{\tau} y\},
\end{aligned} \tag{1}$$

the Γ-lower and, respectively, the Γ-upper limit at the point y. If they turn out to be equal and $F(y)$ denotes their common value then we say that the sequence $\Gamma(q, \tau Y)$-converges to $F(y)$ and we write $\Gamma(q, \tau Y) \lim_{n \to \infty} F_n(y) = F(y)$. Given a family $F_\varepsilon : X \to \overline{\mathbb{R}}$ we say that it $\Gamma(q, \tau Y)$-converges to $F : Y \to \overline{\mathbb{R}}$ at a point $y \in Y$, and we write $\Gamma(q, \tau Y) \lim_{\varepsilon \to 0} F_\varepsilon(y) = F(y)$, if for any sequence ε_n of positive reals converging to 0 we have that $\Gamma(q, \tau Y) \lim_{n \to \infty} F_{\varepsilon_n}(y) = F(y)$.

3 Young Measures

Let, in the current section, Ω be an open bounded subset of $\mathbb{R}^n$. Let $M(\mathbb{R}^m) = C_0(\mathbb{R}^m)^*$ denote the space of $\mathbb{R}$-valued Borel measures on $\mathbb{R}^m$ and

$L_w^\infty(\Omega; M(\mathbb{R}^m)) = L^1(\Omega; C_0(\mathbb{R}^m))^*$ the dual of L^1. An element $\mu \in L_w^\infty(\Omega; M(\mathbb{R}^m))$ can be viewed as a *parameterized measure*, that is a map $x \mapsto \mu_x$ between Ω and $M(\mathbb{R}^m)$, which is essentially bounded and *weakly* measurable* in the sense that the functions $x \mapsto \langle \mu_x, \varphi \rangle$ are measurable for every $\varphi \in C_0(\mathbb{R}^m)$. The subscript w in the notation L_w^∞ refers to this weak* measurability. If μ_x is a probability for a.e. $x \in \Omega$ then μ is called a *Young measure* and $\mathcal{Y}(\Omega; \mathbb{R}^m)$ will denote the space of such Young measures. For instance, if $u : \Omega \to \mathbb{R}^m$ is a measurable function then $\delta_{u(\cdot)} \in \mathcal{Y}(\Omega; \mathbb{R}^m)$.

The space $L_w^\infty(\Omega; M(\mathbb{R}^m))$ will be endowed with the weak* convergence induced by the duality with L^1. Hence $\mu^n \to \mu$ weakly* in $L_w^\infty(\Omega; M(\mathbb{R}^m))$ iff

$$\int_\Omega <\mu_x^n, \varphi> g(x)\,dx \to \int_\Omega <\mu_x, \varphi> g(x)\,dx, \quad \forall \varphi \in C_0(\mathbb{R}^m), \ \forall g \in L^1(\Omega)$$

where $\langle , \rangle$ stays for integration of φ with respect to the involved measure.

Finally, we say that a sequence of measurable functions (u_n) *generates* μ if $\delta_{u_n(\cdot)} \to \mu$ weakly* in $L_w^\infty(\Omega; M(\mathbb{R}^m))$.

4 A 3D–1D Reduction Problem

Let ω be an open, bounded subset of $\mathbb{R}^2$ and let, for every $\varepsilon > 0$,

$$\Omega_\varepsilon = \{x = (x_\alpha, x_3) \in \mathbb{R}^2 \times \mathbb{R} : x_\alpha \in \varepsilon\omega, \ x_3 \in (0, \ell)\},$$

a three dimensional cylinder that we consider as the reference configuration of a hyperelastic body, which reduces to a 1D region as ε goes to zero. Our aim is to obtain the energy of an elastic string as limit of the total energy of 3D bodies occupying the regions Ω_ε. Without loss of generality we can assume that ω contains the origin and that $|\omega| = 1$.

The stored energy density $f : \mathbb{R}^{3\times3} \to (-\infty, +\infty]$ in the reference configuration Ω_ε is assumed to be continuous and to satisfy the following growth assumptions which include the whole class of Antman materials

$$\det F \leq 0 \ \Rightarrow \ f(F) \equiv +\infty,$$

$$\det F > 0 \ \Rightarrow \ \text{there exists two constants } C \geq c > 0 \text{ such that}$$
$$c\left(\frac{1}{|\det F|^s} + |F|^p - 1\right) \leq f(F) \leq C\left(\frac{1}{|\det F|^s} + |F|^p + 1\right), \tag{2}$$

for suitable $p \in [1, +\infty)$ and $s \in (0, +\infty)$.

The hypothesis that the material is homogeneous, which relays in the assumption that f be independent of the point in the reference configuration, is not essential and can be easily dropped. We refer to [13] for a treatment of the non-homogeneous case where moreover also the diameter of the cross-section is allowed to change from point to point.

Up to the scaling factor $1/\varepsilon^2$, the total energy I_ε of the body is given by

$$I_\varepsilon(y) = \frac{1}{\varepsilon^2} \int_{\Omega_\varepsilon} f(Dy(x)) \, dx - \int_{\Omega_\varepsilon} \hat{g}^\varepsilon(x) \cdot y(x) \, dx,$$

where the body force densities $\hat{g}^\varepsilon$ are taken in $L^{p'}(\Omega_\varepsilon; \mathbb{R}^3)$, with $1/p + 1/p' = 1$. Assuming, for instance, the body to be clamped on $\varepsilon\omega \times \{0\}$, the equilibrium configurations will be found by minimizing the energy I_ε over all $y \in W^{1,p}(\Omega_\varepsilon; \mathbb{R}^3)$ such that $y(x_1, x_2, 0) = (x_1, x_2, 0)$.

4.1 Scaling energies and passing to a fixed domain

The scaling factor $1/\varepsilon^2$ in front of the energy functionals serves to avoid the trivial case where the Γ-limit is identically zero. The choice of different scaling exponents would provide other limit models with their own physical meaning. For instance, $1/\varepsilon^3$ and $1/\varepsilon^4$ lead to rod theories; these cases has been studied in a quite different setting by Mora and Müller [18, 19].

To perform our analysis, it is convenient to put all the energy integrals on the same domain $\Omega := \Omega_1$, which is independent of ε, by the change of variables $x'_\alpha = \varepsilon x_\alpha$ $\alpha = 1, 2$. This gives to the energy functionals the following form

$$I_\varepsilon^\Omega(y) := \int_\Omega f\left(\frac{D_\alpha y}{\varepsilon} \Big| D_3 y\right) dx - \int_\Omega g^\varepsilon \cdot y \, dx,$$

where $D_\alpha y$ denotes the first two columns of the deformation gradient, while $D_3 y$ is the third column. Hereafter, for simplicity, we assume that the body force densities g^ε do not depend on ε and set $g := g^\varepsilon$. The total energy I_ε^Ω has to be minimized over all $y \in W^{1,p}(\Omega; \mathbb{R}^3)$ such that $y(x_1, x_2, 0) = \varepsilon(x_1, x_2, 0)$.

4.2 Previous results and some remarks

Acerbi, Buttazzo and Percivale [1] studied the problem under the following growth assumptions

$$\det F \leq 0 \Rightarrow f(F) \equiv +\infty,$$
$$\forall \delta > 0 \ \exists C_\delta > 0 \text{ s.t. } \det F \geq \delta \Rightarrow f(F) \leq C_\delta(|F|^p + 1), \qquad (3)$$
$$\exists c > 0 \text{ s.t. } c(|F|^p - 1) \leq f(F),$$

for a suitable $p \in [1, +\infty)$. The Γ-limit of the functionals I_ε is taken by them under the norm convergence in $L^p((0, \ell); \mathbb{R}^3)$ of the average of the deformation over the cross section and the resulting $\Gamma(L^p)$-limit is

$$G_0(v) = \int_0^\ell f_0^{**}(v') \, dx_3 - \int_0^\ell A v^\alpha g \cdot v \, dx_3$$

if $v \in W^{1,p}((0,\ell);\mathbb{R}^3)$ with $v(0) = 0$. f_0^{**} is the convex envelope of the function

$$f_0(z) := \min\{f(\overline{F}|z) \; : \; \overline{F} \in \mathbb{R}^{3\times 2}\}, \quad z \in \mathbb{R}^3$$

and Av^α denotes the integral mean value with respect to x_1 and x_2.

Under the assumptions (2), conditions (3) are satisfied and a simple computation shows that f_0 satisfies the following estimates

$$c\Big(\frac{1}{|z|^q} + |z|^p - 1\Big) \le f_0(z) \le C\Big(\frac{1}{|z|^q} + |z|^p + 1\Big) \quad \text{for every } z \neq 0, \qquad (4)$$

with $q = \dfrac{ps}{p+2s}$, for suitable positive constants C and c.

In fact, the convexification of the energy density completely disregards the behaviour of the energy near vanishing-volume deformations. On the other hand, in the Young measure setting, where no convexification appears, the growth near such small deformations is as much important as the growth for large deformations.

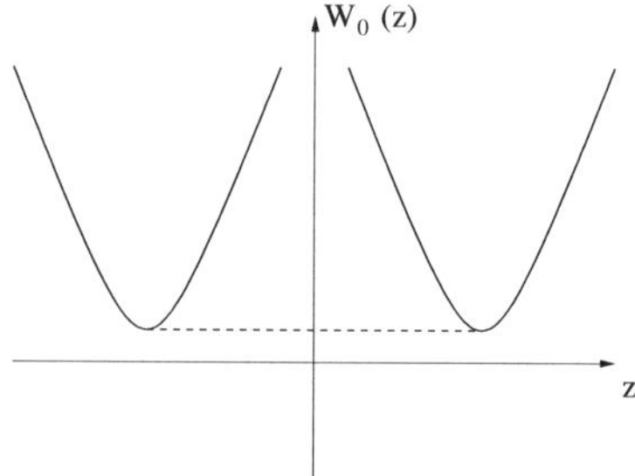

Then we have to expect that conditions (3) be not precise enough to characterize the domain of the limit problem in terms of Young measures. Indeed, we shall see that two different choices of the exponents s and p in (2), will lead to limit problems defined on different spaces.

4.3 The Young measure setting

Acerbi, Buttazzo and Percivale, in taking the Γ-limit, used the norm topology of L^p. Nevertheless, there are other topologies which ensure compactness to the minimizing sequences which can choose to compute the Γ-limit. Limit problems provided by different topologies show different ability in describing the behaviour of the minimizing sequences. As already explained, we shall work within the framework of Young measures.

Via the Dirac mass supported on the gradient, the space of functions $W^{1,p}$ with some prescribed boundary conditions can be identified with a subspace of L_w^∞. Thus we can extend the functionals I_ε^Ω to $L_w^\infty(\Omega; M(\mathbb{R}^{3\times 3}))$ by setting

$$I_\varepsilon^\infty(\mu) = \begin{cases} I_\varepsilon^\Omega(y) & \text{if } \exists\, y \in \mathcal{A}_\varepsilon^\Omega \text{ s.t. } \mu = \delta_{Dy(\cdot)} \\ +\infty & \text{otherwise in } L_w^\infty(\Omega; M(\mathbb{R}^{3\times 3})). \end{cases} \qquad (5)$$

Due to the growth constraints (2), the effective domain of the limit functional will consist of those Young measures which are characterized by a certain kind of growth, as precised by the following definition.

Definition 1. *Let $p \geq 1$ and $q > 0$. With $\mathcal{Y}^{-q,p}((0,\ell);\mathbb{R}^m)$, $m \in \mathbb{N}$, we denote the set of Young measures $\nu \in \mathcal{Y}((0,\ell);\mathbb{R}^m)$ such that*

$$\int_0^\ell \int_{\mathbb{R}^m} \left(|z|^{-q} + |z|^p \right) d\nu_t(z)\, dt < +\infty.$$

The Young measures in this spaces can be characterized as those generated by sequences (z_j) such that $\left(|z_j|^{-q} + |z_j|^p \right)$ is equi-integrable (see [13]).

4.4 Compactness properties of bounded sequences

In the L^p setting, if y^ε are deformations with equi-bounded energy, the growth conditions imply that, up to subsequences, $y^\varepsilon \to y$ and $D_\alpha y^\varepsilon \to 0$ in L^p, hence $D_\alpha y = 0$ and $y = y(x_3)$. On the contrary, in our case

$$\delta_{\left(\frac{D_\alpha y^\varepsilon}{\varepsilon} | D_3 y^\varepsilon \right)} \to \mu \ \text{ weakly* in } L_w^\infty \ \Rightarrow \ \mu = \delta_0 \otimes \nu_x$$

with $\delta_0 \in M(\mathbb{R}^{3\times 2})$ and $\nu_x \in M(\mathbb{R}^3)$, but now ν may depend also on x_1 and x_2. Moreover the only relevant part of the limit parametrized measure is given by the projection on the third column.

4.5 The average-projection mapping and the Γ-convergence result

The considerations above motivate the introduction of the following mapping

$$\rho : L_w^\infty(\Omega; M(\mathbb{R}^{3\times 3})) \to L_w^\infty((0,\ell); M(\mathbb{R}^3)),$$

defined by $\rho = \pi_\#^3 \circ \mathrm{Av}^\alpha = \mathrm{Av}^\alpha \circ \pi_\#^3$, where Av^α denotes average with respect to the first two variables and $\pi_\#^3$ is the image measure under the projection on the third column. The commutativity of composition follows directly from the definitions. The following Γ-convergence theorem is stated with respect to the weak* convergence in $L_w^\infty(\omega; M(\mathbb{R}^{3\times 2}))$ of $\rho(\mu^\varepsilon) \to \nu$, under which the sequences with bounded energy are relatively compact (see [13], Lemma 5.1).

Theorem 1. *Let f be a real extended valued continuous function which satisfies (2). Then $\Gamma(\rho, w^* L_w^\infty((0,\ell); M(\mathbb{R}^3))) \lim_{\varepsilon \to 0^+} I_\varepsilon^\infty(\nu) = I_S(\nu)$ were*

$$I_S(\nu) = \int_0^\ell \langle \nu_t, f_0 \rangle\, dt - \int_0^\ell \mathrm{Av}^\alpha g \cdot y\, dt$$

if $\nu \in \mathcal{Y}^{-q,p}((0,\ell);\mathbb{R}^3)$ with $q = ps/(p+2s)$, and holds $+\infty$ otherwise in $L_w^\infty((0,\ell); M(\mathbb{R}^3))$. The function $y \in W^{1,p}((0,\ell);\mathbb{R}^3)$ which appears in the expression above stays for the underlying deformation of ν with boundary condition $y(0) = 0$.

From our result we can recover the Γ-limit G_0 of Acerbi, Buttazzo and Percivale as follows. Let $v \in L^p((0,\ell);\mathbb{R}^3)$; then

$$G_0(v) = \inf\{I_S(\nu) \ : \ \nu \in L_w^\infty((0,\ell); M(\mathbb{R}^3)), \ \langle \nu, \mathrm{id} \rangle = v', \ v(0) = 0\}$$

with the usual convention $\inf \varnothing = +\infty$. Moreover the infimum is attained.

4.6 The energy density of the string

The map $f_0 : \mathbb{R}^3 \to \overline{\mathbb{R}}$ is the unique continuous integrand satisfying the growth conditions (4) and such that the Γ-limit I_S is the relaxation of the functional

$$E(y) = \int_0^\ell f_0(y'(x_3))\, dx_3 - \int_0^\ell \mathrm{Av}^\alpha g \cdot y\, dx_3$$

with respect to the weak* topology in $L_w^\infty((0,\ell); M(\mathbb{R}^3))$. Hence f_0 can be considered to be the energy density of the string.

5 A 3D–2D Reduction Problem

Let ω be an open bounded subset of $\mathbb{R}^2$ with a regular boundary, which we assume to be the reference configuration of a membrane. We want to obtain the energy of the membrane as limit, when $\varepsilon \to 0$, of the total energy of 3D hyperelastic bodies occupying the cylindrical regions $\Omega_\varepsilon = \omega \times (-\varepsilon/2, \varepsilon/2)$.

Up to the scaling factor $1/\varepsilon$, the total energy of the 3D body Ω_ε is

$$I_\varepsilon(y) = \frac{1}{\varepsilon} \int_{\Omega_\varepsilon} W(Dy)\, dx - \int_{\Omega_\varepsilon} g^\varepsilon(x) \cdot y\, dx,$$

where the deformation $y : \Omega_\varepsilon \to \mathbb{R}^3$ is subject to a prescribed linear boundary condition on the lateral boundary of the cylinder, that is $y(x) = Bx$ on $\Gamma_\varepsilon = \partial\omega \times (-\varepsilon/2, \varepsilon/2)$, where $B : \mathbb{R}^3 \to \mathbb{R}^3$ is a linear map. The energy density $W : \mathbb{R}^{3\times 3} \to \mathbb{R}$ is continuous, generally non-convex, and satisfies a growth condition of order $p \in (1, +\infty)$, that is

$$C(|F|^p - 1) \le W(F) \le C(|F|^p + 1). \tag{6}$$

The body force densities g^ε are assumed to be in $L^{p'}$, with $1/p + 1/p' = 1$.

5.1 Scaling energies and passing to a fixed domain

The choice of a scaling factor different from $1/\varepsilon$ in front of the 3D energies would lead to other limit problems; $1/\varepsilon^3$, for example, corresponds to a plate model (see for instance Friesecke, James and Müller [14, 15]).

Under the change of variable $x_3' = \varepsilon x_3$ (still called x_3) the energy functionals become

$$I_\varepsilon(y) = \int_\Omega W\left(D_\alpha y \Big| \frac{D_3 y}{\varepsilon}\right) dx - \int_\Omega g^\varepsilon(x) \cdot y\, dx$$

where $\Omega := \Omega_1$, and $y \in W^{1,p}(\Omega; \mathbb{R}^3)$ has to satisfy $y(x) = B(x_1, x_2, \varepsilon x_3)$ on $\Gamma = \partial\omega \times (-1/2, 1/2)$. Hereafter, for the sake of simplicity, we assume the scaled body force densities to be independent of ε.

5.2 Previous results and some remarks

Le Dret and Raoult [17] proved that $\Gamma(L^p)\lim_{\varepsilon\to 0} I_\varepsilon(y) = I_{LDR}(y)$ where

$$I_{LDR}(y) = 2\int_\omega QW_0(D_\alpha y)\,dx_\alpha - \int_\omega \mathrm{Av}^3 g(x_\alpha)\cdot y(x_\alpha)\,dx_\alpha$$

if $y \in W^{1,p}(\Omega;\mathbb{R}^3)$, $y(x) = (x_1, x_2, 0)$ on Γ and $D_3 y = 0$. Here QW_0 denotes the quasi-convex envelope of the function

$$W_0(\overline{F}) := \min\{W(\overline{F}|z) \,:\, z \in \mathbb{R}^3\}, \quad \overline{F} \in \mathbb{R}^{3\times 2}$$

and Av^3 denotes the integral mean value with respect to x_3.

As explained in the introduction, quasi-convex integrands cannot describe any microstructure. In order to overcome this difficulty, Bhattacharya and James [8] considered a 3D body characterized also by an interfacial energy which is taken to be proportional to the square of the Hessian of the deformation; in fact they consider the functionals

$$J_{\varepsilon,\kappa}(y) = \int_{\Omega_\varepsilon} W(Dy) + \kappa|D^2 y|^2\,dx$$

where the positive constant $\kappa > 0$ is fixed and the limit is taken as $\varepsilon \to 0$. The introduction of the extra term leads to a Γ-limit in which no quasi-convex envelope appears. The limit energy obtained, like the three dimensional, has an interfacial energy which makes the minimization quite hard to perform.

About 1 year later Shu [21] has shown that if ε and $\kappa(\varepsilon)$ go to 0 then the total energy considered by Bhattacharya and James Γ-converges, in a suitable topology, to the one obtained by Le Dret and Raoult.

5.3 Young measures generated by gradients

Le Dret and Raoult, in taking the Γ-limit, used the norm topology of L^p. We shall work instead within the framework of Young measures and use a topology which provides a richer description of the microstructure.

To this aim we extend the energies I_ε to $L^\infty_w(\Omega; M(\mathbb{R}^{3\times 3}))$ by setting

$$I_\varepsilon^\infty(\mu) = \begin{cases} I_\varepsilon(y) & \text{if } \exists y \in W^{1,p},\ y(x) = B(x_1, x_2, \varepsilon x_3) \text{ on } \Gamma \,:\, \mu = \delta_{Dy(\cdot)} \\ +\infty & \text{otherwise in } L^\infty_w(\Omega; M(\mathbb{R}^{3\times 3})), \end{cases}$$

and such extension is well defined thanks to the boundary conditions.

The effective domain of the limit problem will turn out to be the space $\mathcal{Y}^{1,p}(\omega;\mathbb{R}^3)$ of Young measures generated by gradients of functions in $W^{1,p}(\omega; \mathbb{R}^3)$. From the characterization of Kinderlehrer and Pedregal [16,20] the center of mass of such a Young measure is a gradient of a function $y \in W^{1,p}$, that is $\langle\mu_x, \mathrm{id}\rangle = Dy(x)$, and y is called an *underlying deformation* of μ. We denote by $\mathcal{Y}^{1,p}_\Gamma(\omega;\mathbb{R}^3)$ the subspace of $\mathcal{Y}^{1,p}(\omega;\mathbb{R}^3)$ whose elements have an underlying deformation satisfying the prescribed boundary condition on $\partial\omega$.

120   L. Freddi and R. Paroni

5.4 Compactness and Γ-convergence results

In the L^p setting, if y^ε are deformations bounded in energy, the growth conditions imply that, up to subsequences, $y^\varepsilon \to y$ and $D_3 y^\varepsilon \to 0$ in L^p, hence $D_3 y = 0$ and $y = y(x_1, x_2)$. On the contrary, in our case

$$\delta_{(D_\alpha y^\varepsilon, \frac{D_3 y^\varepsilon}{\varepsilon})} \overset{*}{\to} \mu \quad \Rightarrow \quad \mu = \nu_x \otimes \delta_0$$

where $\nu_x \in M(\mathbb{R}^{3\times 2})$ and $\delta_0 \in M(\mathbb{R}^3)$, but now ν may depend also on the variable x_3. Moreover the only relevant part of the limit parametrized measure is given by the projection on the first two columns. These facts motivates also in this case the introduction of a suitable average-projection mapping

$$q : L_w^\infty(\Omega; M(\mathbb{R}^{3\times 3})) \to L_w^\infty(\omega; M(\mathbb{R}^{3\times 2})),$$

defined by $q := \bar{\pi}_\# \circ \mathrm{Av}^3 = \mathrm{Av}^3 \circ \bar{\pi}_\#$, where Av^3 denotes average with respect to the third variable and $\bar{\pi}_\#$ is the image measure under the projection on the first two columns. The following Γ-convergence theorem is stated with respect to the weak* convergence in $L_w^\infty(\omega; M(\mathbb{R}^{3\times 2}))$ of $q(\mu^\varepsilon) \to \nu$, under which the sequences with bounded energy are relatively compact (see [12], Theorem 5.6).

Theorem 2. *Let W be a real continuous function which satisfies the growth assumptions* (6). *Then* $\Gamma(q, w^* L_w^\infty(\omega; M(\mathbb{R}^{3\times 2}))) \lim_{\varepsilon \to 0} I_\varepsilon^\infty(\nu) = I_M(\nu)$ *were*

$$I_M(\nu) = \int_\omega \langle \nu_{x_\alpha}, W_0 \rangle \, dx_\alpha - \int_\omega \mathrm{Av}^3 g \cdot y \, dx_\alpha$$

if $\nu \in \mathcal{Y}_\Gamma^{1,p}(\omega; \mathbb{R}^3)$ *and holds* $+\infty$ *otherwise in* $L_w^\infty(\omega; M(\mathbb{R}^{3\times 2}))$. *The function* $y \in W^{1,p}(\omega; \mathbb{R}^3)$ *which appears in the expression above denotes the underlying deformation of ν which satisfies the boundary condition* $y(x) = B(x_1, x_2, 0)$ *on* $\partial\omega$.

Given $y \in W^{1,p}(\Omega; \mathbb{R}^3)$, the Γ-limit of Le Dret and Raoult at y can be obtained from ours by taking the infimum of $I_M(\nu)$ over all $\nu \in L_w^\infty(\omega; M(\mathbb{R}^{3\times 2}))$ which satisfy $\langle \nu, \mathrm{id} \rangle = D_\alpha y$ and $y(x) = (x_1, x_2, 0)$ on $\partial\omega$, with the usual convention $\inf \varnothing = +\infty$. Moreover the infimum is attained.

5.5 The energy density of the thin film

For every matrix B, the Γ-limit I_M is the Young measure relaxation of

$$E_{W_0}(y) = \int_\omega W_0(D_\alpha y) \, dx_\alpha - \int_\omega \mathrm{Av}^3 g \cdot y \, dx_\alpha, \quad y(x) = B(x_1, x_2, 0) \text{ on } \partial\omega.$$

The remarkable fact is that among all continuous integrands with p-growth, W_0 is the unique density such that the relaxation of the corresponding energy E_{W_0} produces the functional I_M, for every linear map B. Hence W_0 can be considered to be the energy density of the thin film.

References

1. E. Acerbi, G. Buttazzo, and D. Percivale, A variational definition of the strain energy for an elastic string, *J. Elasticity*, Vol. 25, pp.137–148 (1991).
2. G. Anzellotti, S. Baldo, and D. Percivale, Dimension reduction in variational problems, asymptotic development in Γ-convergence and thin structures in elasticity, *Asymptot. Anal.*, Vol. 9, pp.61–100 (1994).
3. J.M. Ball and R.D. James, Fine phase mixtures as minimizers of energy, *Arch. Ration. Mech. Anal.*, Vol. 100, pp.13–52 (1987).
4. J.M. Ball and R.D. James, Proposed experimental tests of a theory of fine microstructure and the two well problem, *Phil. Trans. R. Soc. London A.*, Vol. 338, pp.389–450 (1992).
5. P. Bělík and M. Luskin, A computational model for the indentation and phase transformation of a martensitic thin film, *J. Mech. Phys. Solids*, Vol. 50, pp.1789–1815 (2002).
6. P. Bělík and M. Luskin, A total-variation surface energy model for thin films of martensitic cristals, *Interface. Free Bound.*, Vol. 4, pp.71–88 (2002).
7. H. Ben Belgacem, Relaxation of singular functionals defined on Sobolev spaces, ESAIM *Control Optim. Calc. Var.*, Vol. 5, pp.71–85 (2000)
8. K. Bhattacharya and R.D. James, A theory of thin films of martensitic materials with applications to microactuators, *J. Mech. Phys. Solids*, Vol. 47, pp.531–576 (1999).
9. M. Bocea and I. Fonseca, A Young measure approach to a nonlinear membrane model involving the bending moment, *Proc. Roy. Soc. Edinburgh Sect. A,* Vol. 134 no. 5, pp.845–883 (2004).
10. A. Braides, *Γ-convergence for beginners*, Oxford University Press, 2002.
11. G. Dal Maso, *An introduction to Γ-convergence*, Birkhäuser, 1993.
12. L. Freddi and R. Paroni, The energy density of martensitic thin films via dimension reduction, *Interface. Free Bound.*, Vol. 6, pp.439–459 (2004).
13. L. Freddi and R. Paroni, A 3D–1D Young measure theory of an elastic string. *Asymptotic Anal.*, Vol. 39 no. 1, pp.61–89 (2004)
14. G. Friesecke, R.D. James, and S. Müller, A theorem on geometric rigidity and the derivation of nonlinear plate theory from three-dimensional elasticity, *Commun. Pure Appl. Math.*, Vol. 55, pp.1461–1506 (2002).
15. G. Friesecke, R.D. James, and S. Müller, A hierarchy of plate models derived from nonlinear elasticity by Γ-convergence, *Arch. Ration. Mech. Anal.*, Vol. 180, pp.183–236 (2006).
16. D. Kinderlehrer and P. Pedregal, Characterizations of Young measures generated by gradients, *Arch. Ration. Mech. Anal.*, Vol. 115, pp.329–365 (1991).
17. H. Le Dret and A. Raoult, The nonlinear membrane model as variational limit of nonlinear three-dimensional elasticity, *J. Math. Pures Appl.*, Vol. 74, pp.549–578 (1995).
18. M.G. Mora and S. Müller, A nonlinear model for inextensible rods as a low energy Γ-limit of three-dimensional nonlinear elasticity, *Ann. I. H. Poincaré*, Vol. 21, pp.271–293 (2004).
19. M.G. Mora and S. Müller, Derivation of a rod theory for multiphase materials, *Calc. Var. Partial Dif.*, Vol. 28, pp.161–178 (2007).
20. P. Pedregal, *Parametrized measures and variational principles*, Progress in nonlinear differential equations and their applications, Birkhäuser, 1997.

21. Y.C. Shu, Heterogeneous thin films of martensitic materials, *Arch. Ration. Mech. Anal.*, Vol. 153, pp.39–90 (2000).
22. L.C. Young, *Lectures on the calculus of variations and optimal control theory*, W. B. Saunders Co., 1969.

Joint Vibrations of a Rectangular Shell and Gas in It

Elena Gavrilova

Department of Mathematics, St. Ivan Rilski University of Mining & Geology, Studentski Grad, 1700 Sofia, Bulgaria, `elenag@mbox.contact.bg`

Abstract Thin rectangular elastic plates are often used as structural components of parallelepiped cavities filled with gas and subjected to different dynamic loads. Such systems find application in the glass-skin technology of tall buildings, as outside skin plates of supersonic air crafts, as covers of tanks in the chemical industry, as chambers in hydraulic structures, etc. The main problem of the mechanics of these systems is to determine their response to specific dynamic loads or to standard catastrophic ones. Subordinate to this, but important enough for engineering practice, appears to be the problem of the determination of the dynamic characteristics of such systems.

In this study, a closed rigid rectangular parallelepiped tank, filled with gas, is under consideration. A part of one of its walls is a thin linearly elastic rectangular plate. The problem focuses on the stationary forced vibrations of the gas and the elastic plate under the action of a source, being situated in the gas tank. Let the source have a range of sizes which are small in comparison with the lengths of the excited waves: then it is possible to be accepted as a point source. It is supposed that the productivity and the frequency of the source are given and do not show any influence of the earlier excited waves. The problem is considered in a linear approximation without giving an account of the dissipating forces.

A combination of the use of Green's function, the method of the crossed strips of G. Warburton and the method of Bubnov-Galerkin is employed to investigate the dynamic behavior of this gas-structure interaction system in cases of arbitrary supporting conditions of the elastic plate. An approximate solution is arrived at based on ignoring diffraction by the elastic plate waves. Some numerical examples are given which demonstrate the necessity of taking into account which part of the spectrum of the natural frequencies of the elastic plate the forced frequency is located in.

Keywords: closed rigid rectangular parallelepiped tank, thin linearly elastic rectangular plate, gas-structure interaction system, forced stationary vibrations, Green function, method of the crossed strips of G. Warburton, method of Bubnov-Galerkin

G. Jaiani, P. Podio-Guidugli (eds.), *IUTAM Symposium on Relations of Shell, Plate, Beam, and 3D Models*, © Springer Science+Business Media B.V. 2008

1 Introduction

Thin rectangular elastic plates are often used as structural components closing in or covering parallelepiped cavities filled with compressible fluid (gas) and subjected to dynamic loads. Such mechanical systems are employed in the glass-skin technology of tall buildings, as panels of shop windows, as outside skin plates of supersonic aircrafts, as covers of tanks in the chemical industry, as chambers in hydraulic structures, etc. [1].

The main problem of the mechanics of such systems is to determine their response to specific dynamic loads or to standard catastrophic ones. Subordinate to this, but important enough for engineering practice, appears to be the problem of the determination of the dynamic characteristics of such systems.

For the solution of the problem of determining the natural vibrations of a rectangular plate at different supporting conditions, Warburton [2] has suggested the method of crossed strips, using the variational principle of Rayleigh-Ritz. In [1] the method of Bubnov-Galerkin [3] is used and it is elaborated in the form of an easy scheme for application to the dynamic problem of the stationary vibrations of a special fluid-structure interaction system. It consists of a parallelepiped tank that is filled with an acoustic fluid as a thin elastic plate is inserted into a rectangular orifice of its arbitrary wall.

2 The Problem

Thin linearly elastic rectangular plate $EFGH$ with sizes a and b ($0 < a \leq L_1$, $0 < b \leq L_2$) and surface S is inserted into a rectangular orifice of the absolutely rigid wall $ABCD$ of a rectangular parallelepiped tank $ABCDA'B'C'D'$, all of

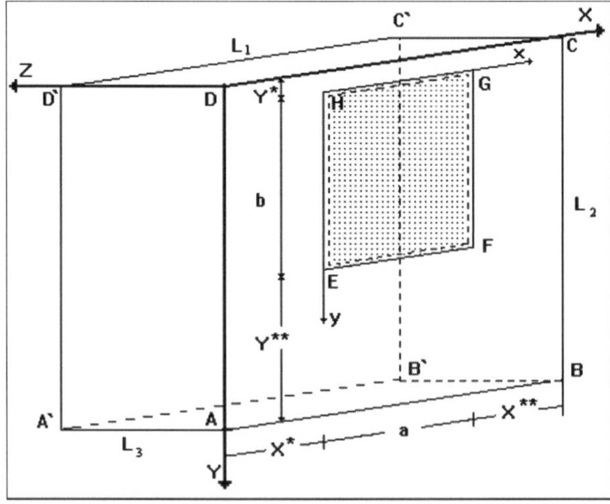

Fig. 1 The geometry of the gas-structure interaction system under consideration

whose other walls are absolutely rigid (Fig. 1). The tank volume is filled with gas with sound velocity c_0 and mass density ρ_0. Let ρ be the mass density per unit area of the plate, $\overline{D} = \frac{Eh^3}{12(1-\nu^2)}$ - the flexural rigidity, h - the thickness, E - Young's modulus of elasticity and ν - Poisson's ratio of the elastic plate material. Two rectangular co-ordinate systems $DXYZ$ and $Hxyz$ are used (see Fig. 1) and they are connected by the transformations

$$X = x + X^*, \quad Y = y + Y^*, \quad Z = z. \tag{1}$$

Let the velocity potential function of the gas motion be $\varphi(X, Y, Z; t)$ and the function describing the middle plate surface vibrations be $w(x, y; t)$. As Kirchoff-Love's assumption is taken to be valid, the motion of the middle plate surface will be adequated with the whole plate motion. The problem we are considering is that of the stationary forced vibrations of the gas and the elastic plate under the action of a source being situated in the gas tank. Let the source have a range of sizes which are small in comparison with the lengths of the excited waves: then it is possible to be accepted as a point source. It is supposed that the productivity Q and the frequency ω of the source are given and are not influenced by the earlier excited waves. Then the velocity potential function of the gas motion satisfies the following wave Equation [4]:

$$\frac{\partial^2 \varphi}{\partial X^2} + \frac{\partial^2 \varphi}{\partial Y^2} + \frac{\partial^2 \varphi}{\partial Z^2} - \frac{1}{c_0^2} \frac{\partial^2 \varphi}{\partial t^2} = q = Q e^{i\omega t} \tag{2}$$

with the boundary conditions

$$\frac{\partial \varphi}{\partial X}|_{X=0,L_1} = 0, \qquad \frac{\partial \varphi}{\partial Y}|_{Y=0,L_2} = 0, \qquad \frac{\partial \varphi}{\partial Z}|_{Z=L_3} = 0. \tag{3}$$

The function $w(x, y; t)$ satisfies the partial differential equation [1]

$$\overline{D}\left(\frac{\partial^2}{\partial x^2} + \frac{\partial^2}{\partial y^2}\right)^2 w + \rho h \frac{\partial^2 w}{\partial t^2} = -\rho_0 \frac{\partial \varphi}{\partial t}|_S, \tag{4}$$

together with the boundary conditions, which describe the way of support of the elastic plate – e.g., in the case of a simply-supported plate:

$$w|_{x=0,a} = \frac{\partial^2 w}{\partial x^2}|_{x=0,a} = 0, \quad w|_{y=0,b} = \frac{\partial^2 w}{\partial y^2}|_{y=0,b} = 0; \tag{5}$$

in the case of a clamped plate:

$$w|_{x=0,a} = \frac{\partial w}{\partial x}|_{x=0,a} = 0, \quad w|_{y=0,b} = \frac{\partial w}{\partial y}|_{y=0,b} = 0. \tag{5a}$$

Both functions φ and w have to satisfy the compatibility condition:

$$\frac{\partial \varphi}{\partial Z}|_{Z=0} = \begin{cases} \frac{\partial w}{\partial t} & (x, y) \in S, \\ 0 & (x, y) \notin S. \end{cases} \tag{6}$$

3 Analytical Solution

Let X_0, Y_0, Z_0 be the coordinates of the source. Then $Q = Q_0 \delta(X - X_0, Y - Y_0, Z - Z_0)$ as the point source is presented by the Dirac delta function $\delta(X - X_0, Y - Y_0, Z - Z_0)$, that is, $\delta = \infty$ when $X = X_0, Y = Y_0, Z = Z_0$ and $\delta = 0$ when $X \neq X_0, Y \neq Y_0, Z \neq Z_0$, except $\delta(X - X_0, Y - Y_0, Z - Z_0) = \delta(X - X_0)\delta(Y - Y_0)\delta(Z - Z_0)$ [5].

From Equation (2) the next partial differential equation is obtained:

$$\frac{\partial^2 \varphi}{\partial X^2} + \frac{\partial^2 \varphi}{\partial Y^2} + \frac{\partial^2 \varphi}{\partial Z^2} - \frac{1}{c_0^2}\frac{\partial^2 \varphi}{\partial t^2} = Q_0 \delta(X - X_0, Y - Y_0, Z - Z_0)e^{i\omega t} \qquad (7)$$

and its solution is [6]

$$\varphi = \left[\iiint_{V_0} q(X_0, Y_0, Z_0)G(X, Y, Z, X_0, Y_0, Z_0)\, dV_0 \right.$$

$$\left. + \iint_{S_0} \left(\frac{\partial \varphi}{\partial Z} \right)_{S_0} G(X, Y, Z, X_0^s, Y_0^s, Z_0^s)\, dS_0 \right] e^{i\omega t}, \qquad (8)$$

where X_0^s, Y_0^s, Z_0^s are coordinates of points on the tank surface; X, Y, Z - the coordinates of the observation point; V_0, S_0 – the volume and the surface of the tank; $G(X, Y, Z, X_0^s, Y_0^s, Z_0^s)$ - Green's function for the observation point in X, Y, Z and the source on the border. The function G is found from the solution of the following Helmholz equation that will be nonhomogeneous only at one point (at the source):

$$\frac{\partial^2 G}{\partial X^2} + \frac{\partial^2 G}{\partial Y^2} + \frac{\partial^2 G}{\partial Z^2} + \frac{\omega^2}{c_0^2}G = \delta(X - X_0, Y - Y_0, Z - Z_0), \qquad (9)$$

and from the boundary conditions:

$$\frac{\partial G}{\partial X}\Big|_{X=0,L_1} = 0, \qquad \frac{\partial G}{\partial Y}\Big|_{Y=0,L_2} = 0, \qquad \frac{\partial G}{\partial Z}\Big|_{Z=0,L_3} = 0. \qquad (10)$$

The function G is presented as a sum of the natural functions of the homogeneous equation, corresponding to Equation (9), satisfying the boundary conditions (10) [1]:

$$G(X, Y, Z, X_0, Y_0, Z_0) = \sum_{q=0}^{\infty}\sum_{r=0}^{\infty}\sum_{s=0}^{\infty} \frac{\varepsilon_{qrs}}{L_1 L_2 L_3} A_{qrs}$$

$$\cos(\alpha_q X)\cos(\beta_r Y)\cos(\gamma_s Z), \qquad (11)$$

where

$$\alpha_q = \frac{q\pi}{L_1}, \quad \beta_r = \frac{r\pi}{L_2}, \quad \gamma_s = \frac{s\pi}{L_3}; \qquad (12)$$

$$\varepsilon_{000} = 1, \varepsilon_{q00} = \varepsilon_{0r0} = \varepsilon_{00s} = 2, \varepsilon_{0rs} = \varepsilon_{q0s} = \varepsilon_{qr0} = 4,$$

$$\varepsilon_{qrs} = 8\ (q, r, s = 1, 2, \ldots). \qquad (13)$$

The substitution of expression (11) in Equation (9) and the integration of the obtained equation, using the method of Bubnov-Galerkin [3] as well as the integration property of the delta-function [5], give:

$$A_{qrs} = -\frac{c_0^2}{\omega_{qrs}^2 - \omega^2}\cos(\alpha_q X_0)\cos(\beta_r Y_0)\cos(\gamma_s Z_0).$$

Here $\omega_{qrs} = c_0\sqrt{\alpha_q^2 + \beta_r^2 + \gamma_s^2}$ are the natural frequencies of the gas vibrations in a tank with absolutely rigid walls. Therefore

$$G(X, Y, Z, X_0, Y_0, Z_0) =$$

$$-\frac{c_0^2}{L_1 L_2 L_3}\sum_{q=0}^{\infty}\sum_{r=0}^{\infty}\sum_{s=0}^{\infty}\frac{\varepsilon_{qrs}\cos(\alpha_q X_0)\cos(\beta_r Y_0)\cos(\gamma_s Z_0)}{\omega_{qrs}^2 - \omega^2} \quad (14)$$

$$\cos(\alpha_q X)\cos(\beta_r Y)\cos(\gamma_s Z).$$

According to conditions (3) and (6) $\frac{\partial\varphi}{\partial Z} = \frac{\partial w}{\partial t}$ on the surface S of the elastic plate (a part of the tank wall $Z_0^s = 0$), but on the other tank walls the first derivatives of the function φ are equal to zero. Then the substitution of Equation (14) in expression (8) gives:

$$\varphi(X, Y, Z; t) = -\frac{c_0^2}{L_1 L_2 L_3}\sum_{q=0}^{\infty}\sum_{r=0}^{\infty}\sum_{s=0}^{\infty}\frac{\varepsilon_{qrs}}{\omega_{qrs}^2 - \omega^2}\cos(\alpha_q X)\cos(\beta_r Y)$$

$$\cos(\gamma_s Z)\left[\iiint_{V_0} q(X_0, Y_0, Z_0)\cos(\alpha_q X_0)\cos(\beta_r Y_0)\cos(\gamma_s Z_0)\,dV_0 \quad (15)\right.$$

$$\left. + \iint_S \frac{\partial w}{\partial t}\cos(\alpha_q X_0^s)\cos(\beta_r Y_0^s)\,dS\right]e^{i\omega t}.$$

The following expansion in the Fourier series is valid [4]:

$$\frac{\text{ch}\,[\gamma_{qr}(L_3 - Z)]}{\gamma_{qr}\,\text{sh}(\gamma_{qr}L_3)} = \frac{c_0^2}{L_3}\sum_{s=0}^{\infty}\frac{e_s}{\omega_{qrs}^2 - \omega^2}\cos(\gamma_s Z),$$

where

$$\gamma_{qr} = i\gamma_s = \sqrt{\alpha_q^2 + \beta_r^2 - \frac{\omega^2}{c_0^2}}, \quad e_0 = 1, \quad e_s = 2\,(s = 1, 2, \ldots).$$

Then

$$\varphi(X, Y, Z; t) = -\frac{1}{L_1 L_2}\sum_{q=0}^{\infty}\sum_{r=0}^{\infty}\frac{\varepsilon_{qr}\,\text{ch}\,[\gamma_{qr}(L_3 - Z)]}{\gamma_{qr}\,\text{sh}(\gamma_{qr}L_3)}\cos(\alpha_q X)\cos(\beta_r Y)$$

$$\left[\iiint_{V_0} q(X_0, Y_0, Z_0)\cos(\alpha_q X_0)\cos(\beta_r Y_0)\,\text{ch}(\gamma_{qr}Z_0)\,dV_0 \quad (16)\right.$$

$$\left. + \iint_S \frac{\partial w}{\partial t}\cos(\alpha_q X_0^s)\cos(\beta_r Y_0^s)\,dS\right]e^{i\omega t},$$

where $\varepsilon_{00} = 1, \varepsilon_{q0} = \varepsilon_{0r} = 2, \varepsilon_{qr} = 4\,(q, r = 1, 2, \ldots)$.

The next denotations of the characteristic functions of the elastic plate

$$\begin{cases} P(k_m x) = P_{mx} = \frac{1}{2}[\mathrm{ch}(k_m x) + \cos(k_m x)], \\ Q(k_m x) = Q_{mx} = \frac{1}{2}[\,\mathrm{sh}\,(k_m x) + \sin(k_m x)], \\ T(k_m x) = T_{mx} = \frac{1}{2}[\,\mathrm{ch}(k_m x) - \cos(k_m x)], \\ U(k_m x) = U_{mx} = \frac{1}{2}[\,\mathrm{sh}\,(k_m x) - \sin(k_m x)], \end{cases} \tag{17}$$

$$\begin{cases} P(k_n y) = P_{ny} = \frac{1}{2}[\mathrm{ch}(k_n y) + \cos(k_n y)], \\ Q(k_n y) = Q_{ny} = \frac{1}{2}[\,\mathrm{sh}\,(k_n y) + \sin(k_n y)], \\ T(k_n y) = T_{ny} = \frac{1}{2}[\,\mathrm{ch}(k_n y) - \cos(k_n y)], \\ U(k_n y) = U_{ny} = \frac{1}{2}[\,\mathrm{sh}\,(k_n y) - \sin(k_n y)] \end{cases} \tag{17a}$$

are used, where k_m and k_n are the plate wave numbers along x and y [1]. These functions will be taken to construct the strip trial functions for the scheme of the Bubnov-Galerkin method. The developed scheme is applied for the solution of the problem that is under consideration at basic combinations of supporting conditions along the four plate edges. The combinations investigated (in the order "East-West-North-South") are: (1) SS-SS-SS-SS; (2) SS-SS-SS-C; (3) SS-SS-C-C; (4) SS-C-C-C; (5) C-C-C-C (SS - simply-supported, C - clamped). It is possible to consider other supporting conditions along the four plate edges of the elastic plate as well.

The function $w(x, y; t)$ is represented in the form

$$w(x, y; t) = W(x, y)e^{i\omega t}, \tag{18}$$

$$W(x, y) = \sum_{m=1}^{\infty} \sum_{n=1}^{\infty} W_{mn} W_m(x) W_n(y), \tag{18a}$$

and the corresponding trial functions $W_m(x)$ and $W_n(y)$ as well as the corresponding wave numbers k_m and k_n (or the dispersion equations) are given in Table 1.

Table 1 Trial functions and wave numbers

Case	$W_m(x)$	$W_n(y)$
1	$\sin(k_m x)$ $k_m = m\pi/a$	$\sin(k_n y)$ $k_n = n\pi/b$
2	$\sin(k_m x)$ $k_m = m\pi/a$	$T_{ny} - G_{nb}U_{ny},\ G_{nb} = T_{nb}/U_{nb}$ $\mathrm{tg}(k_n b) = \mathrm{th}(k_n b)$
3	$\sin(k_m x)$ $k_m = m\pi/a$	$T_{ny} - G_{nb}U_{ny},\ G_{nb} = T_{nb}/U_{nb}$ $\cos(k_n b)\,\mathrm{ch}\,(k_n b) = 1$
4	$T_{mx} - G_{ma}U_{mx},\ G_{ma} = T_{ma}/U_{ma}$ $\mathrm{tg}(k_m a) = \mathrm{th}\,(k_m a)$	$T_{ny} - G_{nb}U_{ny},\ G_{nb} = T_{nb}/U_{nb}$ $\cos(k_n b)\,\mathrm{ch}\,(k_n b) = 1$
5	$T_{mx} - G_{ma}U_{mx},\ G_{ma} = T_{ma}/U_{ma}$ $\cos(k_m a)\,\mathrm{ch}\,(k_m a) = 1$	$T_{ny} - G_{nb}U_{ny},\ G_{nb} = T_{nb}/U_{nb}$ $\cos(k_n b)\,\mathrm{ch}\,(k_n b) = 1$

In a case of a simply-supported plate

$$W_m(x) = \sin(k_m x), \quad W_n(y) = \sin(k_n y), \quad k_m = \frac{m\pi}{a}, \quad k_n = \frac{n\pi}{b}.$$

In the other cases of plate support, the expressions of the function $W(x,y)$ are taken using Table 1.

Using Equations (15) and (18), the expression of the acoustical pressure on the surface of the elastic plate is:

$$p = -\rho_0 \frac{\partial \varphi}{\partial t}\Big|_S = \frac{i\omega \rho_0}{L_1 L_2} \sum_{q=0}^{\infty} \sum_{r=0}^{\infty} \frac{\varepsilon_{qr}}{\gamma_{qr}} \cos(\alpha_q X) \cos(\beta_r Y). \tag{19}$$

$$\left[\begin{array}{l} \frac{1}{\text{sh}(\gamma_{qr} L_3)} \int \int \int_{V_0} Q \cos(\alpha_q X_0) \cos(\beta_r Y_0) \, \text{ch}[\gamma_{qr}(L_3 - Z_0)] \, dV_0 + \\ i\omega \, \text{cth}(\gamma_{qr} L_3) \int \int_S W(X_0^s, Y_0^s) \cos(\alpha_q X_0^s) \cos(\beta_r Y_0^s) \, dS \end{array} \right] e^{i\omega t}.$$

The substitution of $Q = Q_0 \delta(X - X_0, Y - Y_0, Z - Z_0)$ as well as the series (18) in Equation (19) and its integrating on the volume of the tank and on the surface of the elastic plate give the following expression of the acoustical pressure on the surface of the elastic plate:

$$p = -\rho_0 \frac{\partial \varphi}{\partial t}\Big|_S = \frac{i\omega \rho_0}{L_1 L_2} \sum_{q=0}^{\infty} \sum_{r=0}^{\infty} \frac{\varepsilon_{qr}}{\gamma_{qr}} \left\{ i\omega \text{cth}(\gamma_{qr} L_3) \sum_{m=1}^{\infty} \sum_{n=1}^{\infty} W_{mn} I_{mnqr} + \right.$$

$$\left. \frac{Q_0}{\text{sh}(\gamma_{qr} L_3)} \cos(\alpha_q X_0) \cos(\beta_r Y_0) \, \text{ch}[\gamma_{qr}(L_3 - Z_0)] \right\} \tag{20}$$

$$\cos[\alpha_q(x + X^*)] \cos[\beta_r(y + Y^*)] e^{i\omega t},$$

where

$$I_{mnqr} = I_{mq} \cdot I_{nr}, \quad I_{mq} = \int_0^a W_m(x) \cos[\alpha_q(x + X^*)] \, dx,$$

$$I_{nr} = \int_0^b W_n(y) \cos[\beta_r(y + Y^*)] \, dy.$$

In a case of a simply-supported plate

$$I_{mq} = \begin{cases} \frac{(-1)^m k_m}{\alpha_q^2 - k_m^2} \cos[\alpha_q(a + X^*)] & k_m \neq \alpha_q \\ -\frac{a \sin(\alpha_q X^*)}{2} & k_m = \alpha_q \end{cases};$$

$$I_{nr} = \begin{cases} \frac{(-1)^n k_n}{\beta_r^2 - k_n^2} \cos[\beta_r(b + Y^*)] & k_n \neq \beta_r \\ -\frac{b \sin(\beta_r Y^*)}{2} & k_n = \beta_r \end{cases}.$$

The first and the second members of the expression (20) correspond to the contribution of the elastic plate vibrations and to that of the source on the acoustical pressure upon the plate surface. If the source is at a point through which one of the planes $(\cos(\alpha_q X_0) = 0, \cos(\beta_r Y_0) = 0)$ goes, then the corresponding form of the natural gas vibrations does not excite. On the contrary, this or that transverse form becomes predominant when the source is on the loops of the transverse vibrations $(\cos(\alpha_q X_0) = 1, \cos(\beta_r Y_0) = 1)$.

It is evident that the pressure on the plate will be greatest when the source is close to it $(Z_0 = 0,\ \mathrm{ch}\,\gamma_{qr} L_3 > 1)$, and smallest when it is on the opposite wall $(Z_0 = L_3,\ \mathrm{ch}\,0 = 1)$. All this is valid when γ_{qr} is real – then the excitation frequencies are low in comparison with the natural frequencies of the transverse vibrations. If the frequency of the source is high $(\gamma_{qr}$ – imaginary), then the pressure on the plate, according to the source position, changes following the law $\cos|\gamma_{qr}(L_3 - Z_0)|$. The value $\mathrm{sh}(\gamma_{qr} L_3) = 0$ corresponds to the resonance of a gas cavity with all absolutely rigid walls. Then $\gamma_{qr} = 0$ or $\omega_{qr} = c_0\sqrt{\alpha_q^2 + \beta_r^2}$ – these are the natural frequencies of the transverse forms of vibrations. Giving an imaginary value to γ_{qr}, the free natural vibrations of the considered gas-structure interaction system are obtained:

$$\gamma_{qr} L_3 = s\pi \quad (s = 0, 1, 2, \ldots)$$

or $\omega_{qrs} = c_0\sqrt{\alpha_q^2 + \beta_r^2 + \left(\frac{s\pi}{L_3}\right)^2}$. Let

$$F_{mn\overline{mn}} = F_{m\overline{m}} + 2H_{m\overline{m}}H_{n\overline{n}} + F_{n\overline{n}},$$

where

$$F_{m\overline{m}} = \frac{2}{a}\int_0^a W_m^{IV}(x)W_{\overline{m}}(x)\,dx, \quad F_{n\overline{n}} = \frac{2}{b}\int_0^b W_n^{IV}(y)W_{\overline{n}}(y)\,dy,$$

$$H_{m\overline{m}} = \frac{2}{a}\int_0^a W_m^{II}(x)W_{\overline{m}}(x)\,dx, \quad H_{n\overline{n}} = \frac{2}{b}\int_0^b W_n^{II}(y)W_{\overline{n}}(y)\,dy,$$

$$T = \frac{1}{ab}\int_0^a W_m^2(x)\,dx.\int_0^b W_n^2(y)\,dy.$$

After the substitution of Equations (18) and (20) in Equation (4) and its integration with the Bubnov-Galerkin method, the following infinite system of equations for the determination of the unknown coefficients W_{mn} is obtained:

$$[\Omega_{mn}^2 - \omega^2]W_{mn} - \frac{T\rho_0\omega^2}{\rho h a b L_1 L_2}\sum_{q=0}^{\infty}\sum_{r=0}^{\infty}\sum_{\overline{m}=1}^{\infty}\sum_{\overline{n}=1}^{\infty}\frac{\varepsilon_{qr}\mathrm{cth}(\gamma_{qr}L_3)}{\gamma_{qr}}I_{mnqr}$$

$$I_{\overline{mn}qr}W_{\overline{mn}} = -\frac{Ti\omega\rho_0 Q_0}{\rho h a b L_1 L_2}\sum_{q=0}^{\infty}\sum_{r=0}^{\infty}\frac{\varepsilon_{qr}\ \mathrm{cth}(\gamma_{qr}L_3)}{\gamma_{qr}}I_{mnqr} \qquad (21)$$

$$\cos(\alpha_q X_0)\cos(\beta_r Y_0)\ \mathrm{ch}\,[\gamma_{qr}(L_3 - Z_0)],$$

where

$$\Omega^2_{mn} = \frac{\overline{D}}{\rho h} F_{mnmn}.$$

In a case of a simply-supported plate

$$F_{m\overline{m}} = \begin{cases} k^4_m & m = \overline{m} \\ 0 & m \neq \overline{m} \end{cases}, \quad F_{n\overline{n}} = \begin{cases} k^4_n & n = \overline{n} \\ 0 & n \neq \overline{n} \end{cases}, \quad H_{m\overline{m}} = \begin{cases} -k^2_m & m = \overline{m} \\ 0 & m \neq \overline{m} \end{cases},$$

$$H_{n\overline{n}} = \begin{cases} -k^2_n & n = \overline{n} \\ 0 & n \neq \overline{n} \end{cases}, \quad T = \frac{1}{4}, \quad \Omega^2_{mn} = \frac{\overline{D}}{\rho h}(k^2_m + k^2_n)^2.$$

The other cases of plate support may be considered in a similar way.

The real part of the series (18a) gives the displacement of the elastic plate. After substituting $Q_0 = 0$ in Equation (21) and taking the determinant of the corresponding homogeneous system as equal to zero, the equation for the determination of the natural frequencies of the considered gas-structure interaction system is obtained.

4 Numerical Calculations

The obtained theoretical solution is very complicated, that is why an approximate solution is made based on ignoring diffraction by the elastic plate waves. The approximate solution can be used when the frequencies of the source are not close to the resonance frequencies of the gas-structure interaction system and when the cavity is filled with air. The corresponding equations are obtained from Equation (21), ignoring the members with fourth sums. The numerical calculations are made for two-dimensional vibrations in the plane ZX, when the source is placed on the line $X_0 = L_1/2$. It is seen that only symmetrical "steady-state" waves are excited in the cavity and in the elastic plate ($q = 0, 2, 4, \ldots$; $m = 1, 3, 5, \ldots$). The numerical calculations are made using the corresponding exact and approximate theories:

$$[\Omega^2_m - \omega^2]W_m - \frac{2\rho_0\omega^2}{\rho h a L_1} \sum_{q=0,2,\ldots}^{\infty} \sum_{\overline{m}=1,3,\ldots}^{\infty} \frac{\varepsilon_q}{\gamma_q \, \text{th} \, (\gamma_q L_3)} W_{\overline{m}} I_{mq} I_{\overline{m}q} = \tag{22}$$

$$-\frac{2i\omega\rho_0 Q_0}{\rho h a L_1} \sum_{q=0,2,\ldots}^{\infty} \frac{(-1)^{\frac{q}{2}}\varepsilon_q}{\gamma_q \, \text{sh} \, (\gamma_q L_3)} I_{mq} \, \text{ch} \, [\gamma_q(L_3 - Z_0)] \; (m = 1, 3, 5, \ldots);$$

$$[\Omega^2_m - \omega^2]W_m = -\frac{2i\omega\rho_0 Q_0}{\rho h a L_1} \sum_{q=0,2,\ldots}^{\infty} \frac{(-1)^{\frac{q}{2}}\varepsilon_q}{\gamma_q \, \text{sh} \, (\gamma_q L_3)} I_{mq}$$

$$\text{ch} \, [\gamma_q(L_3 - Z_0)] \; (m = 1, 3, \ldots). \tag{23}$$

Some conclusions can be drawn regarding the series congruence and the region of application of the approximate formula (23). If the frequency of the source

$f = \frac{\omega}{2\pi}$ is less than the first natural frequency of the plate, then it is enough to take only one member ($\overline{m} = 1$) in the series on $\overline{m}$ in the formula (22). The congruence of the series on q is very good. The calculations made show that it is necessary to take into account which part of the spectrum of the natural frequencies of the elastic plate the forced frequency is located in, and the same is valid in respect to the summation of the series on q.

The second formula (23) can be used when the excitation frequencies are not close to the resonance ones. When the frequency of the source $f = \frac{\omega}{2\pi} \to 0$, very strong increase of the amplitudes appears except at the resonance points – the explanation of this is because it is considered $Q_0 = $ const. The second formula (23) cannot be used if there is a heavy liquid in the rectangular tank.

References

1. Gavrilova EG (1994) Hydroelasticity of thin-walled prismatic structures with elastic inclusions, PhD Thesis, Bulgarian Academy of Sciences, Sofia
2. Warburton G (1954) Inst Mech Eng 168: 371–381
3. Awrejcewicz J, Krysko VA (2003) Nonclassical thermoelastic problems in nonlinear dynamics of shells. Applications of the Bubnov-Galerkin and finite difference numerical methods. Scientific computation. Springer, Berlin
4. Ilgamov MA (1969) Vibrations of the elastic shells containing liquid and gas. Nauka, Moscow
5. Bracewell RN (1999) The Fourier transform and its applications. McGraw-Hill, New York
6. Polyanin AD (2002) Handbook of linear partial differential equations for engineers and scientists. Chapman & Hall/CRC Press, Boca Raton, ISBN 1-58488-299-9

On Physical and Mathematical Moments and the Setting of Boundary Conditions for Cusped Prismatic Shells and Beams

George Jaiani

I.Vekua Institute of Applied Mathematics of Iv. Javakhishvili Tbilisi State University, 2, University St., 0186, Tbilisi, Georgia, `jaiani@viam.sci.tsu.ge`

Abstract This paper deals with the analysis of the physical and geometrical sense of N-th ($N = 0, 1, \ldots$) order moments and weighted moments of the stress tensor and the displacement vector in the theory of cusped prismatic shells [1,2] and beams [3]. The peculiarities of the setting of boundary conditions at cusped edges in terms of moments and weighted moments are analyzed. The relation of such boundary conditions to the boundary conditions of the three-dimensional theory of elasticity is also discussed.

Keywords: cusped beams, cusped plates, cusped prismatic shells, physical moments, mathematical moments, mathematical modeling, linear elasticity

1 Physical and Mathematical Moments of Stresses

Let X_{ij}, $i, j = 1, 2, 3$, be stress tensor. The k-th order mathematical moment of $X_{ij} \in C(\Omega \cup \overset{(+)}{h} \cup \overset{(-)}{h})$ is defined as follows (see [4]):

$$
\overset{k}{X}_{ij}(x_1, x_2) := \int\limits_{\overset{(-)}{h}(x_1,x_2)}^{\overset{(+)}{h}(x_1,x_2)} X_{ij}(x_1, x_2, x_3) P_k(ax_3 - b) dx_3, \tag{1}
$$

$$
(x_1, x_2) \in \omega, \quad (x_1, x_2, x_3) \in \Omega,
$$

where

$$
P_k(t) := \sum_{l=0}^{\left[\frac{k}{2}\right]} (-1)^l \frac{(2k - 2l)!}{2^k l!(k - l)!(k - 2l)!} t^{k-2l}, \quad k = 0, 1, ..., \tag{2}
$$

are Legendre polynomials (see [5], Section 15.1), $\left[\frac{k}{2}\right]$ is the integer part of $\frac{k}{2}$, Ω is a domain occupied by a prismatic shell of variable thickness, with ω its projection on the plane $x_3 = 0$,

G. Jaiani, P. Podio-Guidugli (eds.), *IUTAM Symposium on Relations of Shell, Plate, Beam, and 3D Models*, © Springer Science+Business Media B.V. 2008

$$\overset{(\pm)}{h} := \{(x_1, x_2, x_3) \in R^3 : (x_1, x_2) \in \omega, \quad x_3 = \overset{(\pm)}{h}(x_1, x_2)\}$$

are the upper and lower faces of the shell, $\Gamma := \partial\Omega \backslash (\overset{(+)}{h} \cup \overset{(-)}{h})$ is the lateral cylindrical surface whose element is parallel to the axis $0x_3$,

$$a := \frac{1}{h}, \quad b := \frac{\tilde{h}}{h},$$

$$2h(x_1, x_2) := \overset{(+)}{h}(x_1, x_2) - \overset{(-)}{h}(x_1, x_2),$$

$$2\tilde{h}(x_1, x_2) := \overset{(+)}{h}(x_1, x_2) + \overset{(-)}{h}(x_1, x_2).$$

Physical moments are defined as follows:

$$S_{ij}(x_1, x_2) := \overset{0}{M}_{ij}(x_1, x_2), \quad i, j = 1, 2, 3,$$

$$\overset{k}{M}_{ij}(x_1, x_2) := \int_{\overset{(-)}{h}(x_1, x_2)}^{\overset{(+)}{h}(x_1, x_2)} X_{ij}(x_1, x_2, x_3) x_3^k dx_3, \tag{3}$$

$$k = 0, 1, ..., \quad i, j = 1, 2, 3,$$

where S_{23}, S_{13} are the so called intersecting forces, $S_{\alpha\beta}$, $\alpha, \beta = 1, 2$ are the so called membrane (or normal and tangent) forces, $\overset{1}{M}_{11}$, $\overset{1}{M}_{22}$ are the bending moments and $\overset{1}{M}_{12}$ is the twisting moment. In what follows, on generalizing these definitions, $\overset{k}{M}_{ij}(x_1, x_2)$ will be called physical moments of the k-th order.

As it is clear (1) and (3) make sense for $2h > 0$.

If a point P belongs to the cusped edge $\Gamma_0 \bar{\subset} \Gamma$ of the shell, i.e., if $2h(P_\omega) = 0$, $P_\omega \in \gamma_0$, (see Fig. 1), then the mathematical and physical moments we define as the following limits:

$$\overset{k}{X}_{ij}(P_\omega) := \lim_{\omega \ni Q_\omega \to P_\omega} \overset{k}{X}_{ij}(Q_\omega), \quad i, j = 1, 2, 3,$$

$$\overset{k}{M}_{ij}(P_\omega) := \lim_{\omega \ni Q_\omega \to P_\omega} \overset{k}{M}_{ij}(Q_\omega), \quad i, j = 1, 2, 3, \tag{4}$$

where $\gamma_0 \subseteq \partial\omega$ is the projection of Γ_0 on the plane $x_3 = 0$, P_ω and Q_ω are the projections of $P \in \Gamma$ and $Q \in \Omega$, respectively. Clearly, Γ_0 is a curve lying on the cylindrical surface bounding the shell from the side. In what follows by the normals n at points of Γ_0 we mean normals at the same points to the above cylindrical surface. When the cusped edge lies on $\partial\omega$, then obviously $P_\omega \equiv P$. (see Figures 1–3, where normal sections of the shell are shown).

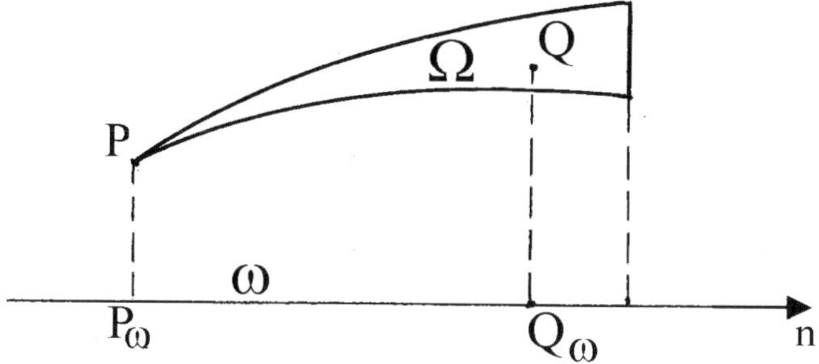

Fig. 1 A prismatic shell profile, when $P \neq P_\omega$

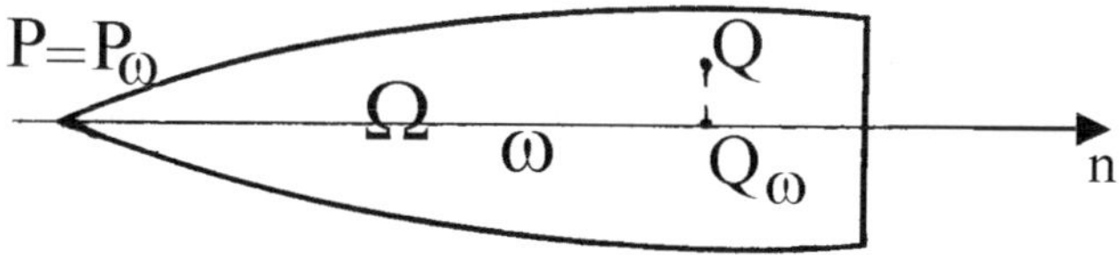

Fig. 2 A prismatic shell profile, when $P = P_\omega$ (symmetric case)

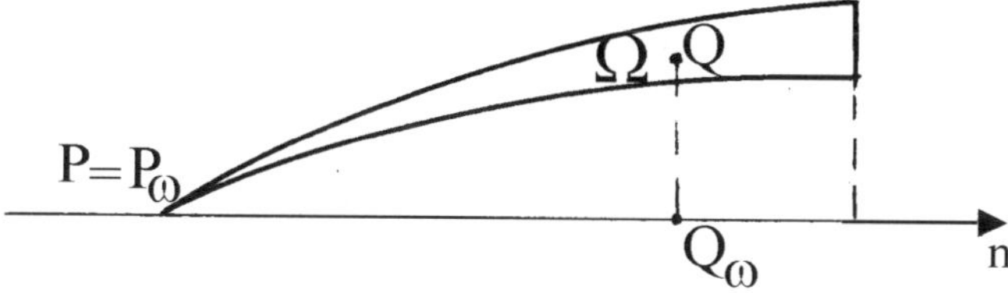

Fig. 3 A prismatic shell profile, when $P = P_\omega$ (non symmetric case)

When X_{ij} are bounded functions on Ω and $P \in \Gamma_0$, then $\overset{k}{X}_{ij}(P_\omega) = 0$ and $\overset{k}{M}_{ij}(P_\omega) = 0$.

It is easy to see that

$$\overset{k}{X}_{ij}(P_\omega) \neq 0 \quad \text{and} \quad \overset{k}{M}_{ij}(P_\omega) \neq 0 \quad \text{only if} \quad \lim_{\Omega \ni Q \to P \in \Gamma_0} X_{ij}(Q) = \infty.$$

Suppose that some neighborhoods of the point $P \in \Gamma_0$ along Γ_0 and on the upper and lower faces are not loaded. Forces and physical moments concentrated at the point P of the cusped edge of the shell are defined as follows (see Fig. 4):

$$\overset{k}{F}_j(P) := \lim_{\rho \to 0} \int_S X_{nj}(Q) x_3^k dS = \lim_{\rho_\omega \to 0} \int_S X_{nj}(Q_\omega, x_3) x_3^k dS$$

$$= \lim_{\rho_\omega \to 0} \int_{S_\omega} dS_\omega \int_{\overset{(-)}{h}(x_1,x_2)}^{\overset{(+)}{h}(x_1,x_2)} X_{nj}(Q_\omega, x_3) x_3^k dx_3 = \lim_{Q_\omega \to P_\omega} \int_{S_\omega} \overset{k}{M}_{nj}(Q_\omega) dS_\omega, \quad j = \overline{1,3},$$

where S is a lying inside the shell arbitrary cylindrical surface with the element parallel to $0x_3$, S_ω is its projection on the plane $x_3 = 0$; ρ is maximum distance between P and $Q \in S$, and ρ_ω is the maximum distance between P_ω and $Q_\omega \in S_\omega$.

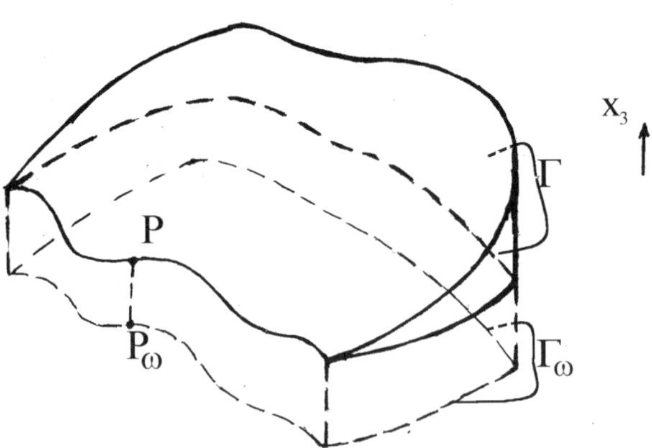

Fig. 4 A force concentrated at a point P of a cusped edge

Suppose that there are no forces and moments concentrated at points. Forces and physical moments concentrated along an arc $d\Gamma_0$ (the upper and lower faces are not loaded) are defined as follows:

$$\overset{k}{E}_{nj}(P)d\Gamma_0 := \lim_{d\to 0} \int\limits_{S_0\cup\Delta'\cup\Delta''} X_{nj}(Q)x_3^k dS$$

$$= \lim_{d\to 0} \int\limits_{\overset{(-)}{h}(x_1,x_2)}^{\overset{(+)}{h}(x_1,x_2)} X_{nj}(Q_\omega,x_3)x_3^k dx_3 dl_0 + \lim_{d\to 0} \int_0^d dl' \int\limits_{\overset{(-)}{h}(x_1,x_2)}^{\overset{(+)}{h}(x_1,x_2)} X_{nj}(Q_\omega,x_3)x_3^k dx_3$$

$$+ \lim_{d\to 0} \int_0^d dl'' \int\limits_{\overset{(-)}{h}(x_1,x_2)}^{\overset{(+)}{h}(x_1,x_2)} X_{nj}(Q_\omega,x_3)x_3^k dx_3$$

$$= \lim_{Q_\omega\to P_\omega} \overset{k}{M}_{nj}(Q_\omega)dl_0 = \overset{k}{M}_{nj}(P_\omega)d\Gamma_0 = \overset{k}{M}_{ij}(P_\omega)n_i(P_\omega)d\Gamma_0,$$

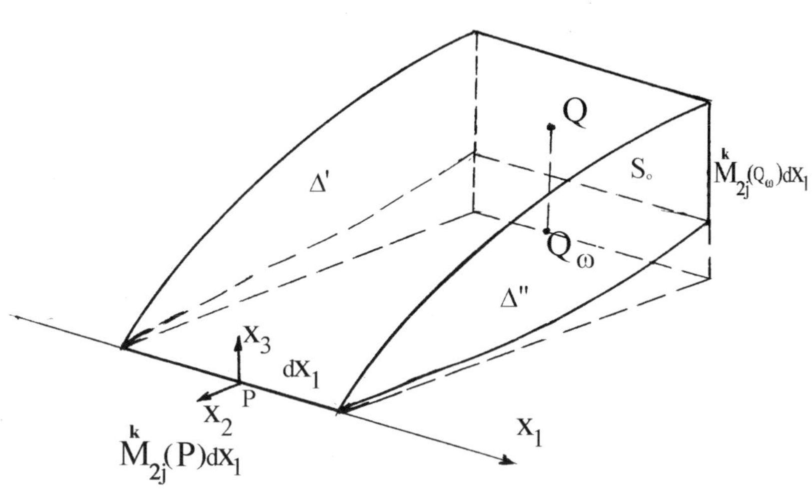

Fig. 5 Moments concentrated along a cusped edge

where S_0 is the part of an arbitrary cylindrical surface parallel to the lateral cylindrical surface passing through the point P cut out by the upper and lower faces and the planes (their parts containing in the body we denote by Δ' and Δ'') passing through the endpoints of $d\Gamma_0$ orthogonal to the lateral cylindrical surface containing Γ_0, d is the distance between the above cylindrical surfaces, dl_0 is the arc lying on S_0 parallel to $d\Gamma_0$ (for the particular case, when $P \equiv P_\omega$,

138   G. Jaiani

$d\Gamma_0 \equiv dx_1$, $dl \equiv dx_1$, see Fig. 5). $\overset{k}{E}_{nj}(P)$, $j = 1, 2, 3$, are the components of the line-concentrated forces ($k = 0$) and moments ($k \geq 1$) at the point P. In the case $P \equiv P_\omega$ we have $\overset{k}{E}_{nj}(P) = \overset{k}{M}_{nj}(P)$.

Remark 1. Since the body is in equilibrium, it follows that

$$\int_{S_0 \cup \Delta' \cup \Delta''} X_{nj}(Q)x_3^k dS = \int_{S_{d\Gamma_0}} X_{nj}(Q)x_3^k dS_{d\Gamma_0},$$

where $S_{d\Gamma_0}$ are arbitrary cylindrical surfaces lying inside the body parallel to the axis $0x_3$ and passing through the endpoints of $d\Gamma_0$.

By virtue of (2), we have

$$P_k(ax_3 - b) = P_k\left(\frac{x_3 - \tilde{h}}{h}\right) =$$

$$\frac{1}{2^k} \sum_{l=0}^{\left[\frac{k}{2}\right]} (-1)^l \frac{1}{l!(k-l)!} \frac{(2k-2l)!}{(k-2l)!}(x_3 - \tilde{h})^{k-2l} h^{2l-k}$$

$$= \frac{1}{2^k} \sum_{l=0}^{\left[\frac{k}{2}\right]} \sum_{r=0}^{k-2l} (-1)^{l+r} \frac{(2k-2l)!}{l!(k-l)!r!(k-2l-r)!} x_3^{k-2l-r} \tilde{h}^r h^{2l-k} \tag{5}$$

$$= \frac{(2k)!}{2^k(k!)^2} \cdot \frac{x_3^k}{h^k} + \frac{1}{2^k}\left\{\sum_{l=1}^{\left[\frac{k}{2}\right]}\sum_{r=0}^{k-2l} + \sum_{\substack{r=1\\l=0}}^{k}\right\}(-1)^{l+r}$$

$$\times \frac{(2k-2l)!}{l!(k-l)!r!(k-2l-r)!} x_3^{k-2l-r} \tilde{h}^r h^{2l-k}, \quad \sum_{l=1}^{0}(\cdot) \equiv 0.$$

Hence,

$$x_3^k = \frac{2^k(k!)^2}{(2k)!} h^k P_k(ax_3 - b) - \frac{(k!)^2}{(2k)!}\left\{\sum_{l=1}^{\left[\frac{k}{2}\right]}\sum_{r=0}^{k-2l} + \sum_{\substack{r=1\\l=0}}^{k}\right\}(-1)^{l+r} \tag{6}$$

$$\times \frac{(2k-2l)!}{l!(k-l)!r!(k-2l-r)!} \tilde{h}^r h^{2l} x_3^{k-2l-r}.$$

From (3), (6), (1) we get

$$\overset{k}{M}_{ij}(Q_\omega) = \int\limits_{\overset{(-)}{h}}^{\overset{(+)}{h}} x_3^k X_{ij}(Q_\omega, x_3)dx_3 = \frac{2^k(k!)^2}{(2k)!}h^k \overset{k}{X}_{ij}(Q_\omega)$$

$$-\frac{(k!)^2}{(2k)!}\left\{\sum_{l=1}^{\left[\frac{k}{2}\right]}\sum_{r=0}^{k-2l}+\sum_{\substack{r=1 \\ l=0}}^{k}\right\}(-1)^{l+r}\frac{(2k-2l)!}{l!(k-l)!r!(k-2l-r)!}$$

$$\times \tilde{h}^r h^{2l}\,\overset{k-2l-r}{M}_{ij}(Q_\omega), \quad k = 0,1,2,\cdots. \tag{7}$$

This result yields the recurrence formulae that allow calculating $\overset{k}{X}_{ij}$ by means of $\overset{s}{M}_{ij}$, $s = 0,1,...,k$, and $\overset{k}{M}_{ij}$ by means of $\overset{s}{X}_{ij}$, $s = 0,1,...,k$.

E.g., for $k = 0,1$,

$$\overset{0}{M}_{ij}(Q_\omega) = \overset{0}{X}_{ij}(Q_\omega), \tag{8}$$

$$\overset{1}{M}_{ij}(Q_\omega) = h\,\overset{1}{X}_{ij}(Q_\omega) + \tilde{h}\,\overset{0}{X}_{ij}(Q_\omega), \tag{9}$$

$$h\,\overset{1}{X}_{ij}(Q_\omega) = \overset{1}{M}_{ij}(Q_\omega) - \tilde{h}\,\overset{0}{M}_{ij}(Q_\omega). \tag{10}$$

Now, letting Q tend to P, i.e., Q_ω to P_ω, we obtain from (7) (see Fig. 1):

$$\overset{k}{E}_{ij}(P) := \overset{k}{M}_{ij}(P_\omega) = \frac{2^k(k!)^2}{(2k)!}\lim_{\omega \ni Q_\omega \to P_\omega} h^k(Q_\omega)\,\overset{k}{X}_{ij}(Q_\omega)$$

$$-k!\sum_{r=1}^{k}(-1)^r\frac{1}{(k-r)!r!}\tilde{h}^r(P_\omega)\,\overset{k-r}{M}_{ij}(P_\omega), \tag{11}$$

since $l > 0$ in the double sum of (7) and

$$h(P_\omega) = 0.$$

Note that $\tilde{h}(P_\omega)$ equals the x_3-coordinate of point P. Hence, when $P = P_\omega$, we have $\tilde{h}(P_\omega) = 0$ and, therefore, from (11) we get:

$$\overset{k}{M}_{ij}(P_\omega) = \frac{2^k(k!)^2}{(2k)!}\lim_{\omega \ni Q_\omega \to P} h^k(Q_\omega)\,\overset{k}{X}_{ij}(Q_\omega). \tag{12}$$

Thus, when we consider boundary conditions in terms of stresses, i.e., when forces and moments

$$\overset{k}{M}_{nj}(P_\omega) = \overset{k}{M}_{ij}(P_\omega)n_i(P_\omega)$$

concentrated along the cusped edge are given, for the mathematical moments at cusped edges from (11) in the N-th approximation we get the following boundary conditions:

$$\lim_{\omega \ni Q_\omega \to P_\omega} h^k(Q_\omega) \overset{k}{X}_{nj}(Q_\omega)$$

$$= \frac{(2k)!}{2^k k!} \left[\frac{1}{k!} \overset{k}{M}_{ij}(P_\omega) + \sum_{r=1}^{k} (-1)^r \frac{1}{(k-r)! r!} \tilde{h}^r(p_\omega) \overset{k-r}{M}_{ij}(P_\omega) \right] n_i(P_\omega), \quad (13)$$

$$j = 1, 2, 3, \quad \text{are prescribed for} \quad k = \overline{0, N},$$

which are weighted boundary conditions for $k \geq 1$.

The homogeneous boundary conditions (13) at cusped edges of the two-dimensional model correspond, for the three-dimensional model, to the cases when on the faces and on the lateral non-cusped edge (boundary) $\Gamma \backslash \bar{\Gamma}_0$ either the displacements or the stresses are prescribed. In this instance, the homogenous boundary conditions (13) are automatically satisfied for the bounded stresses or for $u_i \in H^1(\Omega)$, since in the last case $X_{ij} \in L_2(\Omega)$ and by an application of Fubini theorem the summability of X_{ij} along x_3 in (1) can be proved. Therefore,

$$\lim_{\omega \ni Q_\omega \to P_\omega} \overset{k}{X}_{nj}(Q_\omega) = \lim_{\omega \ni Q_\omega \to P_\omega} \overset{k}{X}_{ij}(Q_\omega) n_i(Q_\omega) = 0,$$

since the integration limits in (1) tend to 0. So, the homogeneous boundary conditions (13) at cusped edges are not real boundary conditions; instead they are replaced by the requirement that the desired quantities belong to certain spaces on ω, where we are looking for solutions without any boundary conditions at cusped edge. The last case arises in [2].

The nonhomogeneous boundary conditions (13) at cusped edges of the two-dimensional model correspond to the three-dimensional model, when at above-mentioned cusped edges Γ_0 forces and physical moments concentrated along the cusped edges are applied, while on the other parts $\overset{(+)}{h}, \overset{(-)}{h}, \Gamma \backslash \bar{\Gamma}_0$ of the body's boundary the same conditions as in the above-formulated case of homogeneous boundary conditions (13) are given.

In the N-th approximation the stress tensor is given in the form [4]:

$$X_{ij}(Q) \cong \sum_{k=0}^{N} a\left(k + \frac{1}{2}\right) \overset{k}{X}_{ij}(Q_\omega) P_k(a x_3 - b), \quad i, j = 1, 2, 3. \quad (14)$$

In this case, the relation (1) between X_{ij} and $\overset{k}{X}_{ij}$ is correct when the symbol of approximate equality in (14) is replaced by the exact equality symbol. This occurs if either X_{ij} are N-th order polynomials with respect to x_3 or $N = +\infty$ (i.e., we consider the Fourier-Legendre series representation for

$$X_{ij}(Q_\omega, \cdot) \in C^2\left(\left[\overset{(+)}{h}(Q_\omega), \overset{(-)}{h}(Q_\omega)\right]\right).$$

Surface forces X_{ni} can be considered only at points of blunt cusped edges (in this case the union of the upper and lower surfaces is a smooth surface and there exist normals to the above surface at points $P \in \Gamma_0$) and, as it follows from (14), (13) they become infinite as $Q \to P$ if the boundary conditions (13) (see also (10)) are inhomogeneous; e.g., for $N = 0$, it follows from (14) that

$$X_{ij}(Q) = \frac{1}{2h} \overset{0}{X}_{ij}(Q_\omega), \quad \overset{0}{X}_{ij}(P_\omega) \neq 0.$$

Let us note that the inhomogeneous boundary conditions (13), i.e., (11) and (12), mean that along the cusped edge forces and moments concentrated along the edge are given (see Figures 6, 7, where the plane sections of the corresponding three-dimensional problems are given in the case of symmetric prismatic shclls (platcs)). Somc such problems in two- and one-dimensional formulations are solved in [6], [7] (see also [3]). A similar analysis can be carried out for cusped beams.

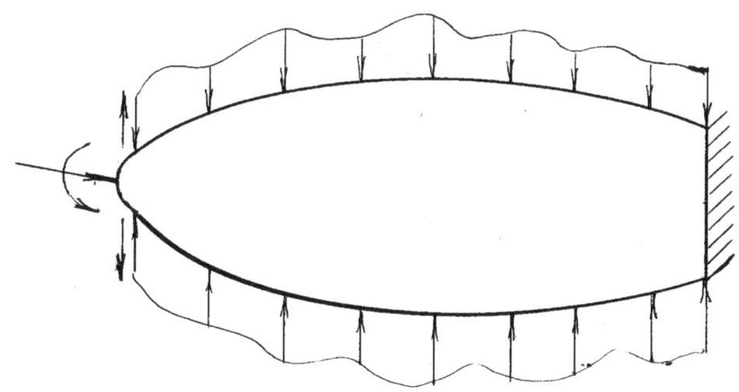

Fig. 6 A concentrated force and a concentrated moment applied at a blunt cusped edge

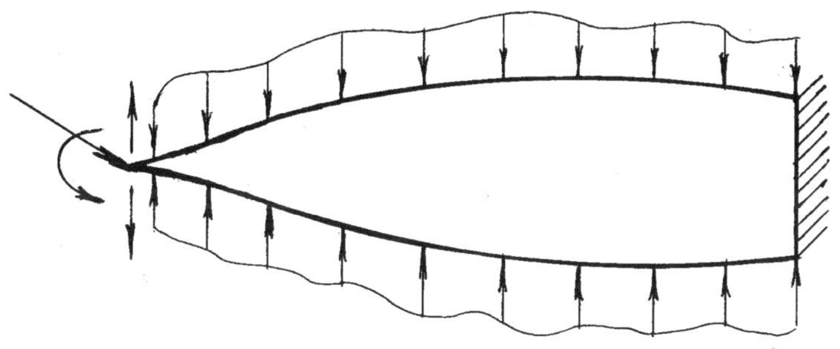

Fig. 7 A concentrated force and a concentrated moment applied at a sharp cusped edge

2 Displacement Vector. Weighted Moments

The k-th order moments of the displacement vector components are defined analogously to (1):

$$\overset{k}{u_i}(x_1, x_2) := \int\limits_{\overset{(-)}{h}(x_1,x_2)}^{\overset{(+)}{h}(x_1,x_2)} u_i(x_1, x_2, x_3) P_k(ax_3 - b)dx_3,$$

$$i = 1, 2, 3, \quad k = 0, 1, 2, \cdots .$$

If u_i, $i = 1, 2, 3$, are bounded on Ω functions, then, for $P \in \Gamma_0$, $\overset{k}{u_i}(P_\omega) = 0$.

In the N-th approximation, by virtue of (5), the components u_i are expressed by $\overset{k}{u_i}$ as follows:

$$
\begin{aligned}
u_i(x_1, x_2, x_3) &\cong \sum_{k=0}^{N} a\left(k + \frac{1}{2}\right) \overset{k}{u_i}(x_1, x_2) P_k(ax_3 - b) \\
&= \sum_{k=0}^{N} \left(k + \frac{1}{2}\right) h^k \overset{k}{v_i}(x_1, x_2) P_k(ax_3 - b) \\
&= \sum_{k=0}^{N} \frac{1}{2^k}\left(k + \frac{1}{2}\right) \sum_{l=0}^{\left[\frac{k}{2}\right]} (-1)^l \frac{(2k - 2l)!}{l!(k-l)!(k-2l)!} \\
&\quad \times \overset{k}{v_i}(x_1, x_2)(x_3 - \tilde{h})^{k-2l} h^{2l}, \quad i = 1, 2, 3,
\end{aligned}
$$

$$(15)$$

where

$$\overset{k}{v_i}(x_1, x_2) := \frac{\overset{k}{u_i}(x_1, x_2)}{h^{k+1}}, \quad i = 1, 2, 3, ..., \quad k = \overline{0, N},$$

are weighted moments of the displacement vector components.

In particular, in the $N = 0$ and $N = 1$ approximations,

$$u(Q) \cong \frac{1}{2} \overset{0}{v_i}(Q_\omega), \quad i = 1, 2, 3,$$

and

$$u_i(Q) \cong \frac{1}{2} \overset{0}{v_i}(Q_\omega) + \frac{3}{2} \overset{1}{v_i}(Q_\omega)(x_3 - \tilde{h}), \quad i = 1, 2, 3,$$

respectively.

Let $\overset{k}{v_i}$ be bounded; then, for

$$k - 2l + 2l = k > 0,$$

and for $P \neq P_\omega$, since on the one hand, $x_3 \to x_3^0 \neq 0$ as $Q \to P$ (if $P \equiv P_\omega$, evidently $x_3^0 = 0$) and on the other hand $\lim\limits_{Q_\omega \to P_\omega} \tilde{h}(Q_\omega) = x_3^0$ (see Fig.1), the limits on the right-hand side of (15) are zero as $\Omega \ni Q \to P$, i.e., as $\omega \ni Q_\omega \to P_\omega$. Hence, only the summand for $k = 0$ remains, i.e.,

$$\lim_{\omega \ni Q \to P} u_i(Q) = \frac{1}{2} \overset{0}{v}_i(P_\omega) \ \text{ if } \ I_0 := \int\limits_{P_\omega}^{Q_\omega} \frac{dn}{h} < +\infty, \tag{16}$$

where n is the inward normal to $\partial \omega$ at the point P_ω, since for $I_0 = +\infty$ the limit $\lim\limits_{\omega \ni Q_\omega \to P_\omega} \overset{0}{v}_i(Q_\omega)$, in general, cannot be assigned (see [1], [3], [8–11] and the references therein).

From (15), we have that

$$\frac{\partial^j u_i(Q)}{\partial x_3^j} = \sum_{k=j}^{N} \frac{1}{2^k} \left(k + \frac{1}{2} \right) \sum_{l=0}^{\left[\frac{k-j}{2} \right]} (-1)^l \frac{(2k - 2l)!}{l!(k - l)!(k - 2l)!}$$

$$\times \overset{k}{v}_i(Q_\omega)(k - 2l)(k - 2l - 1) \cdots (k - 2l - j + 1)$$

$$\times (x_3 - \tilde{h})^{k-2l-j} h^{2l}, \quad j = 1, 2, ..., N, \quad i = 1, 2, 3. \tag{17}$$

From (17), if additionally $\overset{k}{v}_i$ are bounded, we get

$$\lim_{\Omega \ni Q \to P} \frac{\partial^j u_i(Q)}{\partial x_3^j} = \frac{1}{2^j} \left(j + \frac{1}{2} \right) \frac{(2j)!}{j!} \overset{j}{v}_i(P_\omega), \tag{18}$$

$$\forall j = 1, 2, \cdots, N, \quad i = 1, 2, 3, \quad when \ \ I_N := \int\limits_{P_\omega}^{Q_\omega} \frac{dn}{h^{2N+1}} < +\infty.$$

For a fixed j the last inequality should be replaced by the condition

$$I_j := \int\limits_{P_\omega}^{Q_\omega} \frac{dn}{h^{2j+1}} < +\infty, \tag{19}$$

since for $I_j = +\infty$ the limit $\lim\limits_{\omega \ni Q_\omega \to P_\omega} \overset{j}{v}_i(Q_\omega)$, in general, can not be assigned (see [8–11]).

Thus, from (16) and (18) we can define $\overset{j}{v}_i$ $(j = 0, 1, \cdots, N, i = 1, 2, 3)$, provided the left-hand sides of (16) and (18), i.e., the displacements u_i ($i = 1, 2, 3$) and their derivatives with respect to x_3 up to the N-th order at point $P \in \Gamma_0$ are known.

Remark 2. (16) and (18), with (19), signify that the boundary conditions of two-dimensional models, when on γ_0

$$\overset{j}{v}_i(P), \ j = 0, 1, ..., N, \ i = 1, 2, 3, \quad \text{are prescribed,}$$

correspond to the boundary conditions of the three-dimensional model, when on Γ_0

$$\frac{\partial^j u_i(P)}{\partial x_3^j} \quad \text{if} \quad \int_{P_\omega}^{Q_\omega} \frac{dn}{h^{2j+1}} < +\infty, \ j = 0, 1, ..., N, \ i = 1, 2, 3, \quad \text{are prescribed.}$$

Clearly, all the $\overset{j}{v}_i(P)$ $(j = 0, 1, ..., N, \ i = 1, 2, 3)$ can be prescribed on γ_0 at the same time only if

$$\int_{P_\omega}^{Q_\omega} \frac{dn}{h^{2N+1}} < +\infty, \tag{20}$$

which corresponds to the boundary conditions of the three-dimensional model, when on Γ_0

$$\frac{\partial^j u_i(P)}{\partial x_3^j}, \quad \forall j = 0, 1, ..., N, \quad i = 1, 2, 3, \quad \text{if} \quad \int_{P_\omega}^{Q_\omega} \frac{dn}{h^{2N+1}} < +\infty,$$

are prescribed at the same time. Such boundary conditions fall outside the limits of the classical three-dimensional theory of elasticity. When $N = +\infty$ (i.e., when we actually have a three-dimensional model), then

$$\lim_{N \to +\infty} \int_{P_\omega}^{Q_\omega} \frac{dn}{h^{2N+1}} = +\infty \tag{21}$$

e.g., if $P_\omega \equiv (x_1, 0)$ and $h = h_0 x_2^k$, $h_0, k = const > 0$, then

$$\int_{P_\omega}^{Q_\omega} \frac{dn}{h^{2N+1}} = \frac{1}{h_0^{2N+1}} \int_0^\varepsilon \frac{dx_2}{x_2^{k(2N+1)}} < +\infty,$$

when

$$k(2N + 1) < 1.$$

Since the last condition is violated for $N \to +\infty$, (21) becomes clear]; therefore, the above boundary conditions disappear and, when on the faces and non-cusped edge of the shell surface forces are applied, we arrive at the ordinary boundary value problem in stresses for the classical three-dimensional model.

Remark 3. Whenever $\overset{0}{v}_i(P_\omega) = 0$ (see (16)) along the cusped edge (because of $I_0 < +\infty$ this cusped edge is a blunt one [1]; in other words, the upper and lower faces of the cusped prismatic shell both form jointly a smooth surface passing smoothly through the edge, where the thickness becomes zero), then in the corresponding three-dimensional problem the cusped edge is fixed; on the faces the stresses- and on the lateral non-cusped edge either the displacements or the stresses- are prescribed (see Fig. 8). The physical sense of $\overset{0}{v_i}(P) \neq 0$ is evident. Such a formulation of the three-dimensional boundary value problem is not usual and falls outside the limits of the classical three-dimensional theory of elasticity.

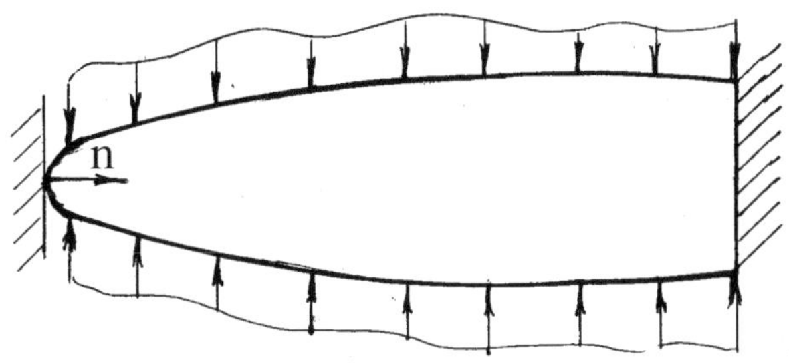

Fig. 8 A fixed cusped edge

Remark 4. If we consider admissible (i.e., correct) boundary value problems posed in weighted Sobolev spaces, then, on generalizing known results [12], [8] (see also [13] and the references therein and [14]) for the N-th approximation at a cusped edge, we get the following boundary conditions in weighted moments of displacements in the sense of traces [9]:

$$\overset{k}{v_i}(P) \text{ if } \int_P^{Q_\omega} \frac{dn}{h^{2k+1}} < +\infty, \ k = 0, 1, ..., N, \ i = 1, 2, 3, \text{ are given.}$$

Whence, by virtue of (15), for displacements we obtain that there exist traces

$$u_i(P), \quad i = 1, 2, 3, \text{ if } \int_P^{Q_\omega} \frac{dn}{h^{2N+1}} < +\infty.$$

3 General Conclusion

Summarizing the assertions of two previous sections, we arrive to the conclusion that within the framework of the concrete ($N = 0, 1, ...$) models we are able to study certain physical problems that do not find their solution within the framework of the 3D theory of elasticity (e.g., see Fig. 8).

Acknowledgements

Research supported by the INTAS-South-Caucasus Programme (project 06-1000017-8886)

References

1. Jaiani G.V. (1996) Elastic Bodies with Non-smooth Boundaries-Cusped Plates and Shells. ZAMM, 76 Suppl. 2, 117–120
2. Jaiani G.V., Kharibegashvili S.S., Natroshvili D.G., Wendland W.L. (2004) Two-dimensional Hierarchical Models for Prismatic Shells with Thickness Vanishing at the Boundary, Journal of Elasticity, 77(2), 95–122
3. Jaiani G.V. (2001) On a Mathematical Model of Bars with Variable Rectangular Cross-sections, ZAMM, 81(3) 147–173
4. Vekua I.N. (1955) Shell Theory: General Methods of Construction. Pitman Advanced Publishing Program, Boston-London-Melbourne
5. Whittaker E.T., Watson G.N. (1927) A Course of Modern Analysis. Cambridge University Press, Vol. 2
6. Jaiani G.V. (1980) On Some Boundary Value Problems for Cusped Shells, in "Theory of Shells", Koiter W.T. and Mikhailov G.K., Eds., North-Holland Pub. Comp., 339–343
7. Jaiani G.V. (1982) Solution of Some Problems for a Degenerate Elliptic Equation of Higher Order and Their Applications to Prismatic Shells, Tbilisi, University Press (Russian)
8. Devdariani G.G., Jaiani G.V., Kharibegashvili S.S., Natroshvili D.G. (2000) The First Boundary Value Problem for the System of Cusped Prismatic Shells in the First Approximation, Applied Mathematical Information 5(2) 26–46
9. Jaiani G.V. (2001) Application of Vekua's Dimension Reduction Method to Cusped Plates and Bars, Bulletin TICMI, 5, 27–34
10. Devdariani G.G. (2001) The First Boundary Problem for a Degenerate Elliptic System, Bull. TICMI, 5, 23–27
11. Jaiani G.V., Schulze B.-W. (2007) Some Degenerate Elliptic Systems and Applications to Cusped Plates, Mathematische Nachrichten, 280, 4, 388–407
12. Kharibegashvili S.S., Jaiani G.V. (2000) On a Vibration of an Elastic Cusped Bar, Bull. TICMI, 4, 24–28
13. Jaiani G.V., Kufner A. (2002) Oscillation of Cusped Euler-Bernulli Beams and Kirchhoff-Love Plates, Preprint 145, Academy of Sciences of the Czech Republic, Mathematical Institute, Prague
14. Tsiskarishvili G.V., Khomasuridze N.G. (1991) Cylindrical Bending of a Cusped Cylindrical Shells. Proceedings of I.Vekua Institute of Applied Mathematics of Tbilisi State University, 42, 72–79 (Georgian)

Material Conservation Laws Established Within a Consistent Plate Theorie

Reinhold Kienzler and Dipak K. Bose

University of Bremen, Department of Production Engineering, Bremen, Germany,
`rkienzler@uni-bremen.de`

Abstract Conservation laws and related path-independent integrals have been derived within a second-order consistent plate theory considering shear deformations, warping of the cross-section and change-of-thickness effects. By means of path-independent integrals, material forces, i. e., driving-forces for defects may be calculated and used to assess plates with, e. g., cracks. In the absence of defects, the integrals may serve as a validation for the accuracy of numerical calculations.

Keywords: consistent plate theory, second-order approximation, conservation laws, path-independent integrals, defects, cracks

1 Introduction

Whereas Kirchoff's plate theory [12] is rather well established since more than 150 years, various versions of plate theories exist, which take shear deformations and change-of-thickness effects into account cf., e.g., [21]. Most prominently, Mindlin's [16] and Reissner's [17] theories are invoked. The number of engineering (sometimes self-contradictory) a priori assumptions necessary to establish these theories can be reduced considerably by the application of the consistent-approximation technique [14]. The idea is to approximate the complete set of governing equations uniformly, i. e., to the same degree of accuracy. This method has been extended [7, 8] by the proper demand that not only during the derivation of the equations, terms of a certain order of magnitude are to be retained (and terms of higher order are to be neglected) but also during the reduction of the equation system using an elimination process. All theories in common is an integration process with respect to the plate thickness giving rise to a plate parameter $c^2 = h^2 / \left(12a^2\right)$ (h is the characteristic length in thickness direction assumed to be constant throughout the paper, and a is the characteristic in-plane dimension). This plate parameter is usually small $c^2 << 1$. If only terms up to the zeroth order $O[(c^2)^0]$ are considered, only rigid body motions of the plate are admitted. The consistent

G. Jaiani, P. Podio-Guidugli (eds.), *IUTAM Symposium on Relations of Shell,
Plate, Beam, and 3D Models*, © Springer Science+Business Media B.V. 2008

first-order approximation $O[(c^2)^1]$ delivers exactly Kirchhoff's theory including effective shear forces for a proper formulation of the boundary conditions, and the second-order approximation $O[(c^2)^2]$ results in a consistent shear-deformable plate theory. It has been shown that various higher-order plate theories coincide within the second-order approximation [8, 9].

Since in his pioneering paper [5], Eshelby advanced the notion of a force on a defect, a whole edifice of *Configurational Mechanics or Mechanics in Material Space* or most adequately *Eshelbian Mechanics* arose, cf., e. g., [6, 11, 15]. One of the most important and useful ingredients are path-independent integrals characterizing the energy changes due to the motion of defects (cavities, inclusions, dislocations, cracks, etc.) relatively to the elastic material in which they find themselves. The broadest potential of application, especially in fracture mechanics, has the so-called J-integral [18] correlated with a translation of a defect. Two further integrals designated as L and M [13] describe the change of energy of the system due to rotation and self-similar expansion of defects, respectively [3]. Defects may also be present in plates and have to be assessed properly. Path-independent integrals have been used (cf., e.g., [19]) and severe discrepancy have been revealed when Kirchhoff's theory was applied. It is therefore intriguing to formulate path-independent integrals consistent with a uniform second-order approximation and this is being done in this paper.

2 A Consistent Second-order Plate Theory

Point of departure are the governing equations of the linear theory of three-dimensional elasticity with appropriate boundary conditions. The material behavior is supposed to be, for simplicity, homogeneous and isotropic. The equilibrium equations relate the divergence of the stress tensor τ_{ij} to the applied body forces f_i , the kinematic relations link strains γ_{ij} and displacement gradients $u_{i,j}$ and Hooke's law connects stresses and strains.

In a first step, the displacements are expanded in thickness direction x_3 into a power series involving, among others, the transverse displacement of the plate-middle-surface w and the change of the slope of the straight line normal to the undeformed midplane ψ_α ($\alpha = 1, 2$). Comparing coefficients of equal order in x_3 , a series expansion of the strain tensor results from the kinematic relations. With these, the strain-energy density W and the potential of external forces V can be calculated. After integrating both with respect to x_3 and applying a variational principle, a linear system of coupled differential equations is obtained to determine the expansion coefficients of the displacements. This system of equations involve various powers of the plate parameter c^2 mentioned above. In turn, the system of equations is reduced by an elimination process. If during this process, consequently terms up to the order c^4 are retained and higher-order terms are omitted, we arrive finally at two uncoupled differential equations in w and a measure of shear deformation $\Psi = \psi_{2,1} - \psi_{1,2}$ as follows

$$K\Delta\Delta w = a^3 \left(P - (\kappa - 12\varepsilon)c^2 \Delta P \right),$$

$$(1)$$

$$c^2 \left(\Psi - \frac{3}{2}c^2 \Delta\Psi \right) = 0,$$

with the abbreviations

$$\kappa = \frac{3}{10}\frac{8+\nu}{1-\nu}, \qquad \varepsilon = \frac{1}{10}\frac{\nu}{1-\nu}, \tag{2}$$

the external transverse force P per unit of plate surface $\overline{S}$, the Laplacian $\Delta = ()_{,\alpha\alpha}$, the plate stiffness

$$K = \frac{Eh^3}{12(1-\nu^2)}, \tag{3}$$

Young's modulus E and Poisson's ratio ν. A comma followed by an index means partial differentiation with respect to the indicated dimensionless variable $x_\alpha/a, \alpha = 1,2$ and the summation convention is implied for repeated indices. Details of the derivation may be found in [8, 9].

The bending moments $M_{\alpha\beta}$ and the transverse shear forces Q_α are given by

$$\begin{aligned}
M_{\alpha\beta} = -\frac{K}{a} &\left\{ \left(1 + c^2\kappa\,\Delta\right)\left[(1-\nu)\,w_{,\alpha\beta} + \nu\delta_{\alpha\beta}w_{,\gamma\gamma}\right] \right. \\
&\left. +c^2\frac{3}{5}(1-\nu)\left(\varepsilon_{\alpha\gamma}\Psi_{,\beta\gamma} + \varepsilon_{\beta\gamma}\Psi_{,\alpha\gamma}\right) \right\} + 12\delta_{\alpha\beta}c^2 a^2 \varepsilon P, \\
Q_\alpha = -\frac{K}{a^2} &\left\{ \left(1 + c^2\kappa\Delta\right)\Delta w_{,\alpha} + \frac{2}{5}(1-\nu)\varepsilon_{\alpha\beta}\Psi_{,\beta} \right\} + 12c^2 a\varepsilon P_{,\alpha},
\end{aligned} \tag{4}$$

with the Kronecker tensor of unity $\delta_{\alpha\beta}$ and the skew symmetric Levi-Civita tensor of unity $\varepsilon_{\alpha\beta}$. The boundary conditions are

$$M^*_{\alpha\beta}n_\alpha = M_{\alpha\beta}n_\alpha \quad \text{or} \quad \psi^*_\beta = -\left[(1 + c^2\kappa\,\Delta)w_{,\beta} + c^2\frac{6}{5}\varepsilon_{\beta\gamma}\Psi_{,\gamma}\right],$$

$$(5)$$

$$Q^*_\alpha n_\alpha = Q_\alpha n_\alpha \quad \text{or} \quad w^* = (1 + 3c^2\varepsilon\,\Delta)w.$$

Variables with asterix designate prescribed quantities along the boundary of the plate. This completes the set of equations of a self-consistent second-order plate theory without introducing any a priori assumptions and neglections. Further discussion and comparison with other theories available in the literature are given in [8, 9].

3 Material Conservation Laws

We consider a body B with surface S, surface element dA and unit-outward normal vector n_i. The path-independent integrals mentioned in the introduction are given within the three-dimensional theory of elasticity as [3, 13, 18]

$$J_k = \int_S b_{jk} n_j \, dA,$$

$$L_i = \int_S \varepsilon_{ijk} \left[b_{lk} x_j + \tau_{lk} u_j \right] n_l \, dA, \tag{6}$$

$$M = \int_S \left[b_{jk} x_k - \frac{1}{2} \tau_{jk} u_k \right] n_j \, dA,$$

$(i, j, k, l = 1, 2, 3)$ with the completely skew-symmetric permutation tensor in three dimensions ε_{ijk} $(\varepsilon_{3\alpha\beta} \equiv \varepsilon_{\alpha\beta})$. The tensor b_{jk} is the energy-momentum tensor of elastostatics or the Eshelby tensor [5]

$$b_{jk} = (W + V)\delta_{jk} - \tau_{ji} u_{i,k}. \tag{7}$$

If the material is homogeneous, i.e., no defects are present within the closed surface S, the integral J_k vanishes for constant body forces f_i. Otherwise $J_k \neq 0$ and calculates the material force on the defect. If the defect is a crack, J_k is designated as crack-driving force. To render L_i and M zero, the material must be isotropic and homogeneous of grade 2, respectively, both in the absence of body forces. The integrals (6) are path-independent, i.e., integration over any surface S enclosing one and the same defect or defect configuration will lead to the same value of J_k, L_i and M. In the absence of defects ($J_k = L_i = M = 0$) the integrals vanish and may serve, e. g., as a validation of the accuracy of numerical results.

4 Conservation Laws of a Second-Order Plate Theory

In this section, the integrals (6) will now be specified for the consistent plate theory. First, we choose for S a surface of a "plate continuum" as depicted in Fig. 1, consisting of two cover surfaces S^+ and S^- at $x_3 = +h/2$ and $x_3 = -h/2$, respectively, and a cylindrical surface S^Γ. For any integral I_k, say, we have

$$I_k = \int_S d_{jk} n_j \, dA = \int_{S^\Gamma + S^- + S^+} d_{jk} n_j \, dA. \tag{8}$$

Due to symmetry and antisymmetry conditions of the stresses and displacements it turns out during the analysis that the integrals over the cover

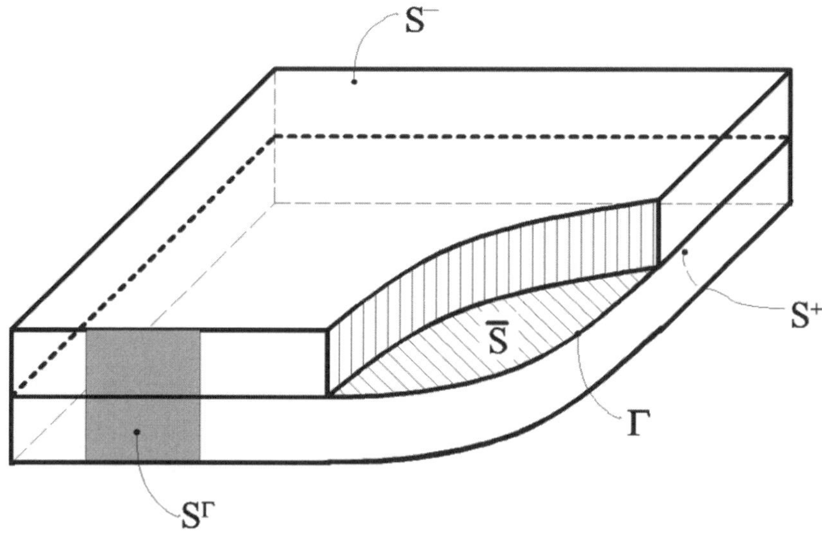

Fig. 1 Plate continuum

surfaces S^+ and S^- do not contribute to the value of neither J_k, L_i or M. The integral S^Γ is changed to a contour integral along Γ (are length s) by integration with respect to the thickness coordinate x_3 $\left(n_j^\Gamma = [n_1, n_2, 0]^T\right)$. Thus I_k reduces to

$$I_k = \oint_\Gamma \bar{d}_{\alpha k} n_\alpha ds, \qquad \text{with} \qquad \bar{d}_{\alpha k} = \int_{-\frac{h}{2}}^{+\frac{h}{2}} d_{\alpha k}\, dx_3.$$

The integration in thickness direction is now applied to the Eshelby tensor (7). Details of the calculation may be found in [1, 2]. The result is

$$\bar{b}_{\alpha\beta} = (\bar{W} + \bar{V})\delta_{\alpha\beta} + \frac{1}{a}M_{\alpha\gamma}\left\{\left(1 + \kappa\, c^2 \Delta\right)w_{,\gamma\beta} + \frac{6}{5}c^2\varepsilon_{\gamma\delta}\Psi_{,\delta\beta}\right\}$$
$$- Q_\alpha\left(1 + 3\varepsilon c^2 \Delta\right)w_{,\beta}, \tag{9}$$
$$\bar{b}_{\alpha 3} = 0,$$

with

$$\overline{W} + \overline{V} = -\frac{1}{2}\left[\frac{1}{a}M_{\alpha\gamma}\left\{(1 + \kappa c^2 \Delta)w_{,\gamma\alpha} + \frac{6}{5}c^2\varepsilon_{\gamma\delta}\Psi_{,\delta\alpha}\right\}\right.$$
$$\left. + Q_\alpha\frac{6}{5}c^2\left\{\frac{2}{1-\nu}\Delta w_{,\alpha} + \varepsilon_{\alpha\beta}\Psi_{,\beta}\right\}\right] \tag{10}$$
$$- aP\left\{1 + 3\varepsilon c^2 \Delta\right\}w.$$

From (6a) and (7–11), the J-integral of the consistent second-order plate theory follows to be

$$J_\beta = \oint_\Gamma \bar{b}_{\alpha\beta} n_\alpha ds,$$

$$J_3 = 0. \tag{11}$$

The divergence theorem applied to (12a) leads in the absence of inhomogeneities to the conservation law

$$\bar{b}_{\alpha\beta,\alpha} = 0, \tag{12}$$

i.e., the Esheby tensor is divergence-free. If in (10), (11), and also in (4), c^2 is set to zero, the J_β-integral of the Kirchhoff-plate theory is recovered [11]. Considering material rotations within a plate theory, only the rotation with respect to the x_3-axis is admissible. Therefore, the L_i-integral has only one relevant component, namely $L := L_3$. In a similar way we obtain the L-integral for a plate as

$$L = \oint_\Gamma \varepsilon_{\alpha\beta} \left[x_\alpha \bar{b}_{\gamma\beta} - M_{\gamma\beta} \left\{ (1 + \kappa c^2 \Delta) w_{,\alpha} + \frac{6}{5} c^2 \varepsilon_{\alpha\delta} \Psi_{,\delta} \right\} \right] n_\gamma ds. \tag{13}$$

Finally, the integral for a plate follows to be as

$$M = \oint_\Gamma \left[x_\beta \bar{b}_{\alpha\beta} + a Q_\alpha \left(1 + \kappa c^2 \Delta \right) w \right] n_\alpha ds, \tag{14}$$

and the accompanying conservation laws in the absence of inhomogeneities are to be found by application of the divergence theorem. A detailed derivation of the path-independent integrals, the adjoined conservation laws and its proof may be found in [2] and, at length, in [1].

5 Concluding Remarks

Conservation laws for a consistent second-order plate theory have been derived by integration of the well-established conservation laws of the three-dimensional theory of elasticity over a plate continuum. Alternatively, one could apply the vector operations gradient, curl and divergence to the Lagrangian $\bar{L}$ per unit of plate middle surface and its moments $x_\alpha \bar{L}$. It has been shown by [20] that differences may occur, especially when dealing with the M-integral. Chien et al. [4] applied Noether's theorem involving geometric symmetry. An overview of different approaches has been given in [11]. It turns out

[1] that differences in the result for the M-integral are due to inconsistencies in the various theories and can be avoided by a consequent application of the uniform-approximation technique. The theory is formulated within the easiest framework possible. The extension of the theory, however, is straight forward to treat, e. g., plates of variable thickness, temperature effects, anisotropic elastic materials, laminated structures and materials with a non-linear characteristic. The coupling between disk and plate terms, however, has to be observed. The application of the uniform-approximation technique to shell theories has already been shown in [7, 10]. Finally, it may be mentioned that the conservation laws presented are useful in the numerical analysis of cracked plates, cf., e. g., [20].

References

1. Bose DK (2004) Erhaltungssätze der Kontinuumsmechanik für eine konsistente Plattentheorie. Diss Univ Bremen, Germany. Shaker, Aachen
2. Bose DK, Kienzler R (2006) On material conservation laws for a consistent plate theory. Arch Appl Mech 75: 607–617
3. Budiansky B, Rice JR (1973) Conservation laws and energy release rates. ASME J Appl Mech 40: 201–203
4. Chien N, Honein T, Herrmann G (1994) Conservation laws for non-homogeneous Mindlin plates. Int J Engng Sci 32: 1125–1136
5. Eshelby J (1951) The force on an elastic singularity. Phil Trans Roy Soc London A244:87–112
6. Gurtin ME (2000) Configurational forces as basic concepts of continuum physics. Springer, New York
7. Kienzler R (1982) Eine Erweiterung der klassischen Schalentheorie: der Einfluss von Dickenverzerrungen und Querschnittsverwölbungen. Ing Arch 52: 311–322
8. Kienzler R (2002 a) On consistent shell theories. Arch Appl Mech 72: 229–247
9. Kienzler R (2002 b) On consistent second-order plate theories. In: Kienzler R, Altenbach H, Ott I (eds) Theories of plates and shells, critical review and new applications. Springer, Berlin, pp. 85–96
10. Kienzler R, Golebiewska-Herrmann A (1985) Material conservation laws in higher-order shell theories. Int J Solids Struct 21: 1035–1045
11. Kienzler R, Herrmann G (2000) Mechanics in material space. Springer, Berlin
12. Kirchhoff GR (1850) Über das Gleichgewicht und die Bewegung einer elastischen Scheibe. Crelles J reine angew Math 40: 51–88
13. Knowles JK, Sternberg E (1972) On a class of conservation laws in linearized and finite elastostatics. Arch Rat Mech Anal 44: 187–211
14. Krätzig WB (1980) On the structure of consistent linear shell theories. In: Koiter WB, Mikhailov GK (eds) Prod. 3rd IUTAM Symp Shell Theory. North-Holland, Amsterdam pp. 353–368
15. Maugin GA (1993) Material inhomogeneities in elasticity. Chapman & Hall, London
16. Mindlin RD (1951) An introduction to the theory of vibrations of elastic plates. ASME J Appl Mech 73: 31–38
17. Reissner E (1944) On the bending of elastic plates. J Math Phys 23: 184–191

18. Rice JR (1968) A path independent integral and the approximate analysis of strain concentration by notches and cracks. ASME J Appl Mech 27: 379–386
19. Sosa HA, Eischen JW (1986) Computation of stress intensity factors for plate bending via a path-independent integral. Engng Fract Mech 25: 451–462
20. Sosa HA, Rafalski P, Herrmann G (1988) Conservation laws in plate theories. Ing Arch 88: 305–320
21. Wang CM, Reddy JN, Lee KH (2000) Shear deformable beams and plates. Elsevier, Oxford

A Small-Parameter Method for I. Vekua's Nonlinear and Nonshallow Shells

Tengiz Meunargia

I. Vekua Institute of Applied Mathematics of Faculty of Exact and Natural Sciences of I. Javakhishvili Tbilisi State University, `tmeun@viam.sci.tsu.ge`

Abstract In the present paper the system of differential equations for the nonlinear theory of non-shallow shells is obtained. Some basic boundary value problems are solved.

Keywords: non-shallow and nonlinear shells

I. Vekua constructed several versions of a refined linear theory of thin and shallow shells, consisting of an original method of reduction of three-dimensional problems of elasticity to two-dimensional ones.

In the present paper the system of differential equations for the nonlinear theory of nonshallow shells is obtained by means of the I. Vekua method. Then those equations are treated by the small-parameter method and some basic boundary value problems are solved.

By thin and shallow shells I. Vekua meant three-dimensional shell-type elastic bodies satisfying the conditions

$$a_\alpha^\beta - x_3 b_\alpha^\beta \cong a_\beta^\alpha, \quad -h(x^1, x^2) \leq x_3 \leq h(x^1, x^2) \ (\alpha, \beta = 1, 2), \qquad (*)$$

where a_α^β and b_α^β are the mixed components of the metric tensor and the curvature tensor of the shell's midsurface, x_3 is the thickness coordinate and h is the semi-thickness, depending on curvilinear coordinates x^1, x^2.

In the sequel, by nonshallow shells we mean elastic bodies not subject to assumption $(*)$, i. e., such that

$$a_\alpha^\beta - x_3 b_\alpha^\beta \neq a_\alpha^\beta \Rightarrow \left| x_3 b_\alpha^\beta \right| \leq q < 1.$$

1. To construct the theory of shells, we use the coordinate system which is normally connected with the midsurface S. This means that the radius-vector of any point of the domain Ω can be represented in the form [1]

$$\boldsymbol{R}(x^1, x^2, x^3) = \boldsymbol{r}(x^1, x^2) + x^3 \boldsymbol{n}(x^1, x^2), \quad (x^3 = x_3),$$

G. Jaiani, P. Podio-Guidugli (eds.), *IUTAM Symposium on Relations of Shell, Plate, Beam, and 3D Models*, © Springer Science+Business Media B.V. 2008

where $\boldsymbol{r}$ and $\boldsymbol{n}$ are the radius vector and the unit normal to the midsurface $S(x_3 = 0)$; x^1, x^2 are the Gaussian parameters of S.

Covariant and contravariant basis vectors $\boldsymbol{R}_i$ and $\boldsymbol{R}^i$ of the surface $\hat{S}(x_3 = \text{const})$ and the corresponding basis vectors $\boldsymbol{r}_i$ and $\boldsymbol{r}^i$ of the midsurface $S(x_3 = 0)$ are connected by the following relations [1]:

$$\boldsymbol{R}_i = A_{i.}^{\cdot j}\boldsymbol{r}_j = A_{ij}\boldsymbol{r}^j, \quad \boldsymbol{R}^i = A_{\cdot j}^{i\cdot}\boldsymbol{r}^j = A^{ij}\boldsymbol{r}_j, \quad (i,j = 1,2,3),$$

where

$$A_{\alpha.}^{\cdot\beta} = a_\alpha^\beta - x_3 b_\alpha^\beta, \quad A_{\cdot\beta}^{\alpha\cdot} = \vartheta^{-1}[(1 - 2Hx_3)a_\beta^\alpha + x_3 b_\beta^\alpha], \quad A_3^i = A_i^3 = \delta_i^3,$$

$$\vartheta = 1 - 2Hx_3 + Kx_3^2, \quad \boldsymbol{R}_3 = \boldsymbol{R}^3 = \boldsymbol{r}_3 = \boldsymbol{r}^3 = \boldsymbol{n} \quad (\alpha,\beta = 1,2). \tag{1}$$

Here $(a_{\alpha\beta}, a^{\alpha\beta}, a_\beta^\alpha)$ and $(b_{\alpha\beta}, b^{\alpha\beta}, b_\beta^\alpha)$ are the components (co, contra, mixed) of the metric tensor and curvature tensor of the midsurface S. By H and K we denote a middle and Gaussian curvature of the surface S, where

$$2H = b_\alpha^\alpha = b_1^1 + b_2^2, \quad K = b_1^1 b_2^2 - b_2^1 b_1^2.$$

The main quadratic forms of the midsurface S have the form

$$\mathrm{I} = ds^2 = a_{\alpha\beta}dx^\alpha dx^\beta, \quad \mathrm{II} = k_s ds^2 = b_{\alpha\beta}dx^\alpha dx^\beta, \tag{2}$$

where k_s is the normal curvature of the surface S, and

$$a_{\alpha\beta} = \boldsymbol{r}_\alpha \boldsymbol{r}_\beta, \quad b_{\alpha\beta} = -\boldsymbol{r}_\alpha \boldsymbol{n}_\beta, \quad k_s = b_{\alpha\beta}s^\alpha s^\beta, s^\alpha = \frac{dx^\alpha}{ds}.$$

Here and in the sequel, under a repeated indices we mean summation; note that the Greek indices range over 1, 2, while Latin indices range over 1, 2, 3.

To construct the theory of non- shallow shells, it is necessary to obtain formulas for a family of surfaces $\hat{S}(x_3 = \text{const})$, analogous to (2) of the midsurface $S(x_3 = 0)$ which have the form [1]

$$\mathrm{I} = d\hat{s}^2 = g_{\alpha\beta}dx^\alpha dx^\beta, \quad \mathrm{II} = k_{\hat{s}}d\hat{s}^2 = \hat{b}_{\alpha\beta}dx^\alpha dx^\beta, \tag{3}$$

where

$$g_{\alpha\beta} = a_{\alpha\beta} - 2x_3 b_{\alpha\beta} + x_3^2(2Hb_{\alpha\beta} - Ka_{\alpha\beta}), \quad \hat{b}_{\alpha\beta} = (1 - 2Hx_3)b_{\alpha\beta} + x_3 Ka_{\alpha\beta},$$

and $k_{\hat{s}}$ the normal curvature of the surface $\hat{S}$.

It is not now difficult to get the expression for the tangential normal $\hat{\boldsymbol{l}}$ of the surface $\hat{S}$ directed to $\hat{\boldsymbol{s}}$ [3]:

$$\hat{\boldsymbol{l}} = \hat{\boldsymbol{s}} \times \boldsymbol{n} = [(1 - x_3 k_s)\boldsymbol{l} - x_3 \tau_s \boldsymbol{s}]\frac{ds}{d\hat{s}}, \quad d\hat{s} = \sqrt{1 - 2x_3 k_s + x_3^2(k_s^2 + \tau_s^2)}ds,$$

where s and l are the unit vectors of the tangent and tangential normal on S, $d\hat{s}$ and ds are the linear elements of the surfaces $\hat{S}$ and S, and τ_s is the geodesic torsion of the surface S.

2. We write the equation of equilibrium of elastic shell-type bodies in a vector form

$$\frac{1}{\sqrt{g}}\frac{\partial\sqrt{g}\boldsymbol{\sigma}^i}{\partial x^i} + \boldsymbol{\Phi} = 0, \Rightarrow \nabla_i\boldsymbol{\sigma}^i + \boldsymbol{\Phi} = 0, \tag{4}$$

where g is the discriminant of the metric quadratic form of the three-dimensional domain Ω, ∇_i are covariant derivatives with respect to the space coordinates x^i, $\boldsymbol{\Phi}$ is on external force, $\boldsymbol{\sigma}^i$ are the contravariant constituents of the stress vector $\boldsymbol{\sigma}_{(l)}^{*}$ acting on the area with the normal $\overset{*}{\boldsymbol{l}}$ and representable as the Cauchy formula as follows:

$$\boldsymbol{\sigma}_{(l)}^{*} = \boldsymbol{\sigma}^i \overset{*}{l}_i, \quad \left(\overset{*}{l}_i = \overset{*}{\boldsymbol{l}}\,\boldsymbol{R}_i, \right).$$

For the stress vector acting on the area with normal $\hat{\boldsymbol{l}}$, we obtain

$$\boldsymbol{\sigma}_{(\hat{l})} = \boldsymbol{\sigma}^\alpha(\hat{\boldsymbol{l}}\boldsymbol{R}_\alpha) = \vartheta\boldsymbol{\sigma}^\alpha(\boldsymbol{l}\boldsymbol{r}_\alpha)\frac{ds}{d\hat{s}}. \tag{5}$$

The stress-strain relation for the geometrically nonlinear theory of elasticity has the form

$$\boldsymbol{\sigma}^i = \sigma^{ij}(\boldsymbol{R}_j + \partial_j\boldsymbol{U}) = E^{ijpq}e_{pq}(\boldsymbol{R}_j + \partial_j\boldsymbol{U}), \tag{6}$$

where σ^{ij} are contravariant components of the stress tensor, e_{ij} are covariant components of the strain tensor, $\boldsymbol{U}$ is the displacement vector, E^{ijpq} and e_{ij} are defined by the formulas:

$$E^{ijpq} = \lambda g^{ij}g^{pq} + \mu(g^{ip}g^{jq} + g^{iq}g^{jp}), \; e_{ij} = \frac{1}{2}(\boldsymbol{R}_i\partial_j\boldsymbol{U} + \boldsymbol{R}_j\partial_i\boldsymbol{U} + \partial_i\boldsymbol{U}\partial_j\boldsymbol{U}). \tag{7}$$

To reduce the three-dimensional problems of the theory of elasticity to the two-dimensional problems, it is necessary to rewrite the relation (4–7) in forms of the bases of the midsurface S of the shell Ω.

The relation (4) can be written as

$$\frac{1}{\sqrt{a}}\frac{\partial\sqrt{a}\vartheta\boldsymbol{\sigma}^\alpha}{\partial x^\alpha} + \frac{\partial\vartheta\boldsymbol{\sigma}^3}{\partial x^3} + \vartheta\boldsymbol{\Phi} = 0, \quad (a = a_{11}a_{22} - a_{12}^2). \tag{8}$$

From (1), (6), (7) we obtain

$$\boldsymbol{\sigma}^i = A_{i_1}^i A_{p_1,}^p M^{i_1j_1p_1q_1}[(\boldsymbol{r}_{q_1}\partial_p\boldsymbol{U}) + \frac{1}{2}A_{q_1}^q(\partial_p\boldsymbol{U}\partial_q\boldsymbol{U})](\boldsymbol{r}_{j_1} + A_{j_1}^j\partial_j\boldsymbol{U}), \tag{9}$$

$$M^{i_1j_1p_1q_1} = \lambda a^{i_1j_1}a^{p_1q_1} + \mu(a^{i_1p_1}a^{j_1q_1} + a^{i_1q_1}a^{j_1p_1}) \; (a^{ij} = \boldsymbol{r}^i\boldsymbol{r}^j).$$

3. The isometric system of coordinates on the surface S is of special interest, for in this system we can obtain basic equations of the theory of shells in a complex form which in turn allows one to construct for a rather wide class of problems complex representations of general solutions by means of analytic functions of one variable $z = x^1 + ix^2$.

The main quadratic forms in the system of coordinates are of the type

$$\mathrm{I} = ds^2 = \Lambda(z,\bar{z})dzd\bar{z}, \quad \mathrm{II} = k_s ds^2 = \frac{1}{2}\Lambda\left[\bar{Q}dz^2 + 2Hdzd\bar{z} + Qd\bar{z}^2\right],$$

$$Q = \frac{1}{2}(b_1^1 - b_2^2 + 2ib_2^1), \quad \Lambda(z,\bar{z}) > 0.$$

Introducing the well-known differential operators $2\partial z = \partial_1 - i\partial_2$, $2\partial_{\bar{z}} = \partial_1 + i\partial_2$ and the notations $\tau^{i}_{.j} = \vartheta\boldsymbol{\sigma}^i\boldsymbol{r}_j$, $X_i = \vartheta\boldsymbol{\Phi}\boldsymbol{r}_i$, for the geometrically nonlinear theory of nonshallow shells from (8) and (9) we obtain the following complex form both for the system of equations of equilibrium and Hooke's law:

$$\frac{1}{\Lambda}\frac{\partial\Lambda\boldsymbol{\tau}^+\boldsymbol{r}_+}{\partial z} + \frac{\partial\bar{\boldsymbol{\tau}}^+\boldsymbol{r}_+}{\partial\bar{z}} - \Lambda(H\tau_3^{\cdot+} + Q\bar{\tau}_3^+) + \frac{\partial\tau_{.+}^3}{\partial x_3} + X_+ = 0,$$

$$\frac{1}{\Lambda}\left(\frac{\partial\Lambda\tau_3^+}{\partial z} + \frac{\partial\Lambda\bar{\tau}_3^+}{\partial\bar{z}}\right) + H(\tau_1^1 + \tau_2^2) + Re\left(\bar{Q}\boldsymbol{\tau}^+\boldsymbol{r}_+\right) + \frac{\partial\tau_3^3}{\partial x_3} + X_3 = 0,$$

(10)

$$\boldsymbol{\tau}^+ = \vartheta\left\{\lambda\Theta + \mu\left[\boldsymbol{R}^+\partial_z\boldsymbol{U} + \left(\bar{\boldsymbol{R}}_+ + 2\partial_z\boldsymbol{U}\right)\partial^{\bar{z}}\boldsymbol{U}\right]\right\}\left(\boldsymbol{R}^+ + 2\partial^{\bar{z}}\boldsymbol{U}\right)$$

$$+ \mu\vartheta\left\{\left[\boldsymbol{R}^+\partial_{\bar{z}}\boldsymbol{U} + (\boldsymbol{R}_+ + 2\partial_{\bar{z}}\boldsymbol{U})\partial^{\bar{z}}\boldsymbol{U}\right]\left(\bar{\boldsymbol{R}}^+ + 2\partial^z\boldsymbol{U}\right)\right.$$

$$\left. + \left[\boldsymbol{R}^+\partial_3\boldsymbol{U} + (\boldsymbol{n} + \partial_3\boldsymbol{U})\partial^{\bar{z}}\boldsymbol{U}\right](\boldsymbol{n} + \partial_3\boldsymbol{U})\right\},$$

$$\boldsymbol{\tau}^3 = \vartheta\left\{\left[\lambda\boldsymbol{\Theta} + 2\mu(\boldsymbol{n}\partial^3\boldsymbol{U} + \frac{1}{2}\partial_3\boldsymbol{U}\partial^3\boldsymbol{U})\right](\boldsymbol{n} + \partial_3\boldsymbol{U})\right.$$

(11)

$$+ \mu\left[(\frac{1}{2}\bar{\boldsymbol{R}}_+\partial_3\boldsymbol{U} + \boldsymbol{n}\partial_z\boldsymbol{U} + \partial_z\boldsymbol{U}\partial_3\boldsymbol{U})(\boldsymbol{R}^+ + 2\partial^{\bar{z}}\boldsymbol{U})\right.$$

$$\left.\left. + (\frac{1}{2}\boldsymbol{R}_+\partial_3\boldsymbol{U} + \boldsymbol{n}\partial_{\bar{z}}\boldsymbol{U} + \partial_{\bar{z}}\boldsymbol{U}\partial_3\boldsymbol{U})(\bar{\boldsymbol{R}}^+ + 2\partial^z\boldsymbol{U})\right]\right\},$$

$$(\partial^3\boldsymbol{U} = \partial_3\boldsymbol{U}).$$

Here

$$\boldsymbol{\tau}^+\boldsymbol{r}_+ = (\boldsymbol{\tau}^1 + i\boldsymbol{\tau}^2)(\boldsymbol{r}_1 + i\boldsymbol{r}_2), \ \bar{\boldsymbol{\tau}}^+\boldsymbol{r}_+ = (\boldsymbol{\tau}^1 - i\boldsymbol{\tau}^2)(\boldsymbol{r}_1 + i\boldsymbol{r}_2), \ \tau_{.3}^{+\cdot} = \boldsymbol{\tau}^+\boldsymbol{n},$$

$$\tau_{.+}^{3\cdot} = \boldsymbol{\tau}^3\boldsymbol{r}_+, \ \tau_3^3 = \boldsymbol{\tau}^3\boldsymbol{n}, \ 2\partial^{\bar{z}}\boldsymbol{U} = (\boldsymbol{R}^+\boldsymbol{R}^+)\partial_z\boldsymbol{U} + \left(\boldsymbol{R}^+\bar{\boldsymbol{R}}^+\right)\partial_{\bar{z}}\boldsymbol{U}, \ \tau_{.\beta}^{\alpha\cdot} = \boldsymbol{\tau}^\alpha\boldsymbol{r}_\beta,$$

$$\Theta = 2Re\left[(\boldsymbol{R}^+ + \partial^{\bar{z}}\boldsymbol{U}]\right)\partial_z\boldsymbol{U} + \partial_3\boldsymbol{U}_3 + 0.5\,(\partial_3\boldsymbol{U})^2, \ \boldsymbol{R}^+ = \vartheta^{-1}\left[(1 - Hx_3)\,\boldsymbol{r}^+ +\right.$$

$$\left. + x_3Q\bar{\boldsymbol{r}}^+\right], \ \boldsymbol{R}_+ = (1 - Hx_3)\boldsymbol{r}_+ - x_3Q\bar{\boldsymbol{r}}_+, \ \boldsymbol{R}_+ = \boldsymbol{R}_1 + i\boldsymbol{R}_2, \ \boldsymbol{R}^+ = \boldsymbol{R}^1 + i\boldsymbol{R}^2,$$

$$\boldsymbol{R}^{+}\boldsymbol{R}^{+} = 4x_3(\Lambda\vartheta^2)^{-1}(1-Hx_3)Q,\ \boldsymbol{R}^{+}\bar{\boldsymbol{R}}^{+} = 2(\Lambda\vartheta^2)^{-1}(\vartheta+2x_3^2 Q\bar{Q}),$$
$$\boldsymbol{R}^{+}\boldsymbol{r}_{+} = 2\vartheta^{-1}Qx_3,\ \bar{\boldsymbol{R}}^{+}\boldsymbol{r}_{+} = 2\vartheta^{-1}(1-Hx_3),\ \boldsymbol{r}^{+}\bar{\boldsymbol{r}}^{+} = 2\Lambda^{-1},\ \boldsymbol{r}^{+} = \boldsymbol{r}^1 + i\boldsymbol{r}^2.$$

We have the formulas

$$\boldsymbol{r}^{+}\partial_z\boldsymbol{U} = \Lambda^{-1}\partial_z U_{+} - HU_3,\ \boldsymbol{r}^{+}\partial_{\bar{z}}\boldsymbol{U} = \partial_{\bar{z}}U^{+} - QU_3,\ X_{+} = X_1 + iX_2,$$
$$\boldsymbol{n}\partial_z\boldsymbol{U} = \partial_z U_3 + 0.5\left(\bar{Q}U_{+} + H\bar{U}_{+}\right),\ \left(U^{+} = \boldsymbol{U}\boldsymbol{r}^{+},\ U_{+} = \boldsymbol{U}\boldsymbol{r}_{+},\ U_3 = \boldsymbol{U}\boldsymbol{n}\right).$$

4. In the present paper the three-dimensional problems of the theory of elasticity are reduced to the two-dimensional ones by the method suggested by I. Vekua. Since the system of Legendre polynomials $\left\{P_m\left(\dfrac{x_3}{h}\right)\right\}$ is complete in the interval [-h,h], for Equation (8) we obtain the infinite system of two-dimensional equations

$$\int_{-h}^{h}\left[\frac{1}{\sqrt{a}}\frac{\partial\sqrt{a}\vartheta\boldsymbol{\sigma}^{\alpha}}{\partial x^{\alpha}} + \frac{\partial\vartheta\boldsymbol{\sigma}^3}{\partial x^3} + \vartheta\boldsymbol{\Phi}\right]P_m\left(\tfrac{x_3}{h}\right)dx_3 = 0,\quad (m = 0,1,...)$$

or in a form

$$\nabla_\alpha\overset{(m)}{\boldsymbol{\sigma}}{}^{\alpha} - \frac{2m+1}{h}\left(\overset{(m-1)}{\boldsymbol{\sigma}}{}^3 + \overset{(m-3)}{\boldsymbol{\sigma}}{}^3 + ...\right) + \overset{(m)}{\boldsymbol{F}} = 0,\qquad (12)$$

where

$$\left(\overset{(m)}{\boldsymbol{\sigma}}{}^i, \overset{(m)}{\boldsymbol{\Phi}}\right) = \frac{2m+1}{2h}\int_{-h}^{h}\left(\vartheta\boldsymbol{\sigma}^i, \vartheta\boldsymbol{\Phi}\right)P_m\left(\tfrac{x_3}{h}\right)dx_3 = \frac{2m+1}{2h}\int_{-h}^{h}\left(\boldsymbol{\tau}^i, \boldsymbol{X}\right)P_m dx_3,$$

$$\overset{(m)}{\boldsymbol{F}} = \overset{(m)}{\boldsymbol{\Phi}} + \frac{2m+1}{2h}\left(\overset{(+)}{\vartheta}\overset{(+)}{\boldsymbol{\sigma}}{}^3 - (-1)^m\overset{(-)}{\vartheta}\overset{(-)}{\boldsymbol{\sigma}}{}^3\right),\ \left(\overset{(\pm)}{\vartheta} = \vartheta(\pm h)\right),$$

∇_α are covariant derivatives on the midsurface S.

The equation of state (9) may be written as

$$\overset{(m)}{\boldsymbol{\sigma}}{}^i = M^{i_1 j_1 p_1 q_1}\sum_{m_1=0}^{\infty}\left[\left(\overset{(m)}{\underset{(m_1)}{A}}{}^{i\ p}_{i_1 p_1}\boldsymbol{r}_{j_1}\right.\right.$$

$$+ \sum_{m_2=0}^{\infty}\overset{(m)}{\underset{(m_1,m_2)}{A}}{}^{i\ j\ p}_{i_1 j_1 p_1}D_j\overset{(m_2)}{\boldsymbol{U}}\left)\left(\boldsymbol{r}_{q_1}\cdot D_p\overset{(m_1)}{\boldsymbol{U}}\right)\right.$$

$$+ \frac{1}{2}\sum_{m_2=0}^{\infty}\left(\overset{(m)}{\underset{(m_1,m_2)}{A}}{}^{i\ p\ q}_{i_1 p_1 q_1}\boldsymbol{r}_{j_1}\right.$$

$$\left.\left.+ \sum_{m_3=0}^{\infty}\overset{(m)}{\underset{(m_1,m_2,m_3)}{A}}{}^{i\ j\ p\ q}_{i_1 j_1 p_1 q_1}D_j\overset{(m_2)}{\boldsymbol{U}}\right)\left(D_p\overset{(m_1)}{\boldsymbol{U}}D_q\overset{(m_3)}{\boldsymbol{U}}\right)\right],$$

$$(13)$$

160   T. Meunargia

where

$$D_i \overset{(m)}{\boldsymbol{U}} = \delta_i^\beta \partial_\beta \overset{(m)}{\boldsymbol{U}} + \delta_i^3 \overset{(m)}{\boldsymbol{U}}{}'; \quad \overset{(m)}{\boldsymbol{U}}{}' = \frac{2m+1}{h} \left(\overset{(m+1)}{\boldsymbol{U}} + \overset{(m+3)}{\boldsymbol{U}} + ... \right), \qquad (14)$$

$$\overset{(m)}{\underset{(m_1)}{A}}{}^{ij}_{i_1 j_1} = \frac{2m+1}{2h} \int\limits_{-h}^{h} \vartheta A^i_{i_1} A^j_{j_1} P_{m_1}\left(\frac{x_3}{h}\right) P_m\left(\frac{x_3}{h}\right) dx_3$$

$$\overset{(m)}{\underset{(m_1,m_2)}{A}}{}^{ijp}_{i_1 j_1 p_1} = \frac{2m+1}{2h} \int\limits_{-h}^{h} \vartheta A^i_{i_1} A^j_{j_1} A^p_{p_1} P_{m_1} P_{m_2} P_m dx_3, \qquad (15)$$

$$\overset{(m)}{\underset{(m_1,m_2,m_3)}{A}}{}^{ijpq}_{i_1 j_1 p_1 q_1} = \frac{2m+1}{2h} \int\limits_{-h}^{h} \vartheta A^i_{i_1}, A^j_{j_1}, A^p_{p_1} A^q_{q_1} P_{m_1} P_{m_2} P_{m_3} P_m dx_3.$$

The boundary conditions on the lateral contour ∂S take the form:
a) for the stresses

$$\overset{(m)}{\boldsymbol{\sigma}}{}_{(l)} = \overset{(m)}{\boldsymbol{\sigma}}{}_{(ll)} \boldsymbol{l} + \overset{(m)}{\boldsymbol{\sigma}}{}_{(ls)} \boldsymbol{s} + \overset{(m)}{\boldsymbol{\sigma}}{}_{(ln)} \boldsymbol{n} = \frac{2m+1}{2h} \int\limits_{h}^{h} \boldsymbol{\sigma}_{(l)} \frac{d\hat{s}}{ds} P_m\left(\frac{x_3}{h}\right) dx_3, \quad (16)$$

b) for the displacements

$$\overset{m}{\boldsymbol{U}} = \frac{2m+1}{2h} \int\limits_{h}^{h} \boldsymbol{U} P_m\left(\frac{x_3}{h}\right) dx_3 = \overset{(m)}{\boldsymbol{U}}{}_{(l)} \boldsymbol{l} + \overset{(m)}{\boldsymbol{U}}{}_{(s)} (\boldsymbol{s}) + \overset{(m)}{\boldsymbol{U}}{}_{3} \boldsymbol{n}. \qquad (17)$$

Thus we have constructed an infinite system of two-dimensional equations of geometrically non-linear and nonshallow shells (12–17), which is consistent with the boundary conditions on the face surfaces, i.e. $\overset{(\pm)}{\boldsymbol{\sigma}}{}^3 = \sigma^3(x^1, x^2, \pm h)$.

The passage to finite systems can be realized by various methods one of which consists in considering of a finite series, i.e.

$$(\vartheta\boldsymbol{\sigma}^i, \boldsymbol{U}, \vartheta\boldsymbol{\Phi}) = \sum_{m=0}^{N} \left(\overset{(m)}{\boldsymbol{\sigma}}{}^i, \overset{(m)}{\boldsymbol{U}}, \overset{(m)}{\boldsymbol{\Phi}} \right) P_m\left(\frac{x_3}{h}\right) = (\boldsymbol{\tau}^i, \boldsymbol{U}, \boldsymbol{X}),$$

where N is a fixed nonnegative number. In other words, it is assumed that

$$\overset{(m)}{\boldsymbol{U}} = 0, \quad \overset{(m)}{\boldsymbol{\sigma}}{}^i = 0, \quad if \quad m > N.$$

Approximation of this type will be called approximation of order N.

The integrals of type (15) can be calculated [3], for example,

$$
\overset{(m)}{\underset{(m_1)}{A}}{}^{\alpha\beta}_{\alpha_1\beta_1} = \frac{2m+1}{2h} \int\limits_{-h}^{h} \vartheta^{-1} B^{\alpha}_{\alpha_1}(x_3) B^{\beta}_{\beta_1}(x_3) P_{m_1}\left(\frac{x_3}{h}\right) P_m\left(\frac{x_3}{h}\right) dx_3 =
$$

$$
\begin{cases}
\dfrac{2m+1}{2\sqrt{E}h}\left[B^{\alpha}_{\alpha_1}(hy) B^{\beta}_{\beta_1}(hy)\begin{pmatrix} P_{m_1}(y)Q_m(y), \ m_1 \le m \\ Q_{m_1}(y)P_m(y), \ m_1 \le m \end{pmatrix}\right]^{y_2}_{y_1} \\[3mm]
+\dfrac{L^{\alpha}_{\alpha_1} L^{\beta}_{\beta_1}}{K}\delta^m_{m_1}, \quad \text{if } E \ne 0 \ K \ne 0 \\[3mm]
a^{\alpha}_{\alpha_1} a^{\beta}_{\beta_1}\delta^m_{m_1}, \qquad \text{if } E = H^2 - K = 0,
\end{cases}
\tag{18}
$$

where $Q_m(y)$ is the Legendre function of the second kind, E is the Euler difference, $B^{\alpha}_{\beta}(x) = a^{\alpha}_{\beta} + x L^{\alpha}_{\beta}$, $L^{\alpha}_{\beta} = b^{\alpha}_{\beta} - 2H a^{\alpha}_{\beta}$. Under the square brackets we mean the following:

$$
[f(y)]^{y_2}_{y_1} = f(y_2) - f(y_1), \quad y_{1,2} = [(H \mp \sqrt{E})h]^{-1}.
$$

For the integrals containing the product of three Legendre polynomials we have

$$
\overset{(m)}{\underset{(m_1,m_2)}{A}}{}^{\alpha_1\alpha_2\alpha_3}_{\beta_1\beta_2\beta_3} = \frac{2m+1}{2h} \int\limits_{-h}^{h} \frac{B^{\alpha_1}_{\beta_1} B^{\alpha_2}_{\beta_2} B^{\alpha_3}_{\beta_3}}{(1 - 2Hx_3 + Kx_3)^2} P_{m_1} P_{m_2} P_m dx_3 = \frac{2m+1}{K^2 h^4} \times
$$

$$
\times \sum_{r=0}^{min(m_1,m_2)} a_{m_1 m_2 r} \sum_{n=0}^{3} {}^n\mathbb{C}^{\alpha_1\alpha_2\alpha_3}_{\beta_1\beta_2\beta_3} h^n \frac{\partial^2}{\partial y_1 \partial y_2}\left[\frac{y^n}{y_1 - y_2}\begin{pmatrix} P_s(y)Q_m(y), s \le m \\ Q_s(y)P_m(y), s \ge m \end{pmatrix}\right]^{y_2}_{y_1},
$$

where $s = m_1 + m_2 - 2r$,

$$
a_{pqr} = \frac{A_{p-r} A_r A_{q-r}}{A_{p+q-r}} \frac{2(p+q) - 4r + 1}{2(p+q) - 2r + 1}, \quad A_p = \frac{1.3 \cdots 2p - 1}{p!},
$$

$$
B^{\alpha_1}_{\beta_1}(x) B^{\alpha_2}_{\beta_2}(x) B^{\alpha_3}_{\beta_3}(x) = \sum_{n=0}^{3} {}^n\mathbb{C}^{\alpha_1\alpha_2\alpha_3}_{\beta_1\beta_2\beta_3} x^n.
$$

For the integrals containing the product of four Legendre polynomials the corresponding presentation can be written similarly.

5. Three-dimensional shell-shaped bodies are characterized by inequalities

$$
|h b^{\alpha}_{\beta}| \le q < 1, \quad (\alpha, \beta = 1, 2).
$$

Therefore they can be represented as follows

$$
|\varepsilon b^{\alpha}_{\beta} R| \le q < 1,
$$

where $\varepsilon = hR^{-1}$ is a small parameter.

162   T. Meunargia

Here h is semi-thickness of the shell, R is a certain characteristic radius of curvature of the midsurface S [3].

Now, following Signorini [2] we assume the validity of the expansions

$$\left(\overset{(m)}{\sigma}{}^i,\ \overset{(m)}{\mathbf{U}},\ \overset{(m)}{\mathbf{F}} \right) = \sum_{n=1}^{\infty} \left(\overset{(m,n)}{\sigma^i},\ \overset{(m,n)}{\mathbf{U}},\ \overset{(m,n)}{\mathbf{F}} \right) \varepsilon^n.$$

Substituting the above expansions into the relations (12–13) and considering (10–11) we obtain the following system of two-dimensional equations in a complex form for the approximation of order N

$$\frac{h}{\varLambda}\frac{\partial \overset{(m,n)}{\sigma_+} \mathbf{r}_+}{\partial z} + h\frac{\partial \overset{(m,n)}{\bar{\sigma}}{}^+\mathbf{r}_+}{\partial \bar z} - (2m+1)\left(\overset{(m-1,n)}{\sigma}{}^3_+ + \cdots \right) = \overset{(m,n-1)}{G}{}_+,$$

$$\frac{h}{\varLambda}\left(\frac{\partial \overset{(m,n)}{\sigma}{}^3_{+.}}{\partial z} + \frac{\partial \overset{(m,n)}{\bar\sigma}{}^3_{+.}}{\partial \bar z} \right) - (2m+1)\left(\overset{(m-1,n)}{\sigma}{}^3_3 + \cdots \right) = \overset{(m,n-1)}{G}{}_3,$$

$$(\boldsymbol\sigma_+ = \boldsymbol\sigma_1 + i\boldsymbol\sigma_2; \quad m = 0, 1, \cdots N; \quad n = 1, 2 \cdots).$$

For this system of equations, in terms of components of the displacement vector

$$\overset{(m)}{U} = \frac{1}{2}\left(\overset{(m)}{U}{}_+\bar{\mathbf{r}}^+ + \overset{(m)}{\bar U}{}_+\mathbf{r}^+ \right) + \overset{(m)}{U}{}_3\mathbf{n}, \text{ we have}$$

$$4\mu\partial_{\bar z}\left(\varLambda^{-1}\partial_z \overset{(m,n)}{U_+} \right) + 2(\lambda + \mu)\partial_{\bar z} \overset{(m,n)}{\Theta}$$

$$- \mu\tfrac{2m+1}{h}\left(\partial_{\bar z} \overset{(m-1,n)}{U}{}_3 + \cdots \right) = \overset{(m,n-1)}{M}{}_+,$$

$$\mu\left(\nabla^2 \overset{(m,n)}{U}{}_3 + \overset{(m,n)}{\Theta'} \right) \tag{19}$$

$$- \frac{2m+1}{h}\left[\left(\lambda \overset{(m-1,n)}{\Theta} + (\lambda + 2\mu) \overset{(m-1,n)}{U'}{}_3 \right) + \cdots \right] = \overset{(m,n-1)}{M}{}_3,$$

where $\overset{(m,n-1)}{M}{}_+$ and $\overset{(m,n-1)}{M}{}_3$ are expressed by means of solutions of the previous approximations, and hence are assumed to be known,

$$\overset{(m,n)}{\Theta} = \varLambda^{-1}\left(\partial_z \overset{(m,n)}{U}{}_+ + \partial_{\bar z} \overset{(m,n)}{\bar U}{}_+ \right), \quad \nabla^2 = 4\varLambda^{-1}\partial_z\partial_{\bar z},$$

$$\overset{(m,n)}{U'} = (2m+1)h^{-1}\left(\overset{(m+1,n)}{U} + \overset{(m+3,n)}{U} + \cdots \right).$$

6. For the approximation of order $N = 0$ general solution of the system (19) can be represented by means of three analytic functions of z in the form [3]:

$$\overset{(0,n)}{U}_{+} = -\frac{\text{æ}}{\pi}\iint_{D}\frac{\Lambda\varphi'(\zeta)d\xi d\eta}{\bar{\zeta} - \bar{z}} + \left(\frac{1}{\pi}\iint_{D}\frac{\Lambda d\xi d\eta}{\bar{\zeta} - \bar{z}}\right)\overline{\varphi'(z)} - \overline{\psi(z)} - \frac{1}{8\mu}\frac{\lambda + \mu}{\lambda + 2\mu}\frac{1}{\pi}\iint_{D}\frac{\Lambda \overset{(n)}{F}_{+}d\xi d\eta}{\bar{\zeta} - \bar{z}}, \tag{20}$$

$$\overset{(0,n)}{U}_{3} = f(z) + \overline{f(z)} + \frac{1}{2\mu}\frac{1}{\pi}\iint_{D}\overset{(n)}{X}_{3}(\zeta,\bar{\zeta})\ln|\zeta - z|d\xi d\eta,$$

$$\left(\overset{(n)}{F}_{+}(z,\bar{z}) = \frac{1}{\pi}\iint_{D}\left(\frac{\overset{(n)}{\bar{X}}_{+}}{\bar{\zeta} - \bar{z}} - \frac{\text{æ}\,\overset{(n)}{X}_{+}}{\zeta - z}\right)d\xi d\eta, \quad \overset{(n)}{X} = \Lambda\,\overset{(m,n-1)}{M}\right), \tag{21}$$

where D is the domain of the plane $\zeta = \xi + i\eta$ onto which the midsurface S is mapped topologically, $\Lambda = \Lambda(\zeta,\bar{\zeta})$, $\text{æ} = (\lambda + 3\mu)(\lambda + \mu)^{-1}$.

The basic boundary conditions ($N = 0$) for any n have the form: (a) for the first boundary problem (in displacements)

$$\overset{(0,n)}{U}_{(\ell)} + i\,\overset{(0,n)}{U}_{(s)} = i\,\overset{(0,n)}{U}_{+}\frac{d\bar{z}}{ds} = \overset{(n)}{d}_{+}, \quad \overset{(0,n)}{U}_{3} = \overset{(n)}{d}_{3} \ (\text{on}\ \partial D), \tag{22}$$

b) for the second boundary problem (in stresses)

$$\overset{(0,n)}{\sigma}_{(\ell\ell)} + i\,\overset{(0,n)}{\sigma}_{(\ell s)} = \frac{1}{2}\left[\overset{(0,n)}{\bar{\sigma}}{}_{+}r_{+} - \left(\overset{(0,n)}{\sigma}{}_{+}r_{+}\right)\frac{dz}{d\bar{z}}\right] = \overset{(n)}{e}_{+},$$

$$\overset{(0,n)}{\sigma}_{(\ell n)} = -Im\left[\left(\overset{(0,n)}{\sigma}{}_{+}n\right)\frac{d\bar{z}}{ds}\right] = \overset{(n)}{e}_{3} \ (\text{on}\ \partial D). \tag{23}$$

Here we present a general scheme of solution of boundary problems when the domain D is a circle of radius r_0.

The first boundary problem for any n takes the form (on $|z| = z_0$)

$$\overset{(0,n)}{U}_{+} = -\frac{\text{æ}}{\pi}\iint_{D}\frac{\Lambda\varphi'(\zeta)d\xi d\eta}{\bar{\zeta} - \bar{z}} + \left(\frac{1}{\pi}\iint_{D}\frac{\Lambda d\xi d\eta}{\bar{\zeta} - \bar{z}}\right)\overline{\varphi'(z)} - \overline{\psi(z)} = \overset{(n)}{G}_{+}, \tag{24}$$

$$\overset{(0,n)}{U}_{3} = f(z) + \overline{f(z)} = \overset{(n)}{G}_{3} \ (z = re^{i\varphi}, \zeta = \rho e^{i\psi}), \tag{25}$$

where $\overset{(n)}{G}_{+}$ and $\overset{(n)}{G}_{3}$ are the known values containing solutions $\overset{(0,1)}{U}_{i}, \cdots, \overset{(0,n-1)}{U}_{i}$ of the previous approximations.

Let $\Lambda(z,\bar{z})$ depend only on $r = |z|$, next $\varphi'(z)$, $\psi(z)$ and $\overset{(n)}{G}_{+}$ are expanded in power series of the type

164   T. Meunargia

$$\varphi'(z) = \sum_{k=0}^{\infty} a_k z^k, \quad \Psi(z) = \sum_{k=0}^{\infty} b_k z^k, \quad \overset{(n)}{G}_+ = \sum_{k=-\infty}^{\infty} A_k e^{ik\vartheta}.$$

Substituting these expansions into (24), we obtain

$$a_0 = \frac{r_0}{\alpha_0} \frac{\text{æ}A_1 + \bar{A}_1}{\text{æ}^2 - 1}, \quad a_k = \frac{r_0^{k+1} A_{k+1}}{\text{æ}\alpha_k} \ (k \geq 1),$$

$$b_k = -\frac{\bar{A}_k}{r_0^k} - \frac{\alpha_0 r_0^{k+2}}{\text{æ}\alpha_{k+1}} A_{k+2}, \ (k \geq 0), \quad \alpha_k = 2 \int_0^{r_0} \rho^{2k+1} \Lambda(\rho) d\rho.$$

$\overset{(0,n)}{U}_3$ is representable in the form of the Poisson integral,

$$\overset{(0,n)}{U}_3(r, \vartheta) = \frac{1}{2\pi} \int_0^{2\pi} \overset{(n)}{G}_3(\psi) \frac{r_0^2 - r^2}{r^2 - 2r_0 r \cos(\psi - \vartheta) + r_0^2} d\psi. \tag{26}$$

Thus for any n we can construct formal solutions of the problem (22), when $N = 0$.

From the second boundary condition (23), we obtain (on ∂D)

$$\overset{(0,n)}{\sigma}_{(\ell\ell)} + i \overset{(0,n)}{\sigma}_{(\ell s)} = \overset{(n)}{\ell}_+ \Rightarrow (\lambda + \mu) \overset{(0,n)}{\Theta} - 2\mu \left(\frac{1}{\Lambda} \overset{(0,n)}{U}_+\right) \frac{d\bar{z}}{dz} = \overset{(n)}{P}_+, \tag{27}$$

$$\overset{(0,n)}{\sigma}_{(\ell n)} = \overset{(n)}{\ell}_3 \Rightarrow \text{Im} \left(\frac{\partial \overset{(0,n)}{U}_3}{\partial z} \frac{d\bar{z}}{ds}\right) = \overset{(n)}{P}_3,$$

$$\tag{28}$$

$$\left(\overset{(0,n)}{\Theta} = \frac{1}{\Lambda} \left(\frac{\partial \overset{(0,n)}{U}_+}{\partial z} + \frac{\partial \overset{(0,n)}{\bar{U}}_+}{\partial \bar{z}}\right)\right).$$

Consider the case of a spherical shell, whose midsurface is a spherical segment of radius $R_0 \sin \vartheta$, where R_0 is the radius of a sphere. Isometric coordinates on the sphere can be represented in the form

$$z = x^1 + ix^2 = re^{i\varphi}, \quad r = \text{tg}\frac{\vartheta}{2}, \quad \Lambda = 4R^2(1 + z\bar{z})^{-2}, \quad (0 \leq \vartheta \leq \vartheta_0).$$

Let the expressions

$$(\varphi'(z), \Psi'(z), f(z)) = \sum_{k=0}^{\infty} (a_k, b_k, c_k) z^k, \quad \left(\overset{(n)}{g}_+, \overset{(n)}{g}_3\right) = \sum_{k=-\infty}^{\infty} (A_k, B_k) e^{ik\varphi},$$

be valid, where $\overset{(n)}{g}_{+}$ and $\overset{(n)}{g}_{3}$ are known values expressed by $\overset{(0,1)}{U}_{i}, \cdots, \overset{(0,n-1)}{U}_{i}$ of the previous approximations. Substituting these expansions into (27), (28) and taking into account that principal vector and moment of stresses are zero, we obtain

$$a_k = \frac{A_k}{2\mu r_0^k} \frac{1}{1 + 2\text{æ}(1 + r_0^2)\beta_k},$$

$$b_k = \frac{-1}{2\mu r_0^{k-1}} \frac{(1 + r_0^2)^{-1}}{k + 2r_0^2} \left[\frac{((1 + r_0^2)k + 2r_0^2)A_{k+1}}{1 + 2\text{æ}(1 + r_0^2)\beta_{k+1}r_0^2} + \bar{A}_{k-1} \right], \ (k \geq 0).$$

$$c_k = \frac{2}{\mu} \frac{R_0}{1 + r_0^2} \frac{B_k}{kr_0^{k-1}} \ (k \geq 1), \quad B_0 = 0, \ \beta_k(z) = \frac{1}{z^{k+2}} \int\limits_0^z \frac{(z - t)t^k \, dt}{(1 + t\bar{z})^3}.$$

From here we obtain the well-known Dini's formula

$$\overset{(0,n)}{U}_3(r_0, \varphi) = -\frac{r_0}{\pi} \int\limits_0^{2\pi} \overset{(n)}{P}_3(r_0, \varphi) \ln|\sigma - z| \, d\vartheta + \text{const}, \quad (\sigma = r_0 e^{i\vartheta}).$$

For the case $N = 1$ we will have

$$\boldsymbol{U} = \sum_{n=1}^{\infty} \left(\overset{(0,n)}{\boldsymbol{U}} + P_1 \left(\frac{x_3}{h} \right) \overset{(1,n)}{\boldsymbol{U}} \right) \varepsilon^n.$$

The complex representation of a general solution in this case can be written as follows:

$$\overset{(0,n)}{U}_+ = \lambda_1 \frac{\partial \omega}{\partial \bar{z}} - \lambda_2 \frac{1}{\pi} \iint\limits_D \frac{\Lambda \varphi'(\zeta) d\xi d\eta}{\bar{\zeta} - \bar{z}} + \left(\frac{1}{\pi} \iint\limits_D \frac{\Lambda d\xi d\eta}{\bar{\zeta} - \bar{z}} \right) \overline{\varphi'(z)} - \overline{\psi(z)},$$

$$\overset{(1,n)}{U}_+ = i\frac{\partial \chi}{\partial \bar{z}} - 2h\overline{\Psi'} - \frac{1}{\pi} \iint\limits_D \frac{\Lambda \phi' d\xi d\eta}{\bar{\zeta} - \bar{z}} - \left(\frac{1}{\pi} \iint\limits_D \frac{\Lambda d\xi d\eta}{\bar{\zeta} - \bar{z}} \right) \overline{\phi'} + \frac{2(\lambda + 2\mu)h^2}{3\mu} \overline{\phi''},$$

$$\overset{(0,n)}{U}_3 = \Psi(z) + \overline{\Psi(z)} - \frac{1}{\pi h} \iint\limits_D \Lambda[\phi'(z) + \overline{\phi'(z)}] \ln|\zeta - z| d\xi d\eta,$$

$$\overset{(1,n)}{U}_3 = \omega - \frac{2\lambda h}{3\lambda + 2\mu}[\varphi'(z) + \overline{\varphi'(z)}], \quad \left(\lambda_1 = -\frac{\lambda h}{6(\lambda + \mu)}, \ \lambda_2 = \frac{5\lambda + 6\mu}{3\lambda + 2\mu} \right),$$

where $\varphi(z), \psi(z), \phi(z), \Psi(z)$ are analytic functions of z, ω and χ are the general solutions of the following equations

$$\nabla^2 \omega - \eta h^{-2}\omega = 0, \quad \nabla^2 \chi - 3h^{-2}\chi = 0, \quad \eta = 3(\lambda + \mu)/(\lambda + 2\mu).$$

Accordingly they ensure the satisfication of six arbitrary given physical or kinematic conditions.

References

1. I.N. Vekua, Shell theory:general methods of construction. Pitman Advanced Pyblishing Program, Boston-London-Melburne, 1985, 287p.
2. P.G. Ciarlet, Mathematical elasticity. Vol I Three.dimensional elasticity. North-Holland Publishing, Amsterdam-New york-Oxford-Tokio. 1988, 471p.
3. T.V. Meunargia, On the aplication of the method of a small parameter in the theory of non-shallow I.N. Vekua's shells, Proc. A.Razmadze Math. Inst. vol 141 (2006) pp. 87–122.

The Extension and Application
of the Hierarchical Beam Theory
to Piezoelectrically Actuated Beams

DCD Oguamanam[1], C McLean[2], and JS Hansen[3]

[1] Dept. of Mechanical and Industrial Engineering, Ryerson University, 350
   Victoria Street, Toronto, Ontario Canada M5B 2K3, `doguaman@ryerson.ca`
[2] Nuclear Safety Solutions Limited, 700 University Ave., Toronto, Ontario M5G
   1X6, `clayton.mclean@utoronto.ca`
[3] Institute for Aerospace Studies, University of Toronto, 4925 Dufferin Street,
   Downsview, Ontario, Canada M3H 5T6, `hansen@utias.utoronto.ca`

Abstract An accurate, internally consistent modelling process is needed in order
to realize the full compatibility of advanced composite and sandwich structures. The
recently proposed unified hierarchical theory for layered beams which uses the moments of displacements, stress and strains without a priori assumptions on the form
of the displacement field variables is one such effort. The theory relies on the concept of fundamental states and assumes that all load states can be decomposed
as supposition of fundamental load states. The extension of this theory to include
piezoelectric actuation is object of this paper. The through-the-thickness stress and
strain distributions, and the displacement moments are calculated and observed to
be comparable to the corresponding values obtained from two-dimensional finite
element analyses with the commercial finite element software ANSYS.

Keywords: beam theory, hierarchical theory, piezo-actuation, layered beam,
smart beam, composite beam

1 Introduction

This paper is about smart or intelligent structures with piezoelectric components. An excellent compendium on the state-of-the-art is provided in [1]. The
prediction of global responses of these structures, particularly the through-the-
thickness stress and strain distributions, using reduced geometrical models has
limited success. This is a major hindrance to the advancement of the applications of these structures [1].

The common beam theories (Euler-Bernoulli (EBT) and Timoshenko
(TBT)) and plate theories (Kirchoff and Reissner/Mindlin) [2] have inconsistencies in the simultaneous application of plane-strain and plane-stress

assumptions. The use of Layer-wise or Zig-Zag theories [3] is limited by the high number of variables and complex formulation.

The recently proposed hierarchical beam theory (HBT) [4] has been shown to accurately predict through-the-thickness stress and strains. It is internally consistent and assumes that the solution of a beam problem is expressible as a superposition of fundamental states which are numerical experiments that are performed on a self-equilibrating segment of the beam and designed to reflect the load state of the beam. The theory eliminates the problems associated with the discontinuity and differentiability of stress, strains and displacements by using their moments.

The present study extends the HBT to smart structures via the inclusion of piezoelectricity. The efficacy of the extended theory is demonstrated via the application to a cantilevered sandwiched beam with shear load at the free end.

2 Hierarchical Beam Theory - HBT

An important premise in the HBT is that the solution of a beam problem can be developed as a superposition of sets of well chosen fundamental states [4]. Fundamental states (FS) are numerical experiments that are performed on an infinitesimal segment of the beam with known geometry and stacking sequence.

The fundamental states are grouped into two classes (see Fig. 1). The first comprises two states, one pure bending and the other pure shear; these states are always included in the solution space of any problem. The second group of fundamental states describe the given load state of the beam. The

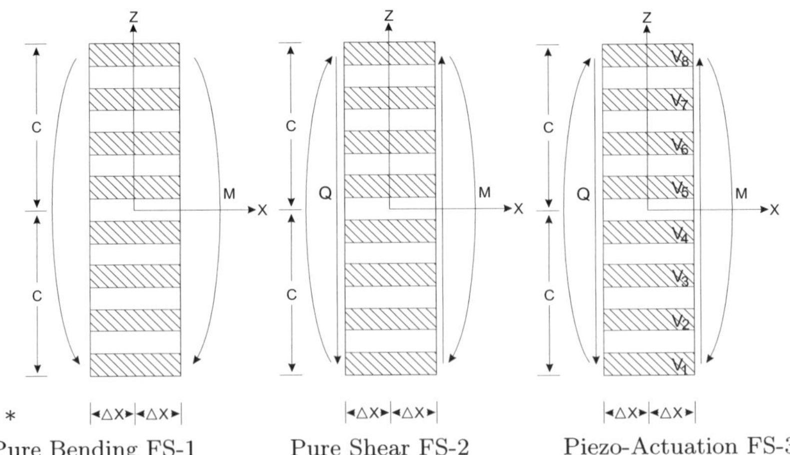

Fig. 1 Schematic of fundamental states - FS

fundamental state variables of this second group are known a priori, hence only those of the first have to determined.

The stress and strain distribution in the beam is expanded as

$$\Xi^T(x,z) = \sum_{i=1}^{n} {}_i\Xi(z)\alpha_i(x) \qquad \text{where } \Xi \in \{\sigma_x, \sigma_z, \sigma_{xz}, \epsilon_x, \epsilon_z, \gamma_{xz}\} \quad (1)$$

The superscript T denotes total. The ith fundamental state variable is denoted by $\alpha_i(x)$ and its coefficient represents the fundamental state component.

Use is made of the moments of stresses and strains, and normalized moments of displacements which eliminates problems of discontinuity or differentiability of the solution field. The stress and strain moments, ${}^T\mathbf{s}^i(x)$ and ${}^T\boldsymbol{\epsilon}^i(x)$, respectively, are defined

$$^T\mathbf{s}^i(x) = \int_{-c}^{c} \boldsymbol{\sigma}^T(x,z)z^i dz, \qquad {}_j\mathbf{s}^i(x) = \int_{-c}^{c} {}_j\boldsymbol{\sigma}(x,z)z^i dz,$$

$$^T\mathbf{e}^i(x) = \int_{-c}^{c} \boldsymbol{\epsilon}^T(x,z)z^i dz \quad \text{and} \quad {}_j\mathbf{e}^i(x) = \int_{-c}^{c} {}_j\boldsymbol{\epsilon}(x,z)z^i dz \quad (2)$$

and the normalized moments of the longitudinal and transverse displacements, $u(x,z)$ and $w(x,z)$, are respectively defined as

$$u_i(x) = \frac{(2i+1)}{2c^{(2i+1)}} \int_{-c}^{c} u(x,z)z^i dz \quad \text{and} \quad w_i(x) = \frac{(2i+1)}{2c^{(2i+1)}} \int_{-c}^{c} w(x,z)z^i dz \quad (3)$$

The homogenized flexural stiffness E_x, homogenized shear stiffness, G_{xz} and the homogenized Poisson's ratio ν_{xz} are defined as

$$E_x = \frac{{}_1 s_x^1}{{}_1 e_x^1}, \quad G_{xz} = \frac{{}_2 s_{xz}^0}{{}_2 e_{xz}^0} \quad \text{and} \quad \nu_{xz} = \frac{{}_1 e_z^1}{{}_1 e_x^1} \quad (4)$$

The exactness of the shear stress and strain moments precludes the introduction of shear-correction factor as employed in the TBT in order to compensate for the assumption of incorrect shear stress distribution. Rather a shear-strain moment correction which is solely based on FS-2 is derived and written as

$$C_{xz} = \left\{ \frac{e_{xz}^0}{2c\left[u_1(x) + \frac{\partial w_0(x)}{\partial x}\right]} \right\}_{state_2} \quad (5)$$

The 2-D equilibrium equation without body forces are valid in each layer of a perfectly bonded K-layered beam. The moment equilibrium equations are valid for all stress moments. They are derived by taking the ith moment of each equilibrium equation with respect to the thickness coordinate z for each layer, k. Summing these moments eliminates inter-ply shear and normal stress components, leaving only the top and bottom surface traction $(T_z^\pm, T_{xz}^\pm)$. Given that the problem is completely specified by s_x^1 and s_{xz}^0, the resulting expressions can be written as

$$\frac{\partial s_x^1}{\partial x} - s_{xz}^0 = -c[T_{xz}^+(x) - T_{xz}^-(x)];$$

$$\frac{\partial s_{xz}^0}{\partial x} = -[T_z^+(x) - T_z^-(x)] \tag{6}$$

The magnitudes of the class 1 fundamental state variables are determined from the solutions of these average moment equilibrium equations.

The HBT enforces interface stress and displacement continuities, and the derivation of the through-the-thickness stress and strains are internally consistent. Internally inconsistency alludes to conventional beam theories where plane strains and stresses are "inadvertently" assumed even though they cannot be simultaneously present. The reader is referred to Ref. [4] for in-depth presentation.

3 Extension of HBT to Include Piezo-Actuation

An extension of HBT requires the development of a new fundamental state that encompasses piezo-actuation (see Fig. 1c). A brief presentation is given here while detailed exposition is provided in [5].

The displacement fields, $u^{(k)}(x, z)$ and $w^{(k)}(x, z)$, for each layer of beam are expanded polynomially as

$$u^{(k)}(x, z) = \sum_{i=0}^{N} \sum_{j=0}^{N-i} u^{(k:i,j)} x^i z^j; \qquad w^{(k)}(x, z) = \sum_{i=0}^{N} \sum_{j=0}^{N-i} w^{(k:i,j)} x^i z^j \tag{7}$$

and the vector of the planar stress and strain fields, $\boldsymbol{\sigma}^{(k)}(x, z)$ and $\boldsymbol{\epsilon}^{(k)}(x, z)$, are similarly expanded polynomially as

$$\boldsymbol{\sigma}^{(k)}(x, z) = \sum_{i=0}^{N-1} \sum_{j=0}^{N-1-i} \sigma^{(k:i,j)} x^i z^j; \qquad \boldsymbol{\epsilon}^{(k)}(x, z) = \sum_{i=0}^{N-1} \sum_{j=0}^{N-1-i} \epsilon^{(k:i,j)} x^i z^j \tag{8}$$

These expansions are used in the linear strain-displacement relations for the kth layer and the expressions that result from equating the coefficients of equal powers of x and z are obtained. These enable the expression of the modified constitutive relation for $i + j \leq N - 1$ and $k = 1, K$ to be written as

$$\left\{ \begin{array}{c} \sigma_x^{(k:i,j)} \\ \sigma_z^{(k:i,j)} \\ \sigma_{xz}^{(k:i,j)} \end{array} \right\} = \begin{bmatrix} C_{11}^{(k)} & C_{12}^{(k)} & 0 \\ C_{12}^{(k)} & C_{22}^{(k)} & 0 \\ 0 & 0 & C_{33}^{(k)} \end{bmatrix} \left\{ \begin{array}{c} u^{(k:i+1,j)}(i+1) \\ w^{(k:i,j+1)}(j+1) \\ u^{(k:i,j+1)}(j+1) + w^{(k:i+1,j)}(i+1) \end{array} \right\} -$$

$$\delta(i)\delta(j) \begin{bmatrix} C_{11}^{(k)} & C_{12}^{(k)} & 0 \\ C_{12}^{(k)} & C_{22}^{(k)} & 0 \\ 0 & 0 & C_{33}^{(k)} \end{bmatrix} \begin{bmatrix} 0 & d_{31}^{(k)} \\ 0 & d_{33}^{(k)} \\ d_{15}^{(k)} & 0 \end{bmatrix} \left\{ \begin{array}{c} 0 \\ E_z^{(k)} \end{array} \right\} \tag{9}$$

where δ denotes the Kronecker delta function, and the piezo-actuation layers are poled in the direction of the thickness under a constant electric field intensity so that

$$E_z^{(k)} = \frac{V^{(k)}}{t^{(k)}} \tag{10}$$

where $V^{(k)}$ is the voltage across the kth layer.

At this juncture the stress polynomial expansions (i.e., Equation (8)) are substituted into the 2-D equilibrium relations and use is made of the modified constitutive relation, Equation (9), to obtain some relations for the components of the mechanical stiffness matrix of each layer of the beam. Further relations among the components of the mechanical stiffness matrix and those of the piezoelectric strain are obtained by satisfying surface traction constraints (each traction is expanded polynomially) and satisfying edge stress constraints, and by enforcing ply stress continuity.

The imposed rigid-body constraints on only one layer within the beam segment and at the origin of the coordinate axis, i.e., $x = z = 0$, are

$$u(x, z) = 0; \qquad w(x, z) = 0; \qquad \frac{\partial u(x, z)}{\partial z} = 0 \tag{11}$$

which, in terms of the displacement polynomials, translate to

$$u^{(1:0,0)} = u^{(1:0,1)} = w^{(1:0,0)} = 0 \tag{12}$$

The enforcement of displacement continuity yields

$$\sum_{j=0}^{N-i} u^{(k:i,j)}(z_k)^j = \sum_{j=0}^{N-i} u^{(k+1:i,j)}(z_k)^j$$

$$\& \sum_{j=0}^{N-i} w^{(k:i,j)}(z_k)^j = \sum_{j=0}^{N-i} w^{(k+1:i,j)}(z_k)^j \tag{13}$$

At this juncture the stress and strain distributions of Equation (1) can be expanded as:

$$\sigma_x^T(x, z) \approx {}_1\sigma_x(z)\alpha_1(x) + {}_2\sigma_x(z)\alpha_2(x) + \cdots + {}_{n_p}\sigma_x(z)\alpha_{n_p}(x)$$
$$+ \cdots + {}_n\sigma_x(z)\alpha_n(x)$$
$$\sigma_z^T(x, z) \approx {}_1\sigma_z(z)\alpha_1(x) + {}_2\sigma_z(z)\alpha_2(x) + \cdots + {}_{n_p}\sigma_z(z)\alpha_{n_p}(x)$$
$$+ \cdots + {}_n\sigma_z(z)\alpha_n(x) \tag{14}$$
$$\sigma_{xz}^T(x, z) \approx {}_1\sigma_{xz}(z)\alpha_1 x + {}_2\sigma_{xz}(z)\alpha_2(x) + \cdots + {}_{n_p}\sigma_{xz}(z)\alpha_{n_p}(x)$$
$$+ \cdots + {}_n\sigma_{xz}(z)\alpha_n(x)$$

and

$$\epsilon_x^T(x,z) \approx {}_1\epsilon_x(z)\alpha_1(x) + {}_2\epsilon_x(z)\alpha_2(x) + \cdots + {}_{n_p}\epsilon_x(z)\alpha_{n_p}$$
$$+ \cdots + {}_n\epsilon_x z\alpha_n(x)$$
$$\epsilon_z^T(x,z) \approx {}_1\epsilon_z(z)\alpha_1(x) + {}_2\epsilon_z(z)\alpha_2(x) + \cdots + {}_{n_p}\epsilon_z(z)\alpha_{n_p}$$
$$+ \cdots + {}_n\epsilon_z(z)\alpha_n(x) \tag{15}$$
$$\gamma_{xz}^T(x,z) \approx {}_1\gamma_{xz}(z)\alpha_1(x) + {}_2\gamma_{xz}(z)\alpha_2(x) + \cdots + {}_{n_p}\gamma_{xz}(z)\alpha_{n_p}(x)$$
$$+ \cdots + {}_n\gamma_{xz}\alpha_n(x)$$

The order of these expansions is governed by the desired degree of the accuracy of the solution. The first and second entries in each expression correspond to the state of pure bending and of the state of pure shear, respectively. Their associated fundamental state variables (i.e., α_1 and α_2) together constitute the first category of fundamental states which are the only unknown fundamental state variables. The remaining fundamental state variables fall into the second category. In this particular representation, α_i for $i = 3, 4, \cdots n_p$ denote the fundamental state variables associated with the piezo-actuation and are the input voltages. The higher order loading terms are captured by the fundamentals states α_i for $i = n_p + 1, n_p + 2, \cdots n$.

The discussion on the determination of the first two fundamental state variables (i.e., α_1 and α_2), and the displacement field moments $u_1(x,z)$ and $w_0(x,z)$ (which are the normalized first and zeroth moments of the respective total displacements) follows in the next section where the focus is on a three-layer cantilevered sandwich beam with an orthotropic elastic core, piezo-actuation faces and carrying a shear load at the free end (see Fig. 2).

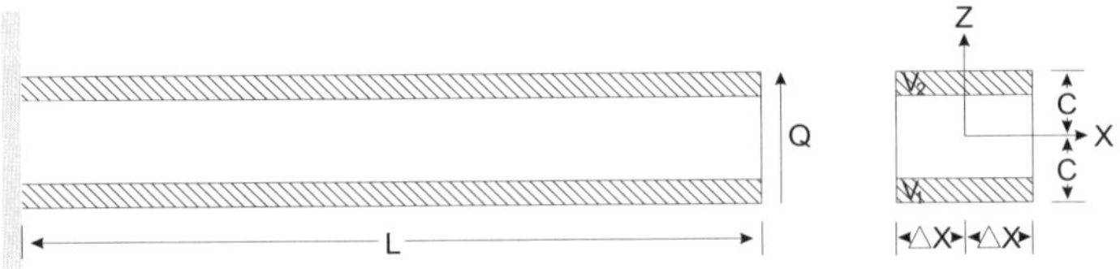

Fig. 2 Schematic of cantilevered sandwich beam

4 Numerical Example

The efficacy of the extended HBT is investigated via the use of the three-layer cantilevered sandwich beam depicted in Fig. 2. The fundamental state variables describing the class 2 states are described by the problem and they must produce a bending state within the beam (i.e., $\alpha_3(x) = -\alpha_4(x)$).

$$\alpha_3(x) = V_1; \quad \text{and} \quad \alpha_4(x) = V_2 \tag{16}$$

The total displacements $u^T(x, z)$ and $w^T(x, z)$ are written as summations of known and unknown displacements. They can also be expressed in terms of correction components, $\tilde{u}(x, z)$ and $\tilde{w}(x, z)$, and the averaged quantities $u_1(x)$ and $w_o(x)$. Specifically,

$$u^T(x, z) = u_1(x)z + \tilde{u}(x, z); \qquad w^T(x, z) = w_0(x) + \tilde{w}(x, z) \tag{17}$$

where the definition implies

$$u_1(x) = \frac{1}{I} \int_{-c}^{c} u^T(x, z)z\,dz; \qquad w_0(x) = \frac{1}{2c} \int_{-c}^{c} w^T(x, z)\,dz \tag{18}$$

and I is second moment of area about the bending axis.

The displacements are used to deduced the moments of the strains which are then used to obtain the stress moments. Specifically,

$$\begin{aligned}
s_x^1(x) &= E_x e_x^1(x) \\
&= E_x \left[I\frac{\partial u_1(x)}{\partial x} - {}_3\bar{e}_x^1(x)\alpha_3(x) - {}_4\bar{e}_x^1(x)\alpha_4(x) \right]
\end{aligned} \tag{19}$$

$$s_{xz}^0(x) = G_{xz}e_{xz}^0(x) = G_{xz}C_{xz}\left\{ 2c\left[u_1(x) + \frac{\partial w_0(x)}{\partial x} \right] - \right.$$
$$\left. 2c\left[{}_3\bar{u}_1(x) + \frac{\partial {}_3\bar{w}_0(x)}{\partial x} \right] - 2c\left[{}_4\bar{u}_1(x) + \frac{\partial {}_4\bar{w}_0(x)}{\partial x} \right] \right\} \tag{20}$$

where, ${}_3\bar{e}_x^1(x)$, ${}_4\bar{e}_x^1(x)$, $\alpha_3(x)$ and $\alpha_4(x)$ are constants.

The above stress moments are substituted into the averaged equilibrium equations, Equation (6), and after some manipulations while observing the boundary conditions yield:

$$u_1(x) = \frac{Q}{2IE_x}(x^2 - 2Lx) + \frac{1}{I}\left[{}_3\bar{e}_x^1(x)\alpha_3(x) + {}_4\bar{e}_x^1(x)\alpha_4(x) \right] x \tag{21}$$

$$w_0(x) = \frac{Q}{2cC_{xz}G_{xz}}x - \frac{Q}{6IE_x}x^3 + \frac{QL}{2IE_x}x^2 + \left[{}_3\bar{u}_1(x) + \frac{\partial {}_3\bar{w}_0(x)}{\partial x} \right]x +$$
$$\left[{}_4\bar{u}_1(x) + \frac{\partial {}_4\bar{w}_0(x)}{\partial x} \right] x - \frac{1}{2I}\left[{}_3\bar{e}_x^1(x)\alpha_3(x) + {}_4\bar{e}_x^1(x)\alpha_4(x) \right] x^2 \tag{22}$$

The fundamental state variables of class 1, $\alpha_1(x)$ and $\alpha_2(x)$ are obtained from the equation

$$\alpha_1(x) = E_x e_x^1(x) \quad \text{and} \quad \alpha_2(x) = G_{xz}e_{xz}^0(x) \tag{23}$$

The distribution of the stresses and strains can now be obtained from Equations (14) and (15) by making the necessary substitutions. The computed homogenized flexural stiffness and the homogenized shear stiffness are exact

Table 1 Summary of the homogenized stiffnesses and the shear correction factors

Theory	100%			10%			1%		
	E_x/E	$G_{xz}A/G_{xz}$	C_{xz}	E_x/E	$G_{xz}A/G_{xz}$	C_{xz}	E_x/E	$G_{xz}A/G_{xz}$	C_{xz}
HBT	169.997	70.135	0.908	132.766	8.473	0.801	129.043	0.871	0.787
TBT	169.997	71.600	0.833	132.766	30.200	0.833	129.043	26.060	0.833

with respect to ANSYS and the hierarchical beam theory, but a 2% error is observed in the shear strain moment correction factor. Table 1 is a tabulation of the prediction of these parameters using both HBT and TBT.

The TBT based homogenized flexural stiffness is defined as the bending moment divided by the beam curvature [2]. Hence the value from HBT is multiplied by the second moment of the cross section, I, to permit a direct/fair comparison. For the same reason, the homogenized shear stiffness predicted by HBT is multiplied by the cross-sectional area of the beam.

The accuracy in TBT predicted shear stiffness value decreases with increasing core flexibility. In fact, HBT and TBT predict identical homogenized shear stiffnesses only when the material is homogenous.

Representative results of the stresses and strains are depicted in Figs. 3 and 5. These results are obtained at the midspan of the beam. The axial stress and strain, σ_x and ϵ_x, show very good agreement with the ANSYS results for both core stiffness.

There is no induced transverse normal stress σ_z and this is reflected in Fig. 4. However a maximum error of 2% is observed between the ANSYS and HBT results within the faces.

The results of the transverse shear stresses and strains are illustrated in Fig. 5. The values predicted by HBT coincide with those of ANSYS, while the profiles of TBT are different and as expected.

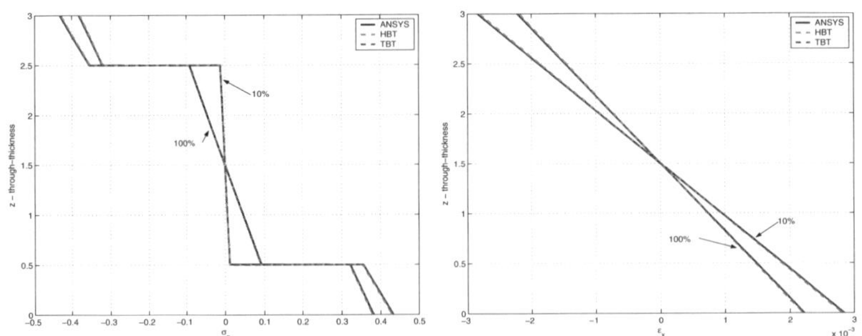

Fig. 3 Through-the-thickness stress and strain, σ_x and ϵ_x, (in Pa) distribution at 50% length for 100% and 10% core stiffness

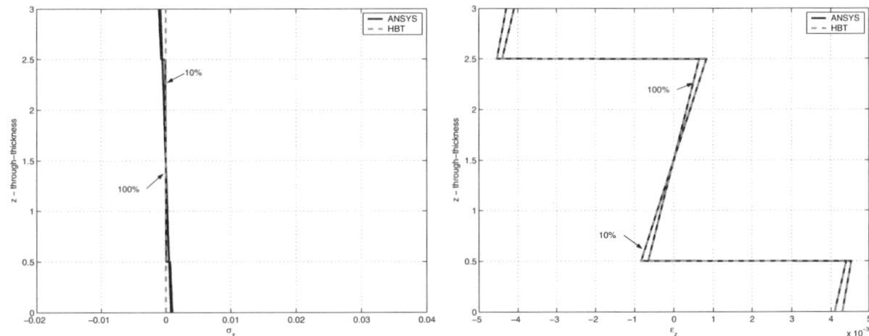

Fig. 4 Through-the-thickness stress and strain, σ_z and ϵ_z, (in Pa) distribution at 50% length for 100% and 10% core stiffness

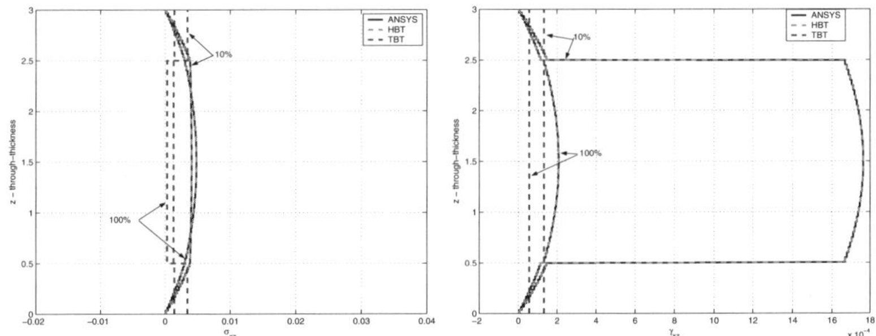

Fig. 5 Through-the-thickness stress and strain, σ_{xz} and γ_{xz}, (in Pa) distribution at 50% length for 100% and 10% core stiffness

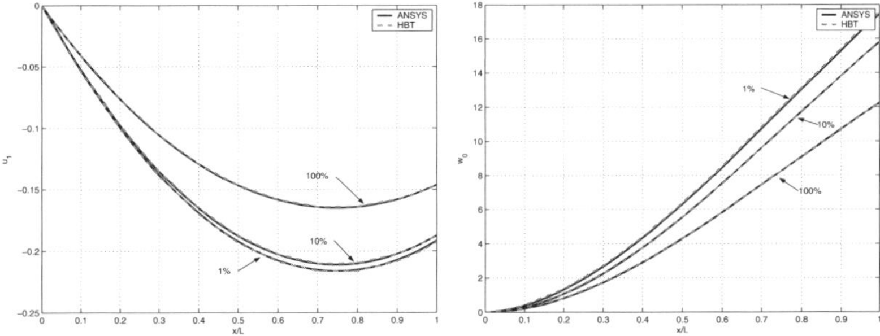

Fig. 6 Distribution of u_1 (in nm) and w_0 (in nm) along the beam length

Finally the plots of the moments of the displacements, u_1 and w_0 are depicted in Fig. 6. An excellent agreement is observed between HBT and ANSYS predictions.

5 Concluding Remarks

The inability of existing beam models to accurately model the through-the-thickness stress and strain distributions is a major hindrance in advancement of the use of advanced piezoelectrically actuated composite and sandwich structures [1]. The hierarchical beam theory (HBT) of Hansen and de Almeida [4] is an internally consistent beam model for accurate prediction of through-the-thickness stresses and strains. The theory is extended in this paper to include piezo-actuation.

The HBT is based on the superposition of fundamental states which are equilibrium solutions that are invariant with respect to the longitudinal axis. The coefficients of the fundamental states are the fundamental variables and they are solely functions of the longitudinal axis. The fundamental variables are not of the same units, but each assumes a unit such that the product with the corresponding fundament state yields the units of stress or strain. The formulation of the theory is such that all but two fundamental variables corresponding to the states of pure bending and pure shear are known a priori from the definition of the problem.

The efficacy of the extended HBT is demonstrated with the analysis of a cantilevered three-layer sandwich beam with piezoelectric outer layers and a shear load at the free end. Excellent agreement is observed with the results obtained using ANSYS.

Acknowledgements

Funding for this work is provided by the Natural Sciences and Engineering Research Council of Canada (NSERC).

References

1. Chopra I (2002) AIAA J 40:2145–2187
2. Shames IH, Dym CL (1985) Energy and finite element methods in structural mechanics. Hemisphere, New York London
3. Icardi U (2003) Applications of zig-zag theories to sandwich beams. Mech Adv Mat Stru 10:77–97
4. Hansen, JS, de Almeida SFM (2001) A theory for laminated composite beams. Technical Report, University of Toronto, Ontario
5. McLean C 2004) A Hierarchical theory for layered beams with piezoelectric actuation. MASc Thesis, University of Toronto, Ontario

Validation of Classical Beam and Plate Models by Variational Convergence

Paolo Podio-Guidugli

Dipartimento di Ingegneria Civile, Università di Roma TorVergata
Viale Politecnico, 1 - 00133 Roma, Italy, ppg@uniroma2.it

Abstract This paper consists of three parts: the first has to do with a method of deduction by scaling of linearly elastic structure models, starting from a displacement formulation of variational equilibrium in three-dimensional linear elasticity; the second part is devoted to elucidating the role of second-gradient elastic energy in the derivation of structure models capable of shearing deformations; in the last part, a validation by Gamma-convergence of the Reissner-Mindlin plate model is offered.

Keywords: Timoshenko beam theory, Reissner-Mindlin plate theory, variational convergence

This writing, expository in nature, consists of three parts. Part I, which is based on joint work with B. Miara [1, 2], is meant to set the stage for the developments in Part III, where the title issue is treated briefly, relaying on joint work with R. Paroni and G. Tomassetti [3, 4]; Part II plays a crucial bridging role. The reader is referred to the cited papers for a more detailed exposition of the matters, including historical remarks and a discussion of motivations.

As is well-known, the most recent techniques of variational convergence have their roots in ideas first put forward by the Italian mathematicians E. De Giorgi and T. Franzoni in 1975 [5]. Roughly speaking, given a parametric family of minimum problems $\min_{u \in X_\varepsilon} F_\varepsilon(u)$, it may happen that a minimum problem $\min_{u \in X} F(u)$ is found, such that minimizers and minima associated with the problems ruled by the functionals F_ε converge to minimizers and minima of the problem ruled by the functional F. If this is the case, the family of functionals F_ε is said to $\Gamma-$converge to the target functional F.

As is also well-known, all classical models from the mechanics of elastic structures can be phrased as minimum problems governed by variational integrals over one- or two-dimensional regions of space. Given anyone of those models, its relationship with three-dimensional elasticity can be established in various ways, one of which is through the notion of Γ−convergence. To do so, one has to find a family of three-dimensional functionals that, as the thickness parameter $\varepsilon \rightarrow 0+$, Γ−converges to a three-dimensional target functional in tight kinship with the lower-dimensional structural model at hand.

The contents of Part I are expedient to suggest what families of functionals and what target functional one should choose to validate variationally all classical beam and plate models, as well as various other models from the theory of linearly elastic structures. The validation process is described in Part III, with reference to a model that has eluded variational hunters so far, the shearable-plate model associated with the names of E. Reissner and R.D. Mindlin. It is suggested in Part II that what makes elusion impossible is to work with an augmented functional including *second-gradient elastic energy*.

PART I. Deduction by Scaling of Linearly Elastic Structure Models

In 1995, I proposed a unified method of deduction of plate and rod theories from three-dimensional linear elasticity, a method based on a formal scaling of both data and unknown in terms of powers of the thickness. That method, first expounded in full in [2], is an evolution of the *method of internal constraints*, that I devised in 1989 to deduce the Germain-Lagrange equation for the bending of a thin plate subject to transverse loadings, and that was later variously generalized. A feature of the *method of formal scaling* presented below is that it yields an exhaustive collection of energy functionals governing the statics of linearly elastic, cylindrical structures.

1 Variational Equilibrium in Three-Dimensional Elasticity

Given the following set of data:

- a *domain* $\mathcal{C}$, a right cylinder $\mathcal{C} \equiv \mathcal{P} \times \mathcal{B}$ (Fig. 1):
- an *elasticity tensor* $\mathbb{C}$, a fourth-order tensor endowed with the standard index symmetries ($\mathbb{C}_{ijhk} = \mathbb{C}_{jihk} = \mathbb{C}_{ijkh} = \mathbb{C}_{hkij}$) and positive definite;
- a field of *distance loads* $\mathbf{d}$, defined over interior part of $\mathcal{C}$;
- a field of *contact loads* $\mathbf{c}$ over $\partial\mathcal{C}$, with $\mathbf{c} = \mathbf{c}^{\pm}$ on top/bottom ends $\mathcal{P}^{\pm}$ of $\mathcal{C}$ and with $\mathbf{c}$ null over $\partial\mathcal{P} \times \mathcal{B}$,

the variational formulation of an equilibrium problem of three-dimensional linear elasticity consists in finding an *elastic state* $(\mathbf{u}, \mathbf{E}, \mathbf{S})$ over the closure of $\mathcal{C}$, with

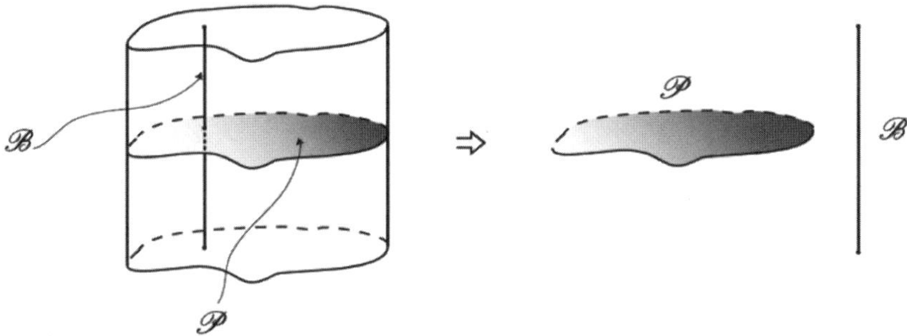

Fig. 1 A right cylinder $\mathcal{C}$ seen as the cartesian product of a *plate-like* region $\mathcal{P}$ and a *beam-like* region $\mathcal{B}$

- $\mathbf{u}$ the *displacement* field,
- $\mathbf{E}(\mathbf{u}) = \operatorname{sym} \nabla \mathbf{u}$ the *strain* field,
- $\mathbf{S}(\mathbf{u}) = \mathbb{C}[\mathbf{E}(\mathbf{u})]$ the *stress* field,

such that

$$\Pi(\mathbf{u}) := \Sigma(\mathbf{u}) - \Delta(\mathbf{u}) \ = \min,$$

where

$$\Sigma(\mathbf{u}) := \frac{1}{2} \int_{\mathcal{C}} \mathbf{S}(\mathbf{u}) \cdot \mathbf{E}(\mathbf{u})$$

is the *energy functional* and

$$\Delta(\mathbf{u}) := \int_{\mathcal{C}} \mathbf{d} \cdot \mathbf{u} + \int_{\mathcal{P}} \mathbf{c}^{\pm} \cdot \mathbf{u}^{\pm}$$

is the *load functional*.

2 Scaling Procedure

A peculiarity of our method is that *the unknown and <u>all</u> data are scaled in terms of powers of the thickness parameter ε, the scaling exponents being relative integers not to be chosen a priori*, but rather to be all determined at the end of the procedure of formal scaling in a manner consistent with one and the same boundedness requirement on the functional ruling the scaled minimum problem.

2.1 Scaling the domain and the displacement

Let *Ref* and *Cur* be two copies of the euclidean point space where the cylinder $\mathcal{C}$ and, respectively, its deformed shape are placed. For $x \in$ *Ref* the referential place of a typical point of $\mathcal{C}$ and for $y \in Cur$ the current place of x, let $\hat{\mathbf{u}}(x) = y - x$ be the relative displacement field. Furthermore, let (x_α, x_3), (y_α, y_3), u_α, u_3 $(\alpha = 1, 2)$ be the cartesian coordinates of x and y and the cartesian components of $\mathbf{u}$.

We scale domain and displacement as follows:

$$x_\alpha = \varepsilon^p \bar{x}_\alpha, \quad x_3 = \varepsilon^q \bar{x}_3;$$
$$u_\alpha = \varepsilon^m \bar{u}_\alpha, \quad u_3 = \varepsilon^n \bar{u}_3. \tag{1}$$

In particular, the fixed cylinder $\mathcal{C}(1)$ is one-to-one mapped onto the cylinder $\mathcal{C}(\varepsilon) = \mathcal{P}(\varepsilon) \times \mathcal{B}(\varepsilon)$.

Remark. For each chosen $\varepsilon > 0$, the (domain,solution) scaling introduced just above may be regarded as a continuous group of linear automorphisms $\mathcal{A}(\varepsilon)$ of $Ref \times Cur$, with ε as the group parameter:

$$\begin{bmatrix} \bar{x} \\ \bar{y} \end{bmatrix} = \mathcal{A}(\varepsilon) \begin{bmatrix} x \\ y \end{bmatrix}, \ \mathcal{A}(\varepsilon) = \begin{bmatrix} \mathbf{A}(\varepsilon) & \mathbf{0} \\ \mathbf{B}(\varepsilon) & \mathbf{C}(\varepsilon) \end{bmatrix}, \ \mathbf{B}(\varepsilon) = \mathbf{A}(\varepsilon) - \mathbf{C}(\varepsilon),$$

with

$$\mathbf{A}(\varepsilon) = diag < \varepsilon^{-p}, \varepsilon^{-p}, \varepsilon^{-q} >, \ \mathbf{C}(\varepsilon) = diag < \varepsilon^{-m}, \varepsilon^{-m}, \varepsilon^{-n} > .$$

2.2 Scaling the strain measure

The strain measure of linear elasticity is

$$\mathbf{E} := \mathrm{sym}\,\mathbf{H}, \quad \mathbf{H} := \nabla \mathbf{u}, \quad H_{ij} = u_{i,j} \ .$$

The (domain,displacement) scaling (1) implies that

$$\mathbf{H} = \varepsilon^{-p+m} \bar{H}_{\alpha\beta} \mathbf{c}_\alpha \otimes \mathbf{c}_\beta + \varepsilon^{-q+n} \bar{H}_{33} \mathbf{c}_3 \otimes \mathbf{c}_3$$
$$+ \varepsilon^{-p+n} \bar{H}_{\alpha3} \mathbf{c}_\alpha \otimes \mathbf{c}_3 + \varepsilon^{-q+m} \bar{H}_{3\alpha} \mathbf{c}_3 \otimes \mathbf{c}_\alpha, \quad \bar{H}_{ij} = \frac{\partial \bar{u}_i}{\partial \bar{x}_j} = \bar{u}_{i,j}.$$

We assume that

$$-p + n = -q + m, \tag{2}$$

whence the scaled strain measure acquires the desired structure:

$$\mathbf{E} = \varepsilon^{\alpha_1} \bar{E}_{\alpha\beta} \mathrm{sym}\,(\mathbf{c}_\alpha \otimes \mathbf{c}_\beta) + \varepsilon^{\alpha_2} \bar{E}_{\alpha3} \mathrm{sym}\,(\mathbf{c}_\alpha \otimes \mathbf{c}_3) + \varepsilon^{\alpha_3} \bar{E}_{33} \mathbf{c}_3 \otimes \mathbf{c}_3,$$

with

$$\alpha_1 = -p + m, \quad \alpha_2 = -p + n, \quad \alpha_3 = -q + n.$$

2.3 Scaling the data

Material moduli. We assume $\mathcal{C}(\varepsilon)$ to be made of a linearly elastic material being *transversely isotropic* with respect to the axial direction:

$$
\begin{aligned}
S_{\alpha\beta} &= 2\mu E_{\alpha\beta} + (\lambda(E_{11} + E_{22}) + \tau_2 E_{33})\delta_{\alpha\beta}, \\
S_{3\alpha} &= 2\gamma E_{3\alpha}, \\
S_{33} &= \tau_1 E_{33} + \tau_2(E_{11} + E_{22}),
\end{aligned}
$$

with

$$
\mu > 0, \ \gamma > 0, \ \tau_1 > 0, \ \tau_1(\lambda + \mu) - \tau_2^2 > 0, \tag{3}
$$

to guarantee positivity of

$$
\sigma = \frac{1}{2} S_{ij} E_{ij},
$$

the *elastic energy* per unit volume of *Ref*.

We scale the material moduli as follows:

$$
\bar{\lambda} = \varepsilon^{-r}\lambda, \ \ \bar{\mu} = \varepsilon^{-r}\mu, \ \ \bar{\gamma} = \varepsilon^{-u}\gamma, \ \ \bar{\tau}_1 = \varepsilon^{-v}\tau_1, \ \ \bar{\tau}_2 = \varepsilon^{-z}\tau_2,
$$

with

$$
r + v - 2z = 0, \tag{4}
$$

so as to give sense to the last of (3) whatever the value of ε.

Loads. We scale the distance loads:

$$
\bar{d}_\alpha = \varepsilon^{-s}b_\alpha, \ \ \bar{d}_3 = \varepsilon^{-t}b_3
$$

and the contact loads:

$$
\bar{c}_\alpha^\pm = \varepsilon^{-y}c_\alpha^\pm, \ \ \bar{c}_3^\pm = \varepsilon^{-w}c_3^\pm \quad \text{(top and bottom)}
$$

under the assumption that

$$
s + q = y, \quad t + q = w \tag{5}
$$

so that, in the *load functional*

$$
\Delta(\mathbf{u}, \varepsilon) := \int_{\mathcal{C}(\varepsilon)} \mathbf{d} \cdot \mathbf{u} + \int_{\mathcal{P}(\varepsilon)} \mathbf{c}^\pm \cdot \mathbf{u}^\pm,
$$

the load measures $d_\alpha u_\alpha d(vol)$ and $c_\alpha^\pm u_\alpha^\pm d(area)$ scale the same, as well as the load measures $d_3 u_3 d(vol)$ and $c_3^\pm u_3^\pm d(area)$.

182   P. Podio-Guidugli

2.4 The scaled energy functional

Summing up, we have introduced 12 scaling exponents:

$$(m, n, p, q; r, u, v, z; s, t, y, w),$$

8 of which are independent, due to (2), (4), and (5). We choose the *scaling list*

$$\ell := (m, n, p; u, v, z; s, t).^{1}$$

Equilibrium of the three-dimensional elastic body occupying the cylinder $\mathcal{C}(\varepsilon)$ is characterized by minimization of the functional

$$\Pi(\mathbf{u}, \varepsilon) = \Sigma(\mathbf{u}, \varepsilon) - \Delta(\mathbf{u}, \varepsilon), \quad \Sigma(\mathbf{u}, \varepsilon) := \int_{\mathcal{C}(\varepsilon)} \frac{1}{2} S_{ij}(\mathbf{E}(\mathbf{u})) E_{ij}(\mathbf{u})$$

Scaling the functional $\Sigma(\mathbf{u}, \varepsilon)$, we obtain the *auxiliary energy functional*

$$\widehat{\Sigma}(\bar{\mathbf{u}}, \varepsilon; \hat{\alpha}, \hat{\beta}_1, \hat{\beta}_2, \hat{\gamma}) = \varepsilon^{\hat{\alpha}} A(\bar{\mathbf{u}}) + \varepsilon^{\hat{\beta}_1} B_1(\bar{\mathbf{u}}) + \varepsilon^{\hat{\beta}_2} B_2(\bar{\mathbf{u}}) + \varepsilon^{\hat{\gamma}} \Gamma(\bar{\mathbf{u}}),$$

where

$$A(\bar{\mathbf{u}}) = \int_{\mathcal{C}(1)} \frac{1}{2} \bar{\tau}_1 \bar{E}_{33}^2 \,,$$

$$B_1(\bar{\mathbf{u}}) = \int_{\mathcal{C}(1)} 2\bar{\gamma}(\bar{E}_{13}^2 + \bar{E}_{23}^2) \,, \quad B_2(\bar{\mathbf{u}}) = \int_{\mathcal{C}(1)} \left(\bar{\tau}_2 \bar{E}_{33} (\bar{E}_{11} + \bar{E}_{22}) \right) ,$$

$$\Gamma(\bar{\mathbf{u}}) = \int_{\mathcal{C}(1)} \frac{1}{2} \left((\bar{\lambda} + 2\bar{\mu})(\bar{E}_{11} + \bar{E}_{22})^2 - 4\bar{\mu}(\bar{E}_{11}\bar{E}_{22} - \bar{E}_{12}^2) \right),$$

and where the list $\hat{\ell} = (\hat{\alpha}, \hat{\beta}_1, \hat{\beta}_2, \hat{\gamma})$ of *energy exponents* is

$$\begin{aligned}
\hat{\alpha} &= -m + 3n + p + v, \\
\hat{\beta}_1 &= m + n + p + u, \\
\hat{\beta}_2 &= m + n + p + z, \\
\hat{\gamma} &= 3m - n + p + 2z - v \,.
\end{aligned}$$

Note that, due to the inequalities in (3), each of the partial energies A, B_1, Γ (but not B_2) is positive. Moreover, note that

$$\hat{\alpha} - 2\hat{\beta}_2 + \hat{\gamma} = 0. \tag{6}$$

Quite similarly, scaling the functional $\Delta(\mathbf{u}, \varepsilon)$, we obtain the *auxiliary load functional*

---

[1] Given ℓ,

$$q = m - n + p, \quad r = 2z - v, \quad y = m - n + p + s, \quad w = y = m - n + p + t.$$

$$\widehat{\Delta}(\bar{\mathbf{u}}, \varepsilon; \hat{\delta}_1, \hat{\delta}_2) = \varepsilon^{\hat{\delta}_1} \Delta_1(\bar{\mathbf{u}}) + \varepsilon^{\hat{\delta}_2} \Delta_2(\bar{\mathbf{u}}),$$

where

$$\Delta_1(\bar{\mathbf{u}}) = \int_{\mathcal{C}(1)} \bar{b}_\alpha \bar{u}_\alpha + \int_{\mathcal{P}(1)} \bar{c}_\alpha^\pm \bar{u}_\alpha^\pm, \quad \Delta_2(\bar{\mathbf{u}}) = \int_{\mathcal{C}(1)} \bar{b}_3 \bar{u}_3 + \int_{\mathcal{P}(1)} \bar{c}_3^\pm \bar{u}_3^\pm,$$

and where the *load exponents* are:

$$\hat{\delta}_1 = m + 2p + q + s, \quad \hat{\delta}_2 = n + 2p + q + t.$$

At this point, we introduce our *central assumption*, the finiteness of the auxiliary functional

$$\widehat{\Pi}(\bar{\mathbf{u}}, \varepsilon; \hat{\alpha}, \ldots, \hat{\delta}_2) = \widehat{\Sigma}(\bar{\mathbf{u}}, \varepsilon; \hat{\alpha}, \hat{\beta}_1, \hat{\beta}_2, \hat{\gamma}) - \widehat{\Delta}(\hat{\mathbf{u}}, \varepsilon; \hat{\delta}_1, \hat{\delta}_2)$$

in the $0-$*thinness limit*. Precisely, we require that *the energy exponents* $\hat{\alpha}, \hat{\beta}_1, \ldots, \hat{\delta}_2$ *be such that the auxiliary functional stay bounded above under the scaling group action*:

$$\lim_{\varepsilon \to 0+} \widehat{\Pi}(\bar{\mathbf{u}}, \varepsilon; \hat{\ell}) = \widehat{\Pi}(\bar{\mathbf{u}}, 0; \hat{\ell}) < +\infty. \tag{A}$$

The consequences of this assumption are detailed in the next section.

3 Taxonomy of Energy Functionals for Linearly Elastic Structures

In the $0-$*thinness limit*,

(i) if anyone of the exponents $\hat{\alpha}, \hat{\beta}_1, \ldots, \hat{\delta}_2$ is chosen positive [null], then the corresponding functional $A, B_1, \ldots, \Delta_2$ drops off [is found in] the de-scaled functional obtained from the limit functional $\widehat{\Pi}(\bar{\mathbf{u}}, 0; \hat{\ell})$;

(ii) if anyone of the energy exponents $\hat{\alpha}, \hat{\beta}_1, \hat{\gamma}$ orderly associated with the constitutively positive functionals A, B_1, Γ is chosen negative, then the corresponding functional must be made to vanish identically by restricting the class of functions on which the relative de-scaled functional is defined.

The reader is referred to [2] for an exhaustive list of limit functionals, each of which, after inverse scaling, is a candidate $\Gamma-$limit for a validation via variational convergence of one or another structure theory. To demonstrate the effectiveness of our approach, it is sufficient to concentrate on the two cases corresponding to the classical theories of *shearable* beams and plates, in both of which $\hat{\beta}_1 = 0$. Our attention will be restricted to the leading quadratic part of the auxiliary functional $\widehat{\Pi}$; as to the linear part accounting for the load potential, it will be safe to think of both $\hat{\delta}_1$ and $\hat{\delta}_2$ equal to null.

184   P. Podio-Guidugli

Axial stretching and flexure of beam-like cylinders. Under the assumption that $\hat{\alpha} = 0$, $\hat{\gamma} < 0$, relation (6) implies that $\hat{\beta}_2 < 0$ as well. Hence, to satisfy assumption (A), the admissible displacement fields must be such that

$$\Gamma(\bar{\mathbf{u}}) \equiv 0, \quad B_2(\bar{\mathbf{u}}) \equiv 0.$$

After inverse scaling, the following *beam functional* is arrived at:

$$\Sigma_B(\mathbf{u}, \varepsilon) = \int_{\mathcal{C}(\varepsilon)} \frac{1}{2} \tau_1 E_{33}^2(\mathbf{u}) + \int_{\mathcal{C}(\varepsilon)} 2\gamma\big(E_{13}^2(\mathbf{u}) + E_{23}^2(\mathbf{u})\big), \tag{7}$$

defined over a class of displacement fields:

$$u_1 = \hat{v}_1(x_3) - x_2\hat{\psi}_3(x_3), \quad u_2 = \hat{v}_2(x_3) + x_1\hat{\psi}_3(x_3), \quad u_3 = \hat{u}_3(x_1, x_2, x_3), \tag{8}$$

whose subcollection

$$u_\alpha = \hat{v}_\alpha(x_3), \quad u_3 = \hat{w}(x_3) + x_2\hat{\psi}_1(x_3) - x_1\hat{\psi}_2(x_3)$$

consists of all *Timoshenko displacement fields*.

In-plane stretching and flexure of plate-like cylinders. By choosing $\hat{\alpha} < 0$, $\hat{\gamma} = 0$, and insisting that assumption (A) is satisfied, the admissible displacement field turn out to be those such that

$$A(\bar{\mathbf{u}}) \equiv 0, \quad B_2(\bar{\mathbf{u}}) \equiv 0.$$

The following *plate functional* is obtained from inverse scaling:

$$\Sigma_P(\mathbf{u}, \varepsilon) = \int_{\mathcal{C}(\varepsilon)} \sigma_P(\mathbf{E}(\mathbf{u})) = \int_{\mathcal{C}(\varepsilon)} 2\gamma\big(E_{13}^2(\mathbf{u}) + E_{23}^2(\mathbf{u})\big)$$

$$+ \int_{\mathcal{C}(\varepsilon)} \frac{1}{2}\Big((\lambda + 2\mu)\big(E_{11}(\mathbf{u}) + E_{22}(\mathbf{u})\big)^2 - 4\mu\big(E_{11}(\mathbf{u})E_{22}(\mathbf{u}) - E_{12}^2(\mathbf{u})\big)\Big),$$

$$\tag{9}$$

defined over the displacement class

$$u_\alpha = \hat{u}_\alpha(x_1, x_2, x_3), \quad u_3 = \hat{u}_3(x_1, x_2), \tag{10}$$

whose subcollection

$$u_\alpha = \hat{v}_\alpha(x_1, x_2) + x_3\hat{\varphi}_\alpha(x_1, x_2), \quad u_3 = \hat{u}_3(x_1, x_2). \tag{11}$$

consists of all *Reissner-Mindlin displacement fields*.

Remark. Taking $\hat{\beta}_1 < 0$ opens the way to theories of *unsherable* beams and plates. In fact, this assumption implies that

$$E_{\alpha3}(\mathbf{u}) = 0 \quad (\alpha = 1, 2) \quad \text{in } \mathcal{C}(\varepsilon).$$

With this, (9) reduces to the functional

$$\widetilde{\Sigma}_P(\mathbf{u},\varepsilon) = \int_{\mathcal{C}(\varepsilon)} \frac{1}{2}\Big((\lambda+2\mu)\big(E_{11}(\mathbf{u})+E_{22}(\mathbf{u})\big)^2 - 4\mu\big(E_{11}(\mathbf{u})E_{22}(\mathbf{u})-E_{12}^2(\mathbf{u})\big)\Big),$$

and (10) reduces to the well-known *Kirchhoff-Love representation* of the displacement field in a plate:

$$\mathbf{u}_{KL} = \mathbf{v} - x_3\nabla u_3 + u_3\mathbf{e}_3, \quad \mathbf{v} = \hat{v}_\alpha(x_1,x_2)\mathbf{e}_\alpha, \quad u_3 = \hat{u}_3(x_1,x_2).$$

Quite similarly, the functional (7) reduces to

$$\widetilde{\Sigma}_B(\mathbf{u},\varepsilon) = \int_{\mathcal{C}(\varepsilon)} \frac{1}{2}\tau_1 E_{33}^2(\mathbf{u}),$$

and (8) reduces to the *Bernoulli-Navier representation* for the displacement field.

PART II. The Role of Second-Gradient Elastic Energy in the Derivation of Shearable Structure Models

Had the limit procedure introduced in Part I produced not only the 'right' functional but also the 'right' displacement class, the way to validation by variational convergence would be paved. Indeed, it is a matter of a straightforward calculation to check that

$$\Sigma_P(\mathbf{u}_{RM},\varepsilon) = \Sigma_{RM}(w,\varphi,\varepsilon),$$

with Σ_{RM} the 'two-dimensional' energy functional on which the Reissner-Mindlin theory of shearable plates is based (see definition (12), Section 5). Luckily, as we are going to see here below, a remedy to this incomplete achievement can be found, of such a nature as to explain why we stated that the method of formal scaling is an evolution of the method of internal constraints.

4 Some Facts and a Conjecture

As early as in 2003, I noticed that the Reissner-Mindlin [Timoshenko] displacement field is the general solution of the following systems of PDEs:

$$u_{3,3} = 0, \quad u_{\alpha,33} = 0 \quad [u_{\alpha,\beta} + u_{\beta,\alpha} = 0, \quad u_{3,\alpha\beta} + u_{\alpha,3\beta} = 0].$$

I reasoned that both PDEs, no matter their order, may be interpreted as internal constraints, i.e., as kinematic restrictions on admissible motions to be maintained by suitable fields of reactive stresses; and that assumption (A) would restitute the 2nd-order constraints as well, provided that the energy functional were made to include the additional elastic energy associated with the 2nd-order displacement gradient $\nabla^{(2)}\mathbf{u}$.

To implement this idea in the simplest manner, I considered the *augmented energy functional*

$$\Sigma_a(\mathbf{u},\varepsilon) = \Sigma(\mathbf{u},\varepsilon) + \int_{\mathcal{C}(\varepsilon)} \frac{1}{2}\Big(\tau_P \sum_\alpha (u_{\alpha,33})^2 + \tau_B \sum_{\alpha,\beta}(u_{3,\alpha\beta} + u_{\alpha,3\beta})^2\Big),$$

and scaled the 2nd-order material moduli τ_P and τ_B (both taken positive) as follows:

$$\bar\tau_P = \varepsilon^{-i_1}\tau_P, \quad \bar\tau_B = \varepsilon^{-i_2}\tau_B;$$

so as to arrive at the auxiliary functional

$$\widehat\Sigma_a(\bar{\mathbf{u}},\varepsilon) := \widehat\Sigma(\bar{\mathbf{u}},\varepsilon) + \varepsilon^{\iota_1} I_1(\bar{\mathbf{u}}) + +\varepsilon^{\iota_2} I_2(\bar{\mathbf{u}}),$$

where

$$\iota_1 = 2(m-p) + q + i_1, \quad \iota_2 = 2(n-p) + q + i_2.$$

Clearly, on assuming that $\iota_1 < 0$, $\iota_2 > 0$ [$\iota_1 > 0$, $\iota_2 < 0$], the $0-$thinness limit yields the Reissner-Mindlin plate theory in the case when $\hat\alpha = \hat\beta_1 = 0$, $\hat\gamma < 0$ [the Timoshenko beam theory in the case when $\hat\alpha < 0$, $\hat\beta_1 = \gamma = 0$] (see [2]).

Prompted by these facts, I conjectured that failure to see the role of second-gradient elastic energy was the reason why, to date, neither asymptotic nor variational convergence methods had been expedient to validate any model of *shearable* structure.

PART III. Validation by Gamma-Convergence of the Reissner-Mindlin Plate Model

That the second-gradient energy conjecture holds true has been proved indirectly in [4], where the Reissner-Mindlin plate theory is shown deducible by $\Gamma-$convergence from linear three-dimensional elasticity in the presence of second-gradient energy terms. We proceed to summarize the main developments of [4].

5 The Target Theory in a Nutshell

For $\mathcal{P}$ a flat region of effective thickness $\varepsilon(2h)$ (here ε is regarded as a bookkeeping parameter), the *energy functional of Reissner-Mindlin plates* is:

$$\Sigma_{RM}(w,\varphi,\varepsilon) =$$
$$\int_{\mathcal{P}} \frac{1}{2}\Big(S(\varepsilon)|\nabla w + \varphi|^2 + B(\varepsilon)\big((1-\nu)|\mathrm{sym}\,\nabla\varphi|^2 + \nu(\mathrm{tr}(\nabla\varphi)^2)\big)\Big), \tag{12}$$

where

$$S(\varepsilon) = \varepsilon(2h\gamma) \quad \text{and} \quad B(\varepsilon) = \varepsilon^3\left(\frac{2}{3}h^3\frac{E}{1-\nu^2}\right)$$

are the *shearing* and *bending stiffness*, respectively, and where E and ν are the Young and Poisson moduli.

Basic to the interpretation and use of the Reissner-Mindlin theory as an approximation of the mechanical behavior of thin plate-like elastic bodies is the *kinematical Ansatz* that constructs the three-dimensional displacement field $\mathbf{u}_{RM}$ corresponding to an equilibrium solution (w, φ) of the minimum problem ruled by the functional (12), namely,

$$\mathbf{u}_{RM} = x_3\varphi + w\mathbf{e}_3, \quad \varphi = \hat{\varphi}_\alpha(x_1, x_2)\mathbf{e}_\alpha, \quad w = \hat{w}(x_1, x_2), \quad \text{in } \mathcal{C}(\varepsilon) = \mathcal{P} \times \varepsilon(2h)$$

(cf. (11)).

6 The Target Theorem

Let

$$\widehat{\Sigma}_{RM}(\bar{\mathbf{u}}, \varepsilon) = \int_{\mathcal{C}} \sigma_\varepsilon^{(1)}(\bar{\mathbf{E}}) + \sigma_\varepsilon^{(2)}(\bar{\mathbf{G}}), \quad \bar{\mathbf{E}} = \operatorname{sym} \nabla\bar{\mathbf{u}}, \quad \bar{\mathbf{G}} = \nabla(\nabla\bar{\mathbf{u}}),$$

where $\mathcal{C} = \mathcal{P} \times (2h)$ and where

$$\sigma_\varepsilon^{(1)}(\bar{\mathbf{E}}) = 2\bar{\gamma}(\bar{E}_{13}^2 + \bar{E}_{23}^2) + \frac{1}{2}(\bar{\lambda} + 2\bar{\mu})[(\bar{E}_{11} + \bar{E}_{22})^2 - \frac{4\bar{\mu}}{\bar{\lambda} + 2\bar{\mu}}(\bar{E}_{11}\bar{E}_{22} - \bar{E}_{12}^2)]$$

$$+ \varepsilon^{-2}\left[\frac{1}{2}\bar{\tau}_1(\bar{E}_{33})^2\right] + \varepsilon^{-1}[\bar{\tau}_2(\bar{E}_{11} + \bar{E}_{22})\bar{E}_{33}],$$

$$\sigma_\varepsilon^{(2)}(\bar{\mathbf{G}}) = \varepsilon^{-2}\left[\frac{1}{2}\bar{\tau}_P \sum_\alpha \bar{G}_{\alpha 33}^2\right], \quad \bar{G}_{\alpha 33} = \bar{u}_{\alpha,33}.$$

Note that

$$\sigma^{(1)}(\mathbf{E}) = \sigma_P(\mathbf{E}),$$

the $\mathbf{e}_3-$*transversely isotropic* energy density in (9), positive for

$$\bar{\mu} > 0, \quad \bar{\tau}_1 > 0, \quad \tilde{\lambda} + \bar{\mu} > 0, \quad \bar{\tau}_3 > 0, \quad \text{where } \tilde{\lambda} := \bar{\lambda} - \bar{\tau}_2^2/\bar{\tau}_1 \,.$$

Note also that the technical moduli

$$\bar{E} = 4\bar{\mu}\,\frac{\tilde{\lambda} + \bar{\mu}}{\tilde{\lambda} + 2\bar{\mu}}, \quad \bar{\nu} = \frac{\tilde{\lambda}}{\tilde{\lambda} + 2\bar{\mu}}$$

are in 1–1 correspondence with $(\tilde{\lambda}, \bar{\mu})$. For

$$\tilde{\sigma}(\bar{\mathbf{E}}) := \min_{\rho \in \mathbf{R}} \sigma_P(\mathbf{E} + \rho\,\mathbf{e}_3 \otimes \mathbf{e}_3),$$

it can be shown that

$$\tilde{\sigma}(\bar{\mathbf{E}}) = 2\bar{\gamma}(\bar{E}_{13}^2 + \bar{E}_{23}^2) + \frac{1}{2}(\tilde{\lambda} + 2\bar{\mu})[(\bar{E}_{11} + \bar{E}_{22})^2 - \frac{4\bar{\mu}}{\tilde{\lambda} + 2\bar{\mu}}(\bar{E}_{11}\bar{E}_{22} - \bar{E}_{12}^2)].$$

In preparation for the convergence theorem, that will be taken *verbatim* from [4], let the functional

$$\widetilde{\Sigma}_{RM}(\bar{\mathbf{u}}) = \int_{\mathcal{C}} \tilde{\sigma}(\mathbf{E}(\bar{\mathbf{u}}))$$

be defined over the Riessner-Mindlin displacement class

$$\mathcal{RM} = \{\bar{\mathbf{u}} \mid \bar{u}_{3,3} = 0 \;\&\; \bar{u}_{\alpha,33} = 0\}.$$

Theorem. *The* $\varepsilon-$*family* $\widehat{\Sigma}_{RM}(\cdot, \varepsilon)$ *sequentially* Γ-*converges to* $\widetilde{\Sigma}_{RM}$ *with respect to the* $L^2(\mathcal{C})$ *topology, i.e.,*

1. [LIMINF INEQUALITY] for every sequence $\{\varepsilon_k > 0\} \to 0$ and for every pair of a sequence $\{\bar{\mathbf{u}}^\varepsilon\} \subset L^2(\mathcal{C}, \mathbb{R}^3)$ and a field $\bar{\mathbf{u}} \in L^2(\mathcal{C}, \mathbb{R}^3)$ such that $\bar{\mathbf{u}}^\varepsilon \to \bar{\mathbf{u}}$ in $L^2(\mathcal{C}, \mathbb{R}^3)$,

$$\liminf_{\varepsilon \to 0} \widehat{\Sigma}_{RM}(\bar{\mathbf{u}}^\varepsilon, \varepsilon) \geq \widetilde{\Sigma}_{RM}(\bar{\mathbf{u}});$$

2. [RECOVERY SEQUENCE] for every sequence $\{\varepsilon_k > 0\} \to 0$ and for every field $\bar{\mathbf{u}} \in L^2(\mathcal{C}, \mathbb{R}^3)$, there is a recovery sequence $\{\bar{\mathbf{u}}^{\varepsilon_k}\} \subset L^2(\mathcal{C}, \mathbb{R}^3)$ such that $\bar{\mathbf{u}}^{\varepsilon_k} \to \bar{\mathbf{u}}$ in $L^2(\mathcal{C}, \mathbb{R}^3)$ and

$$\limsup_{\varepsilon_k \to 0} \widehat{\Sigma}_{RM}(\bar{\mathbf{u}}^{\varepsilon_k}, \varepsilon_k) \leq \widetilde{\Sigma}_{RM}(\bar{\mathbf{u}}).$$

References

1. Miara B, Podio-Guidugli P (2006) C.R. Acad. Sci. Paris, Ser. I 343: 675–678
2. Miara B, Podio-Guidugli P (2007) Asymptotic Analysis 51: 113–131
3. Paroni R, Podio-Guidugli P, Tomassetti G (2006) C.R. Acad. Sci. Paris, Ser. I 343: 437–440
4. Paroni R, Podio-Guidugli P, Tomassetti G (2007) Analysis and Applications 5: 165–182
5. De Giorgi E, Franzoni T (1975) Atti Accad. Naz. Lincei Rend. Cl. Sci. Mat. Fis. Nat. 58: 842–850

On the Simulation of Textile Reinforced Concrete Layers by a Surface-Related Shell Formulation

Rainer Schlebusch and Bernd W. Zastrau

Technische Universität Dresden, Faculty of Civil Engineering,
Institute of Mechanics and Shell Structures, D-01062 Dresden, Germany,
`Rainer.Schlebusch@tu-dresden.de`

Abstract The solution of structural analysis problems, especially of shell structures, demands an efficient numerical solution strategy. Since unilateral contact problems are investigated, the shell model is formulated with respect to one of the outer surfaces, i.e., the shell formulation is surface-related. In particular, the investigation of textile reinforced strengthening layers will be carried out by this approach.

The presented shell formulation assumes linear shell kinematics with six displacement parameters. This low-order shell kinematics produces parasitical strains and stresses, leading to poor approximations of the solution or even useless results. Therewith, extensions and/or adjustments of well-known techniques to prevent or reduce locking like the assumed natural strain (ANS) method, proposed in [14], and the enhanced assumed strain (EAS) method, suggested in [13], have to be performed. The effectiveness of the surface-related solid-shell element is finally demonstrated by a numerical example.

Keywords: surface-related shell formulation, solid-shell element, locking phenomena

1 Introduction

The mechanical simulation of textile reinforced fine-grained concrete strengthening layers is one topic of the Collaborative Research Center (CRC) 528 "Textile Reinforcement for Structural Strengthening and Repair" at the Technische Universität Dresden. Therefore, suitable mechanical models need to be established capable of simulating static and dynamic applications. Basic features of textile reinforced concrete as construction material are presented in [5, 8]. Since the strengthening layer is very thin compared to the existing structure, its mechanical simulation is favorably based on a shell formulation. With respect to the aforementioned framework of application, it is taken advantage

G. Jaiani, P. Podio-Guidugli (eds.), *IUTAM Symposium on Relations of Shell, Plate, Beam, and 3D Models*, © Springer Science+Business Media B.V. 2008

of the, in principle, free choice of the reference surface position by attaching the reference surface to one of the outer surfaces.

The application of the well established degeneration concept only does not permit the use of three-dimensional constitutive relations, if for the semi-discretization of the displacement field a linear series expansion in thickness-direction is used. But in combination with the EAS method as found in [4,13], its application becomes possible. This combination comprises a shell formulation with a minimal number of kinematical degrees of freedom to operate with completely three-dimensional constitutive relations.

2 Theoretical Formulation

In this section, the governing equations, i.e., the strong form of the boundary value problem, and the corresponding weak form represented by the generalized HU–WASHIZU principle are introduced as the three-dimensional foundation of the surface-related shell formulation.

2.1 Governing Equations – Strong Form

Since shells are three-dimensional bodies, the field equations of continuum mechanics are the starting point of the following considerations. They are found in many textbooks and read for the reference configuration $\mathcal{B}_{\underline{t}}$ with boundary $\partial\mathcal{B}_{\underline{t}}$ as follows:

$$\mathrm{Div}(\mathbf{F}^{\mathbf{U}} \cdot \mathbf{S}) + \rho_{\underline{t}}\mathbf{f} = \mathbf{0} \quad \forall \mathbf{X} \in \mathcal{B}_{\underline{t}} \,, \tag{1}$$

$$\mathbf{S} - \mathbf{S}^T = \mathbf{0} \quad \forall \mathbf{X} \in \mathcal{B}_{\underline{t}} \,, \tag{2}$$

$$\mathbf{S} - \mathbf{S}^{\mathbf{E}} = \mathbf{0} \quad \forall \mathbf{X} \in \mathcal{B}_{\underline{t}} \,, \tag{3}$$

$$\mathbf{E} - \mathbf{E}^{\mathbf{U}} = \mathbf{0} \quad \forall \mathbf{X} \in \mathcal{B}_{\underline{t}} \,, \tag{4}$$

wherein $\mathbf{X}$ is the position vector of a material particle in $\mathcal{B}_{\underline{t}}$. The strain-compatible second PIOLA–KIRCHHOFF stress tensor $\mathbf{S}^{\mathbf{E}}$ in (3) is given by the constitutive equation:

$$\mathbf{S}^{\mathbf{E}} := \mathbf{S}(\mathbf{E}) \quad \forall \mathbf{X} \in \mathcal{B}_{\underline{t}} \,, \tag{5}$$

the displacement-compatible GREEN–LAGRANGE strain tensor $\mathbf{E}^{\mathbf{U}}$ in (4) is defined by the kinematical equation:

$$\mathbf{E}^{\mathbf{U}} := \frac{1}{2}\left((\mathbf{F}^{\mathbf{U}})^T \cdot \mathbf{F}^{\mathbf{U}} - \mathbf{G}\right) \quad \forall \mathbf{X} \in \mathcal{B}_{\underline{t}} \,, \tag{6}$$

wherein the displacement-compatible deformation gradient $\mathbf{F}^{\mathbf{U}} := \mathbf{G} + \mathbf{H}^{\mathbf{U}}$ and, finally, the corresponding displacement gradient $\mathbf{H}^{\mathbf{U}} := \mathrm{Grad}\,\mathbf{U}$ are introduced. For sake of completeness the pertinent boundary conditions of the boundary value problem are also stated:

$$\widehat{\mathbf{U}} - \mathbf{U} = \mathbf{0} \quad \forall \mathbf{X} \in \partial_{\mathbf{U}}\mathcal{B}_{\underline{t}} \quad \text{and} \quad \widehat{\mathbf{t}}_0 - \mathbf{F}^{\mathbf{U}} \cdot \mathbf{S}^{\mathbf{E}} \cdot \mathbf{N} = \mathbf{0} \quad \forall \mathbf{X} \in \partial_{\mathbf{t}}\mathcal{B}_{\underline{t}} \,. \tag{7}$$

Herein are: $\partial \mathcal{B}_{\underline{t}} = \partial_{\mathbf{U}}\mathcal{B}_{\underline{t}} \cup \partial_{\mathbf{t}}\mathcal{B}_{\underline{t}}$ and $\partial_{\mathbf{U}}\mathcal{B}_{\underline{t}} \cap \partial_{\mathbf{t}}\mathcal{B}_{\underline{t}} = \varnothing$. In the further treatment, the following quantities are used additionally: $\mathbf{G}$ the metric tensor of the reference configuration, $\mathbf{U}$ the displacement field, $\mathbf{E}$ the GREEN–LAGRANGE strain tensor, $\mathbf{S}$ the second PIOLA–KIRCHHOFF stress tensor, $\rho_{\underline{t}}$ the mass density in the reference configuration, $\mathbf{f}$ prescribed volume forces, $\widehat{\mathbf{t}}_0$ prescribed surface tractions and $\widehat{\mathbf{U}}$ prescribed displacements. Complete solutions of the set of differential equations (1)–(4) with pertinent boundary conditions (7) are difficult to gain. An efficient numerical approximations of the solution of this boundary value problem is achieved more advantageously, if the problem is reformulated using variational formalism.

2.2 Governing Equations – Weak Form

The weak formulation of the governing equations is accomplished by a standard procedure and leads to the following unrestricted abstract variational formulation corresponding to the generalized HU–WASHIZU principle [15]:
Find the primal variables:

$$(\mathbf{U}, \widetilde{\mathbf{E}}) \in \mathcal{X}_1 \times \mathcal{X}_2 = \mathbb{H}^1(\mathcal{B}_{\underline{t}}, \mathcal{E}^3) \times L_2(\mathcal{B}_{\underline{t}}, \mathcal{E}^3 \otimes \mathcal{E}^3) \tag{8}$$

and the dual variables:

$$(\mathbf{t}_0, \mathbf{S}) \in \mathcal{M}_1 \times \mathcal{M}_2 = L_2(\mathcal{B}_{\underline{t}}, \mathcal{E}^3) \times L_2(\mathcal{B}_{\underline{t}}, \mathcal{E}^3 \otimes \mathcal{E}^3) \,, \tag{9}$$

such that:

$$a(\mathbf{U}, \widetilde{\mathbf{E}}; \delta\mathbf{U}, \delta\widetilde{\mathbf{E}}) + b_1(\delta\mathbf{U}, \mathbf{t}_0) + b_2(\delta\widetilde{\mathbf{E}}, \mathbf{S}) = \mathcal{F}(\delta\mathbf{U}, \delta\widetilde{\mathbf{E}}) \quad \forall(\delta\mathbf{U}, \delta\widetilde{\mathbf{E}}) \in \mathcal{X}_1 \times \mathcal{X}_2 \,, \tag{10}$$

$$b_1(\mathbf{U}, \delta\mathbf{t}_0) = \mathcal{G}_1(\delta\mathbf{t}_0) \quad \forall\delta\mathbf{t}_0 \in \mathcal{M}_1 \,, \tag{11}$$

$$b_2(\widetilde{\mathbf{E}}, \delta\mathbf{S}) = \mathcal{G}_2(\delta\mathbf{S}) \quad \forall\delta\mathbf{S} \in \mathcal{M}_2 \tag{12}$$

with the residuum of the kinematical field equations:

$$\widetilde{\mathbf{E}} := \mathbf{E} - \mathbf{E}^{\mathbf{U}} \,, \tag{13}$$

the semi-linear form:

$$a(\mathbf{U}, \widetilde{\mathbf{E}}; \delta\mathbf{U}, \delta\widetilde{\mathbf{E}}) = \int_{\mathcal{B}_{\underline{t}}} \mathbf{S}^{\mathbf{E}} : (\delta\widetilde{\mathbf{E}} + D\mathbf{E}^{\mathbf{U}}(\mathbf{U})[\delta\mathbf{U}])\mathcal{D}V, \tag{14}$$

the bi-linear forms:

$$b_1(\delta\mathbf{U}, \mathbf{t}_0) = -\int_{\partial_{\mathbf{U}}\mathcal{B}_{\underline{t}}} \mathbf{t}_{\underline{t}} \cdot \delta\mathbf{U}\mathcal{D}A \tag{15}$$

$$b_2(\delta\widetilde{\mathbf{E}}, \mathbf{S}) = -\int_{\mathcal{B}_{\underline{t}}} \mathbf{S} : \delta\widetilde{\mathbf{E}}\mathcal{D}V \tag{16}$$

and the linear forms:

$$\mathcal{F}(\delta\mathbf{U}, \delta\widetilde{\mathbf{E}}) = \int_{\mathcal{B}_{\underline{t}}} \rho_{\underline{t}}\mathbf{f} \cdot \delta\mathbf{U}\mathcal{D}V + \int_{\partial_{\mathbf{t}}\mathcal{B}_{\underline{t}}} \widehat{\mathbf{t}}_0 \cdot \delta\mathbf{U}\mathcal{D}A , \tag{17}$$

$$\mathcal{G}_1(\delta\mathbf{t}_0) = -\int_{\partial_{\mathbf{U}}\mathcal{B}_{\underline{t}}} \widehat{\mathbf{U}} \cdot \delta\mathbf{t}_0\mathcal{D}A , \tag{18}$$

$$\mathcal{G}_2(\delta\mathbf{S}) = \quad 0. \tag{19}$$

Details concerning the definitions of the functional spaces $\mathbb{H}^1$ and L_2 used can be found e.g. in [3]. The accomplished re-parametrization using the residuum of the kinematical field equation $\widetilde{\mathbf{E}}$ instead of the tensor $\mathbf{E}$ in the variational formulation was suggested in [13]. Further, following [13], a L_2 orthogonality between the second PIOLA–KIRCHHOFF stress tensor $\mathbf{S}$ and the residuum of the kinematical field equation $\widetilde{\mathbf{E}}$ is enforced. This procedure results in a modified stationarity condition as represented in the following modified abstract variational formulation:

Find:

$$(\mathbf{U}, \widetilde{\mathbf{E}}) \in \mathcal{X}_1 \times \mathcal{X}_2 = \mathbb{H}^1(\mathcal{B}_{\underline{t}}, \mathcal{E}^3) \times L_2(\mathcal{B}_{\underline{t}}, \mathcal{E}^3 \otimes \mathcal{E}^3) , \tag{20}$$

$$(\mathbf{t}_0, \mathbf{S}) \in \mathcal{M}_1 \times \mathcal{M}_2 = L_2(\mathcal{B}_{\underline{t}}, \mathcal{E}^3) \times L_2(\mathcal{B}_{\underline{t}}, \mathcal{E}^3 \otimes \mathcal{E}^3) , \tag{21}$$

such that:

$$a(\mathbf{U}, \widetilde{\mathbf{E}}; \boldsymbol{\eta}, \delta\widetilde{\mathbf{E}}) + b_1(\boldsymbol{\eta}, \mathbf{t}_0) = \mathcal{F}(\boldsymbol{\eta}, \delta\widetilde{\mathbf{E}}) \quad \forall(\boldsymbol{\eta}, \delta\widetilde{\mathbf{E}}) \in \mathcal{X}_1 \times \mathcal{X}_2 , \tag{22}$$

$$b_1(\mathbf{U}, \delta\mathbf{t}_0) = \mathcal{G}_1(\delta\mathbf{t}_0) \quad \forall\delta\mathbf{t}_0 \in \mathcal{M}_1 \tag{23}$$

and the orthogonality condition:

$$\int_{\mathcal{B}_{\underline{t}}} \mathbf{S} : \widetilde{\mathbf{E}}\,\mathcal{D}V = 0 \tag{24}$$

is fulfilled.

Therewith, the stress tensor $\mathbf{S}$ is formally eliminated from the formulation and, consequently, it is not necessary to interpolate $\mathbf{S}$ in the finite element formulation explicitly. Furthermore, the residuum of the kinematical field equation $\widetilde{\mathbf{E}}$ in the form $\mathbf{E} = \mathbf{E}^{\mathbf{U}} + \widetilde{\mathbf{E}}$ is reinterpreted as an additional strain tensor, being used to enhance the displacement-compatible strain tensor $\mathbf{E}^{\mathbf{U}}$ in the discretized problem. Therefore, $\widetilde{\mathbf{E}}$ is called independent enhanced assumed strain tensor. The tensor $\widetilde{\mathbf{E}}$ is used to reduce several artificial stiffening effects stated in Chap. 3. The fulfilment of the orthogonality condition (24) demands specific considerations, published in [12], since the shell formulation is a surface-related one.

2.3 Surface-Related Shell Formulation

Every shell formulation is an approximation of a three-dimensional continuum formulation requiring meaningful assumptions and hypotheses on the involved fields, especially on the displacement field. Here, the displacement field $\mathbf{U}$ is restricted by the following kinematical assumption:

$$\mathbf{U} = \mathbf{V} + \Theta^3 \mathbf{W} . \tag{25}$$

Herein, $\mathbf{V}$ is the displacement field of the reference surface, i.e., of one of the outer surfaces, and $\mathbf{W}$ describes the movement of the so-called shell direc-tor defining the normal expansion of the shell. Corresponding to the particular position of the reference surface, it follows for the thickness coordinate Θ^3 to be in the interval $[0, 1]$. The main disadvantage of this particular shell kine-matics is that it suffers from so-called POISSON thickness locking [4, 11, 16] and is handled by an EAS enhancement.

3 Numerical Solution

The discretization of the variational formulation in the sense of the finite element method is another source of several locking phenomena. In particular, this holds for the low-order interpolation of the developed four-node solid-shell element depicted in Fig. 1. Of course, the element is able to represent several

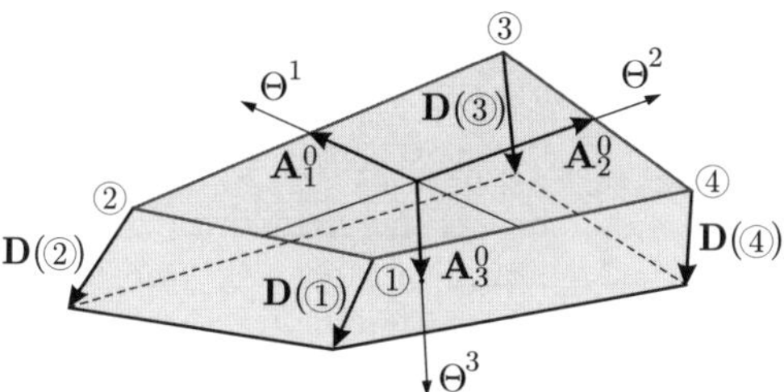

Fig. 1 Illustration of the new four-node surface-related solid-shell element

deformation modes, but the number is always limited the number of degrees of freedom of the finite element.

Probably, the major drawback of the chosen linear shell kinematics (25) for simulating thin structures is the POISSON thickness locking in bending domi-nated problems as stated above, since the kinematics produces only constant transverse strains. The necessary introduction of linear strains is achieved by the aid of the EAS method. This formulation was originally proposed by [4] and has to be adapted to the special position of the reference surface, see [12].

The following locking phenomena have to be expected additionally to POISSON thickness locking:

- transverse shear locking (out-of-plane)
- curvature thickness locking

- membrane locking
- shear locking (in-plane)
- volume locking

An effective concept against transverse shear locking, see [1, 6, 7, 9], and against curvature thickness locking is given by the ANS method [2]. To prevent or reduce membrane, shear and volume locking, the EAS method is also used. All details for the necessary extensions and/or adjustments of these methods can be found in [10].

4 Numerical Example

The following numerical example demonstrates explicitly the effectiveness of the extended and/or adjusted methods. For later reference, several cases of possible enhancements and/or modifications are distinguished and given in Table 1.

Table 1 Definition of distinguished element formulations

case ⓐ:	pure displacement solid-shell element, no modifications or enhancements
case ⓑ:	case ⓐ and ANS modifications against transverse shear locking
case ⓒ:	case ⓑ and ANS modifications against curvature thickness locking
case ⓓ:	case ⓒ and EAS enhancements against POISSON thickness locking

The material properties, the geometrical data and the loads, being distinguished in the numerical example by case ① and ②, are given in Fig. 2. Since only the ratios between the given quantities are important, it is adequate to use only quantities without dimensions. However, the reader may

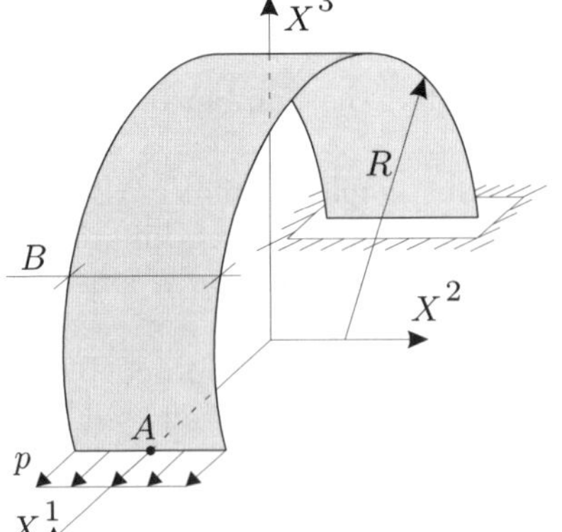

Fig. 2 Semi-circle arc (case ①) or shell (case ②)

assign the dimensions dm and N to all quantities in an appropriate manner. This selection corresponds, e.g., to lead or wood. The investigated arcs have the radii $R = 10.19$ and $R = 101.9$. In this example, linearly elastic, isotropic material behavior is assumed, characterized by YOUNG's modulus $E = 10^8$ and POISSON's ratio $\nu = 0.0$. The system is discretized with a maximum of 32 elements in the circumferential direction and one element in width direction. The displacement of point A in the middle of the free end, see Fig. 2, in vertical X^3-direction is presumed to be a characteristic quantity. The numerical results of the geometrically linear simulations for cases ① and ② are depicted in Figs. 3 and 4, respectively, and for the geometrically nonlinear simulations in Figs. 5 and 6, respectively.

The analytical reference solutions of the geometrically linear simulations are 2.53636 in case ① and 25.3636 in case ②. The reference solutions of the geometrically nonlinear simulations are 2.02682 and 19.8678. They are obtained with a discretization by 128 shell elements in circumferential direction. The reference solutions of the geometrically linear simulations can be obtained exactly by discretization with 128 shell elements. Since POISSON's ratio is set equal to zero in the first instance, POISSON thickness locking does not occur.

At first, the results of the geometrically linear simulations are discussed. The pure displacement finite element behaves, as expected, too stiff in case ① and gives too small displacements with a difference of about -31.6% from

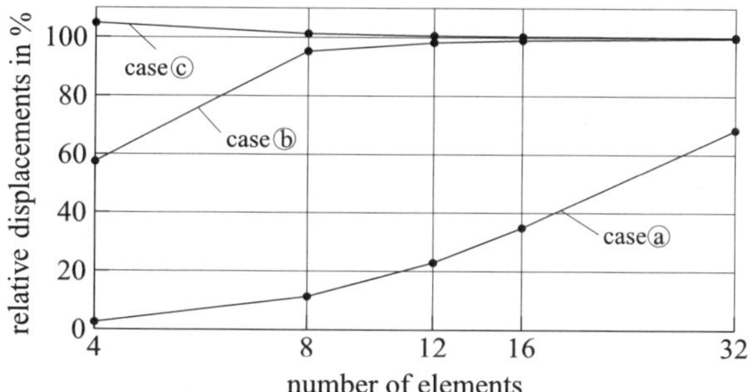

Fig. 3 Results of the geometrically linear simulations in case ①

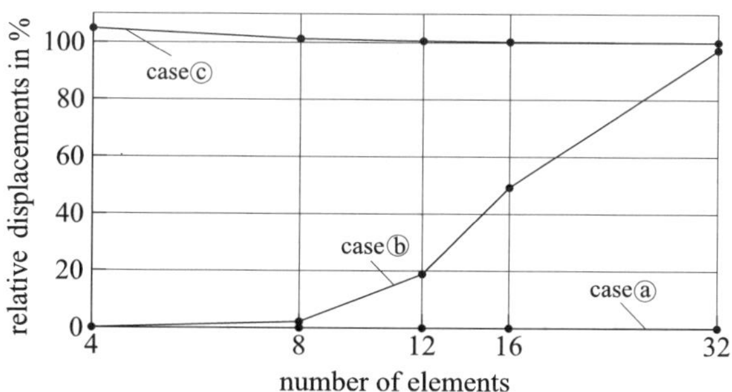

Fig. 4 Results of the geometrically linear simulations in case ②

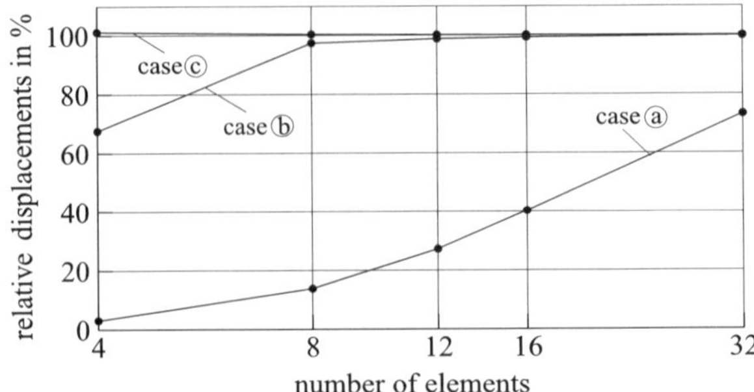

Fig. 5 Results of the geometrically nonlinear simulations in case ①

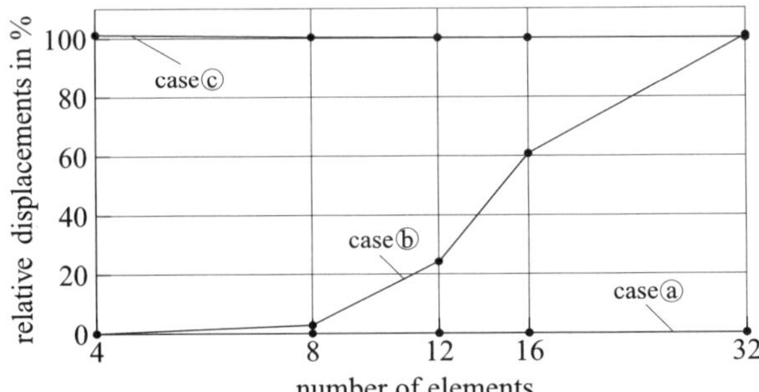

Fig. 6 Results of the geometrically nonlinear simulations in case ②

the analytical reference solution, even if 32 elements are used. In case ②
the simulation produces absolutely useless results. An improvement of the
obtained results becomes possible, if the ANS modifications corresponding
to cases ⓑ and ⓒ in Table 1 are made. The relative error of the results in
both cases decreases close to zero. But the improvement of the results depends
strongly on the number of elements used and on the ANS modifications made.
The so-called "fictional radius of curvature" of all shell elements [10] is in the
same order of magnitude, i.e., about 10.0 in case ① and 100.0 in case ②, and
independent of the number of elements used in the discretization. Therewith,
the fictional radius of curvature of an element is not an adequate measure to
decide whether curvature thickness locking has to be expected or not. The
decisive measure is the angle between the normal vector to the approximated
reference surface and the shell director $\mathbf{D}$ at the nodes, see Fig. 1. This angle is
about 2.8° using a fine discretization by 32 elements and is about 22.5° using
a coarse discretization by four elements. Therefore, it has to be expected
that curvature thickness locking vanishes, if the discretization of the shell
becomes finer. A detailed discussion of the reasons can be found in [10]. This
behavior can principally be seen from the results, if the difference between
the cases ⓑ and ⓒ is assumed to be caused by curvature thickness locking
only. Additionally, it can be stated that the negative influence of curvature

thickness locking is amplified, if the structure becomes thinner. To give reasons for this, the cases ① and ② discretized by four elements are compared. In case ① 57.5% of the analytical reference solution is obtained and in case ②, the thin shell reacts much too stiff and the displacements are close to zero.

The results of the geometrically nonlinear simulations show a similar behavior and are depicted in Figs. 5 and 6. Summarizing, it can be stated that the obtained results are very reliable in this example, since they have practically no relative error, both in cases ① and ② for the fine discretization by 32 elements. Even a relatively coarse discretization by 12 elements is able to produce good results, i.e. the element is efficient as well.

Finally, the element behavior is examined, if POISSON's ratio is not assumed to vanish from the very beginning. If POISSON's ratio ν is set equal to 0.4, POISSON thickness locking has to be expected. This artificial overestimation of the stiffness can be reduced by the EAS enhancement given by case ④ in Table 1. Theoretically, the obtained results remain nearly unchanged by modifying POISSON's ratio. Without going into further details, the results of the simulations remain nearly unchanged [10] and the EAS enhancements work well.

5 Conclusion

This publication states an abstract variational formulation as foundation of a nonlinear, three-dimensional, surface-related shell formulation using linear kinematics. The developed surface-related shell formulation will allow an efficient and reliable numerical simulation of the compound behavior of strengthening layers applied on surfaces. Due to the particular position of the reference surface, well-established techniques against locking phenomena are extended and/or adjusted leading to an efficient and reliable surface-related solid-shell finite element. A numerical example demonstrates finally its efficiency and reliability.

Acknowledgements

The authors gratefully acknowledge financial support of Deutsche Forschungsgemeinschaft DFG within the Sonderforschungsbereich SFB 528 "Textile Reinforcement for Structural Strengthening and Repair" at the Technische Universität Dresden.

References

1. Bathe KJ, Dvorkin EN (1985) Int. J. Num. Meth. Eng. 21:367–383
2. Betsch P, Stein E (1995) Comm. Num. Meth. Eng. 11:899–909

3. Braess D (1997) Finite Elemente: Theorie, schnelle Löser und Anwendungen in der Elastizitätstheorie. Springer, Berlin Heidelberg New York
4. Büchter N, Ramm E (1992) 3d-extension of nonlinear shell equations based on the enhanced assumed strain concept. In: Hirsch C, Périaux J, Oñate E (eds.), Computational Methods in Applied Sciences. Elsevier, Brussels Belgium
5. Curbach M (1998) Sachstandbericht zum Einsatz von Textilien im Massivbau. Beuth, Berlin
6. Dvorkin EN, Bathe KJ (1985) Eng. Comp. 1:77–88
7. Hughes TJR, Tezduyar T (1981) J. Appl. Mech. 48:587–596
8. Jesse F (2005) Tragverhalten von Filamentgarnen in zementgebundener Matrix. Ph.D. Thesis, TU Dresden, Dresden
9. MacNeal RH (1978) Comp. & Struct. 8:175–183
10. Schlebusch R (2005) Theorie und Numerik einer oberflächenorientierten Schalenformulierung. Ph.D. Thesis, TU Dresden, Dresden
11. Schlebusch R, Matheas J, Zastrau B (2003) J. Theor. Appl. Mech. 41(3):623–642
12. Schlebusch R, Zastrau B (2006) Theory and Numerics of a Surface-Related Shell Formulation. In Soares CAM, Martins JAC, Rodrigues HC, Ambrosio JAC, Pina CAB, Pereira EBR, Folgado J (eds.), III European Conference on Computational Mechanics. Springer, Dordrecht
13. Simo JC, Rifai MS (1990) Int. J. Num. Meth. Eng. 29:1595–1638
14. Simo JC, Hughes TJR (1986) J. Appl. Mech. 53:52–54
15. Washizu K (1981) Variational methods in elasticity and plasticity. Pergamon, Frankfurt
16. Zastrau B, Schlebusch R, Matheas J (2003) Special aspects of surface-related shell theories with applications to contact problems. In: Bathe KJ (ed), Proceedings of the Second MIT Conference on Computational Fluid and Solid Mechanics. Elsevier, Amsterdam Boston

The Contact Problems of the Mathematical Theory of Elasticity for Plates with an Elastic Inclusion

Nugzar Shavlakadze

A. Razmadze Mathematical Institute, 1, Aleksidze St., Tbilisi, 0193, Georgia,
nusha@rmi.acnet.ge

Abstract The contact problems of the theory of elasticity and bending theory of plates for finite or infinite plates with an elastic inclusion of variable rigidity are considered. The problems are reduced to integro-differential equations or to systems of integro-differential equations with variable coefficient and singular operator. If such coefficient varies according to power law, we investigate the obtained equations and get exact or approximate solutions and study behavior of unknown contact stresses at the ends of the elastic inclusion.

Keywords: contact problem, integro-differential equation, elastic inclusion, holomorphic functions

1 Introduction

The contact problems on interaction of thin-shell elements (stringers or inclusions) of various geometric forms with massive deformable bodies belong to the extensive field of the theory of contact and mixed problems of mechanics of deformable rigid bodies. Interest in such type of problems is motivated by the fact that investigations in this area make it possible to solve a number of questions connected with problems of engineering industry, shipbuilding and thin-shell constructions.

Stringers and inclusions, like stamps and cuts, concentrate stresses. Therefore, it is of great theoretic and practical importance to investigate the influence exerted by the inclusion on the stress-strain state of deformable bodies, to study concentration of stresses in such problems and to elaborate methods for their lowering. Taking into account thin-shellness in different assumptions and theories, we arrive at new statements of the contact problem of deformable bodies which differ substantially from those of classical contact problems of elasticity. As a result, there arises a class of new problems of solid mechanics with displacement boundary conditions.

G. Jaiani, P. Podio-Guidugli (eds.), *IUTAM Symposium on Relations of Shell, Plate, Beam, and 3D Models*, © Springer Science+Business Media B.V. 2008

Many authors (see [1]) have addressed contact problems for an elastic isotropic or anisotropic plate reinforced with a finite rod or an inclusion of constant section. It has been shown that the contact stresses near the ends of the stiffening member have square-root singularities. The behavior of the tangential contact stress is the same when the cross section of the stiffener is elliptic [2]. For an elastic isotropic wedge stiffened over a finite area by a rod whose cross-sectional area varies linearly, the solution was constructed in [3]. It is proven that the contact stress near the thin end of the stiffening rod has a singularity of order less than $1/2$. If the cross section of the rod is parabolic, then the contact stress near its thin end is finite [4].

Contact problems for isotropic plates stiffened with thin (absolutely rigid or with constant bending rigidity) inclusions and subjected to bending were solved in [5–7]. Those problems are reduced to systems of integral equations with a special characteristic part whose solutions are determined in a class of functions with nonintegrable singularities by using the method of regularization of diverging integrals [8]. Exact solutions are constructed by the method of analytic functions, approximate solutions by the method of orthogonal polynomials [9], [10], [11].

We investigated the contact problems of in-plane tension and bending of plates with inclusion of variable rigidity; for unknown contact stresses we obtained the integral differential equation whose characteristic part is Prandtl's integral differential equation, which under certain conditions has been studied in [1], [12], [13]. In the case when the coefficient of a singular operator tends at the ends to zero with any order at the end of the integration line this equation is equivalent to a singular integral equation of the third kind. We have studied this equation and got exact or approximate solutions [14–18].

2 Statement of the Problems. The Basic Equations

Problem 1

Let an elastic plane with the elastic modulus E and Poisson coefficient ν (Lame parameters λ and μ) be strengthened on the finite segment $[0,1]$ of the OX axis by an inclusion of small thickness $h_0(x)$, with elastic modulus $E_0(x)$ and Poisson coefficient ν_0. The plane is subject at infinity to uniform tensions of intensities p and q directed to the OX and OY axes. In conditions of plane deformation, it is required to determine contact stresses acting on a segment, where inclusion comes in contact with the plane, and also axial stresses at the segment ends. The inclusion will be assumed to be a thin plate free from bending rigidity. Thus we have the following basic equations:

$$E(x)\frac{d^2 u_0(x)}{dx^2} = \tau_-(x) - \tau_+(x), \quad q(x) = q_-(x) = q_+(x), \quad 0 < x < 1,$$

$$E(x) \equiv \frac{E_0(x)h_0(x)}{1 - \nu_0^2}.$$

$$(1.1)$$

Here $q_\pm(x)$ and $\tau_\pm(x)$ are respectively normal and tangential unknown contact stresses on the upper (with the index $''+''$) and lower (with the index $''-''$) contours of the inclusion, $u_0(x)$ is the horizontal displacement of its points.

Thus the normal stress on the inclusion contours has no jump, while the tangential stresses due to $E_0(x) \neq 0$ have jumps everywhere on the inclusion contours.

Denoting $\tau(x) \equiv \tau_+(x) - \tau_-(x)$, from (1.1) we obtain

$$\varepsilon_x^0(x) = \frac{du_0}{dx} = \frac{1}{E(x)}\left[P_1 - \int_0^x \tau(t)dt\right], \quad 0 < x < 1. \tag{1.2}$$

The equilibrium condition for the inclusion has the form

$$\int_0^1 \tau(t)dt = P_1 - P_2, \tag{1.3}$$

where P_1 and P_2 are unknown axial stresses at the ends of the inclusion, $x = 0$ and $x = 1$, respectively.

The condition of compatibility of horizontal deformations of the inclusion and of the elastic homogeneous continuous plane loaded along the segment $[0, 1]$ of the OX axis by tangential stresses and forces at infinity, is taken into account. According to the well-known results [19], the horizontal deformation of points of the OX axis on the plane $\varepsilon_x(x)$ owing to these force factors has the form

$$\varepsilon_x(x) = -\frac{\varkappa}{2\pi\mu(1+\varkappa)} = \int_0^1 \frac{\tau(t)dt}{t-x} + \frac{\varkappa+1}{8\mu}p + \frac{\varkappa-3}{8\mu}q, \tag{1.4}$$
$$-\infty < x < \infty.$$

Taking into account the contact condition

$$\varepsilon_x(x) = \varepsilon_x^0(x), \quad 0 < x < 1$$

from (1.2) and (1.4) we obtain the following integral differential equation:

$$\mu_1(x) - \frac{\gamma E(x)}{\pi}\int_0^1 \frac{\mu_1'(t)dt}{t-x} = \left(\frac{\varkappa-3}{8\mu}q + \frac{\varkappa+1}{8\mu}p\right)E(x), \tag{1.5}$$
$$0 < x < 1,$$

where $\gamma = \frac{\varkappa}{2\mu(1+\varkappa)}$, $\mu_1(x) = P_1 - \int_0^x \tau(t)dt$, $\varkappa = 3 - 4\nu$ and condition (1.3) takes the form

$$\mu_1(1) = P_2. \tag{1.6}$$

Stresses P_1 and P_2 are defined from the relations

$$\int_{-\frac{h_0(0)}{2}}^{\frac{h_0(0)}{2}} \sigma_x(0,y)dy = P_1, \quad \int_{-\frac{h_0(1)}{2}}^{\frac{h_0(1)}{2}} \sigma_x(1,y)dy = P_2, \tag{1.7}$$

where $\sigma_x(x, y)$ is the horizontal component of the stressed field caused in the elastic, homogeneous, continuous plane by forces of intensity $\tau(x)$ which are distributed over the segment $(0, 1)$ and directed along it.

We make use of the well-known formula [19]

$$\sigma_x + i\tau_{xy} = \phi(z) + \overline{\phi(z)} - z\overline{\phi'(z)} - \overline{\Psi(z)}, \tag{1.8}$$

where complex potentials in the case of the elastic homogeneous plane have the form

$$\phi(z) = -\frac{1}{2\pi(1+\varkappa)} \int_0^1 \frac{\tau(t)dt}{t-z} + \frac{p+q}{4},$$

$$\Psi(z) = \frac{\varkappa}{2\pi(1+\varkappa)} \int_0^1 \frac{\tau(t)dt}{t-z} + \frac{1}{2\pi(1+\varkappa)} \int_0^1 t\frac{\tau(t)dt}{(t-z)^2} + \frac{q-p}{2}. \tag{1.9}$$

After simple operations, from (1.8), (1.9) we find that

$$\sigma_x(x, y) = -\frac{3+\varkappa}{4\pi(1+\varkappa)} \int_0^1 \left(\frac{1}{\eta-\xi} + \frac{1}{\eta-\overline{\xi}}\right)\tau(\eta)d\eta -$$

$$\frac{iy}{2\pi(1+\varkappa)} \int_0^1 \left(\frac{1}{(\eta-\xi)^2} - \frac{1}{(\eta-\overline{\xi})^2}\right)\tau(\eta)d\eta + p, \tag{1.10}$$

where $\xi = x + iy$.

Substituting (1.10) in (1.7), we obtain the following relations:

$$-\frac{1}{\pi} \int_0^1 \mathrm{arctg}\left[\frac{h_0(0)}{2\eta}\right]\tau(\eta)d\eta -$$

$$-\frac{4h_0(0)}{\pi(1+\varkappa)} \int_0^1 \frac{\eta}{4\eta^2 + h_0^2(0)}\tau(\eta)d\eta + ph_0(0) = P_1,$$

$$-\frac{1}{\pi} \int_0^1 \mathrm{arctg}\left[\frac{h_0(1)}{2(\eta-1)}\right]\tau(\eta)d\eta - \tag{1.11}$$

$$-\frac{4h_0(1)}{\pi(1+\varkappa)} \int_0^1 \frac{\eta-1}{4(\eta-1)^2 + h_0^2(1)}\tau(\eta)d\eta + ph_0(1) = P_2.$$

Solving integral differential Equation (1.5) for (1.6) and substituting the function $\tau(\eta)$ in (1.1), we arrive at a system of two algebraic equations with respect to P_1 and P_2.

It is easily seen from (1.11) that if the inclusion thickness $h_0(x)$ satisfies the conditions $h_0(0) = h_0(1) = 0$, then $P_1 = P_2 = 0$, i.e., there are no axial stresses at the ends of the inclusion.

Problem 2

Consider the problem of bending of an unbounded plate under the action of bending moment at infinity: $M_x^\infty = M$, $M_y^\infty = 0$. The plate along the line

$y = 0$, $0 < x < 1$, is strengthened by a thin elastic inclusion. It is required to find contact stresses caused by interaction of the inclusion and the plate.

The presence of the strengthening inclusion causes in the plate a jump of generalized transversal force N_y [5–7]. Using the notation $< f >= f(x, -0) - f(x, +0)$, we have

$$< \omega >=< \omega_y' >=< M_y >= 0, \quad < N_y >= \mu_2(x), \quad 0 < x < 0, \tag{2.1}$$

where $\mu_2(x)$ is an unknown contact stress caused by interaction of the inclusion and the plate; note that $\mu_2(x) \equiv 0$ for $x \notin (0, 1)$ and satisfying the following conditions of equilibrium of the inclusion:

$$\int_0^1 \mu_2(t)dt = 0, \quad \int_0^1 t\mu_2(t)dt = -M_1 + M_2, \tag{2.2}$$

where M_1 and M_2 are the unknown moments at the end cross-sections $x = 0$ and $x = 1$, respectively.

As for the inclusion bending $\omega_0(x)$, we obtain the boundary value problem

$$\frac{d^2}{dx^2}D_0(x)\frac{d^2\omega_0(x)}{dx^2} = -\mu_2(x), \quad 0 < x < 1$$

$$D_0(x)\omega_0''(x)\big|_{x=0} = M_1, \quad D_0(x)\omega_0''(x)\big|_{x=1} = M_2, \tag{2.3}$$

$$\left[D_0(x)\omega_0''(x)\right]'_{x=0} = 0, \quad \left[D_0(x)\omega_0''(x)\right]_{x=1} = 0,$$

where $D_0(x) = \frac{E_0(x)h_0^3(x)}{12}$ is the bending rigidity, $h_0(x)$ is the thickness and $E_0(x)$ is the Young modulus of the inclusion material.

The stressed state of the thin isotropic homogeneous plate is defined by a midplane bending $w(x, y)$ which satisfies the biharmonic equation

$$\Delta\Delta\omega = 0. \tag{2.4}$$

When the inclusion and the plate come in contact,

$$\omega(x, 0) = \omega_0(x), \tag{2.5}$$

a solution of the boundary value problem (2.1–2.5) is sought in a class of functions $w(x, y)$ having second derivatives which behave themselves like $r^{-\frac{1}{2}}$ when approach to the points $(0; 0)$, $(1, 0)$ and are bounded at infinity.

The solution of Equation (2.4) can be represented in the form

$$w(x, y) = \text{Re}[\bar{z}\varphi(z) + \chi(z)], \tag{2.6}$$

where $\varphi(z)$ and $\chi(z)$ are functions of a complex variable $z = x + iy$, holomorphic in the plate region. For bending moments M_x and M_y, for torque H_{xy} and cutting forces N_x and N_y we have the following formulas [20]

$$M_y - M_x + 2iH_{xy} = 4(1-\nu)D[\overline{z}\varphi''(z) + \psi'(z)],$$
$$M_x + M_y = -8(1+\nu)D\operatorname{Re}\varphi'(z), \qquad (2.7)$$
$$N_x - iN_y = -8D\varphi''(z),$$

where $\psi(z) = \chi'(z)$, $2h$ is the plate thickness and $D = 2Eh^3/3(1-\nu^2)$ is the cylindrical rigidity of the plate.

Taking into account the conditions of equilibrium of the inclusion, the complex potentials can be expressed by the formulas

$$\varphi(z) = -\frac{M}{8D(1+\nu)}z + \varphi_0(z),$$
$$\psi(z) = -\frac{M}{4D(1-\nu)}z + \psi_0(z), \qquad (2.8)$$

where $\varphi_0(z)$ and $\psi_0(z)$ are single-valued and holomorphic functions in the plate region.

Introducing into consideration the function $\Omega(z)$ by the equality

$$\Omega(z) = z\varphi'(z) + \psi(z),$$

by (2.6) we can easily see that the formula

$$\frac{\partial \omega}{\partial x} + i\frac{\partial \omega}{\partial y} = \varphi(z) + \overline{\Omega(z)} + (z - \overline{z})\overline{\varphi'(z)},$$

is valid. From the first two conditions (2.1) we obtain

$$[\varphi(t) - \overline{\Omega(t)}]^- - [\varphi(t) - \overline{\Omega(t)}]^+ = 0,$$

and by (2.8) we find that

$$\varphi(z) - \overline{\Omega(\overline{z})} = \frac{M}{4D(1-\nu)}z.$$

Hence

$$\psi_0(z) = \overline{\varphi}_0(z) - z\varphi_0'(z). \qquad (2.9)$$

From the last two conditions (2.1), with regard to (2.7) and (2.9) we get

$$[\varphi''(x) + \overline{\varphi''(x)}]^- - [\varphi''(x) + \overline{\varphi''(x)}]^+ = 0,$$
$$[\varphi''(x) - \overline{\varphi''(x)}]^- - [\varphi''(x) - \overline{\varphi''(x)}]^+ = \frac{i\mu_2(x)}{4D},$$

Adding the obtained conditions, we find

$$[\varphi''(x)]^+ - [\varphi''(x)]^- = -\frac{i\mu_2(x)}{8D}, \quad 0 < x < 1 \qquad (2.10)$$

The function $\mu_2(x)$ may have nonintegrable singularities on the segment $[0,1]$. Taking into consideration the proof given in [9] on the transfer of the results of monograph [19] to the regularized values of diverging integrals [6], since $\varphi''(\infty) = 0$, a solution of the boundary value problem (2.10) is given by the formula:

$$\varphi''(z) = -\frac{1}{16\pi D}\int_0^1 \frac{\mu_2(t)dt}{t-z}, \tag{2.11}$$

where z changes in the region, cut along the segment $(0,1)$.

In view of the fact that $\frac{\partial^2\omega(x,0)}{\partial x^2} = 2\varphi'(x) - \frac{M}{4D(1-\nu)}$, by virtue of (2.8), condition (2.3) takes the form

$$\frac{d^2}{dx^2}D_0(x)\left[\frac{1}{8\pi D}\int_0^1 \ln|t-x|\mu_2(t)dt - \frac{M}{2D(1-\nu^2)}\right] = -\mu_2(x).$$

Introducing the notation $\lambda(x) = M_1 - \int_0^x dt\int_0^t \mu_2(\tau)d\tau$ and integrating the last equation twice, we arrive at the equation

$$\lambda(x) - \frac{D_0(x)}{8\pi D}\int_0^1 \frac{\lambda'(t)dt}{t-x} = -\frac{MD_0(x)}{2D(1-\nu^2)}, \quad 0 < x < 1, \tag{2.12}$$

provided

$$\lambda(0) = M_1 \quad\text{and}\quad \lambda(1) = M_2. \tag{2.13}$$

The moments M_1 and M_2 at the end cross-sections of the inclusion are defined from the following relations:

$$M_1 = \int_{-h_0(0)/2}^{h_0(0)/2} M_x(0,y)dy, \quad M_2 = \int_{-h_0(1)/2}^{h_0(1)/2} M_x(1,y)dy,$$

where $M_x(x,y) = \frac{1+\nu}{8\pi}\int_0^1 \left(\ln|t-z| + \ln|t-\bar z|\right)\mu_2(t)dt + M$.

On singularities of contact stress

Using the methods of theory of analytical functions for Equations (1.5, 1.6) and (2.12, 2.13) the obtained results can be formulated in the form of the following theorems

Theorem 1 *In Problem I, if the inclusion rigidity varies according to the rule $E(x) = x^\alpha b(x)$, $(b(x) > 0$, $\alpha \geq 0$, $0 \leq x \leq 1)$, then the jump of the tangential contact stress in the neighbourhood of the point $x = 0$ has the behaviour*

$$\tau(x) = \begin{cases} O(x^{-\frac{1}{2}}), & \text{for} \quad 0 \leq \alpha < 1, \\ O(x^{-1+\delta_0}), & \text{for} \quad \alpha = 1, \quad \delta_0 > \frac{1}{2}. \\ O(1), & \text{for} \quad 1 < \alpha \leq 2, \\ O(x^{\alpha-2}), & \text{for} \quad \alpha > 2. \end{cases}$$

Theorem 2 *In Problem* II, *if the inclusion bending rigidity varies by the rule* $D_0(x) = x^\alpha b(x)$, $(b(x) > 0, \ \alpha \geq 0, \ 0 \leq x \leq 1)$ *then the contact stress of interaction of the inclusion and the plate (i.e., the jump of generalized cross-sectional force) in the neighbourhood of the point* $x = 0$ *has the behaviour*

$$< N_y > = \begin{cases} O(x^{-\frac{3}{2}}), & for \quad 0 \leq \alpha < 1, \\ O(x^{-2+\delta_0}), & for \quad \alpha = 1, \ \delta_0 > \frac{1}{2}, \\ O(x^{-1}), & for \quad 1 < \alpha \leq 2, \\ O(x^{\alpha-3}), & for \quad \alpha > 2. \end{cases}$$

References

1. Alexandrov V, Mkhitaryan S (1983) Contact problems for bodies with thin covers and layers. Nauka, Moscow (in Russian)
2. Morar GA, Popov GYa (1970) Prikl Mat Mekh 34(3):412–422 (in Russian)
3. Nuller BM (1972) Izv AN SSSR Mekh Tv Tela 5:150–155 (in Russian)
4. Shavlakadze N (2003) Izv Ross Acad Nauk Mekh Tv Tela 6:102–108 (in Russian)
5. Popov G (1983) Concentration of elastic stresses near punches, cuts, thin inclusions and supports. Nauka, Moscow (in Russian)
6. Onishchuk O, Popov G (1980) Izv Akad Nauk SSSR Mekh Tv Tela 4:141–150 (in Russian)
7. Onishchuk O, Popov G, Proshcherov Ju (1984) Prikl Mat Mekh 48(2):307–314 (in Russian)
8. Gelfand IM, Shilov GE (1958) Generalized functions and operations over them. Fizmatgiz, Moscow (in Russian)
9. Onishchuk O, Popov G, Farshight P (1986) Prikl Mat Mekh 50(2):393–302 (in Russian)
10. Nuller B (1976) Prikl Mat Mekh 40(2):306–316 (in Russian)
11. Bantsuri R (1975) Dokl Akad Nauk SSSR 222(3):568–571 (in Russian)
12. Vekua I (1945) Prikl Mat Mekh 9(2):112–150 (in Russian)
13. Magnaradze L (1942) Soobshch Acad Nauk GSSR 3(6):503–508
14. Shavlakadze N (1999) Georgian Math J 6(5):489–500
15. Shavlakadze N (1999) Proc A Razmadze Math Inst 120:135–147
16. Shavlakadze N (2001) Izv Ross Acad Nauk Mekh Tv Tela 3:144–155 (in Russian)
17. Shavlakadze N (2002) Prikl Mekh 38(3):114–121 (in Russian) translation in Int Appl Mech 38(3):356–364
18. Bantsuri R, Shavlakadze N (2002) Prikl Mat Mekh 66(4):663–669 (in Russian) translation in J Appl Math Mech 66(4):645–650
19. Muskhelishvili NI (1966) Some basic problems of the mathematic theory of elasticity. Nauka, Moscow (in Russian)
20. Fridman MM (1941) Prikl Mat Mekh 5(1):93–102 (in Russian)

On the Basic Systems of Equations of Continuum Mechanics and Some Mathematical Problems for Anisotropic Thin-Walled Structures

Tamaz Vashakmadze

Iv. Javakhishvili Tbilisi State University, 2, University Str., 0143, Tbilisi, Georgia, `tamazvashakmadze@yahoo.com`

Abstract A dynamic system of partial differential equations (PDEs) which is 3D with respect to spatial coordinates and contains as a particular case both: Navier-Stokes equations and the nonlinear systems of PDEs of the elasticity theory is proposed.

Mathematical models for anisotropic, poroelastic media are created and justified. These models are applied to dynamic and steady-state nonlinear problems for thin-walled structures.

A direct method of constructing von Kármán type equations in dynamical case is proposed.

Keywords: Navier-Stokes equations, von Kármán type equations, mass conservation and continuity conditions, anisotropic poroelastic media

1 On the Basic Systems of Continuum Mechanics

In this part we propose a dynamic system of partial differential equations (PDEs) which is 3D with respect to spatial coordinates and contains as a particular case both: Navier-Stokes equations and the nonlinear systems of PDEs of the elasticity theory. Such a general representation of a dynamic system allows us to prove that the nonlinear phenomena observed in problems of solid mechanics can also be detected in Navier-Stokes type equations, and vice versa.

Following [1] and [2], the basic system of partial differential equations is written in the form

$$\rho \frac{D}{Dt}\left(\frac{\partial u}{\partial t}\right) = f - (1 - \Gamma)\nabla p + \nabla\big[(1 + \nabla u)\tau\big], \tag{1}$$

where ρ is the density, p is the pressure, $u = (u_1, u_2, u_3)^T$ is the displacement vector, and $\partial u / \partial t = v$, $v = (v_1, v_2, v_3)^T$ is the velocity vector, f is the mass force, D/Dt is the total (or convective) derivative, τ is the stress tensor.

It is obvious that Newton's law for a viscous flow and the generalized Hooke's law for a solid can be written in the general form

$$\tau = \left[(1 - \Gamma) \frac{\partial}{\partial t} + \Gamma \right] A_\Gamma \cdot \varepsilon \quad (0 \leq \Gamma \leq 1), \tag{2}$$

where the symmetric matrix A_Γ corresponds to a fluid if $\Gamma = 0$, and to a solid if $\Gamma = 1$. The strain tensor is $\varepsilon = (\varepsilon_{11}, \varepsilon_{22}, \varepsilon_{33}, \varepsilon_{23}, \varepsilon_{13}, \varepsilon_{12})^T$, where $2\varepsilon_{ij} = \partial_i u_j + \partial_j u_i + u_{i,k} u_{j,k}$.

For the mass conservation conditions or equations of continuity we have

$$(1 - \Gamma) B_0[\rho, \varepsilon] + \Gamma B_1[\varepsilon] = 0, \tag{3}$$

$$B_0[\rho, \varepsilon] = \partial_t \rho + \nabla(\rho v), \quad B_1[\varepsilon] = (B_{11}, B_{12}, B_{13}, B_{14}, B_{15}, B_{16})^T,$$

where B_1 describes the Saint-Venant-Beltrami conditions of continuity written in the form

$$\begin{aligned}
B_{1i}(\varepsilon) &= \varepsilon_{ii,kl} + \varepsilon_{kl,ii} - \varepsilon_{li,ki} - \varepsilon_{ki,li} + C_{1i}(u), \\
& i, k, l = 1, 2, 3, \quad i \neq k, \quad i \neq l, \\
B_{17-i}(\varepsilon) &= \varepsilon_{ii,kl} + \varepsilon_{kl,ii} - \varepsilon_{li,ki} - \varepsilon_{ki,li} + C_{17-i}(u), \\
& i = 1, 2, 3, \quad k = l = i + 1, \quad x_1 = x_4,
\end{aligned} \tag{4}$$

$C_{ij}(u)$ are nonlinear homogeneous differential forms of at most third order.

It is obvious that for $\Gamma = 0$, $\nabla u = 0$, $\tau_{ij} = -\frac{2}{3} \mu \delta_{ij} \operatorname{div} v + \mu(v_{i,j} + v_{j,i})$, system (1) implies Euler and Navier-Stokes PDEs (see [1], Chap. 11).

When $\Gamma = 1$ and

$$\nabla(1 + \nabla u)\tau = (\partial_j(\tau_{1j} + \tau_{kj} u_{1,k}), \partial_j(\tau_{2j} + \tau_{kj} u_{2,k}), \partial_j(\tau_{3j} + \tau_{kj} u_{3,k}))^T,$$

from (1) we obtain a system of nonlinear PDEs of the spatial theory of elasticity (see [2]).

If $\Gamma = 0$ and $\nabla u \neq 0$, then (1) is a system of Navier-Stokes type PDEs.

If in system (1) $\Gamma = 1$, then, according to [3], Chap. 1, we have the following nonlinear systems of von Kármán-Mindlin-Reissner (KMR) type PDEs:

$$D\Delta^2 \bar{u}_3 = \left(1 - \frac{h^2(1 + 2\gamma)(2 - \nu)}{3(1 - \nu)} \Delta \right)(g_3^+ - g_3^-) + 2h\left(1 - \frac{h^2(1 + 2\gamma)}{3(1 - \nu)} \Delta \right)[\bar{u}_3, \Phi^*]$$

$$+ h(g_{\alpha,\alpha}^+ + g_{\alpha,\alpha}^-) - \int_{-h}^{h} \left(z f_{\alpha,\alpha} - \left(1 - \frac{1}{1 - \nu} \Delta(h^2 - z^2) f_3 \right) \right) dz + R_3[\bar{u}_3; \gamma], \tag{5}$$

$$\Delta^2 \Phi^* = -\frac{E}{2}[\bar{u}_3, \bar{u}_3] + \frac{\nu}{2} \Delta(g_3^+ + g_3^-) + \frac{1 + \nu}{2h} \bar{f}_{a,a} + R_6[\Phi^*], \tag{6}$$

$$Q_{\alpha3} - \frac{1+2\gamma}{3}\,h^2\Delta Q_{\alpha3} = -D\Delta\bar{u}_{3,\alpha} + \frac{h^2(1+2\gamma)}{3(1-\nu)}\,\partial_\alpha\big(g_3^+ - g_3^- + 2h(1+\nu)\big)\big[\bar{u}_3,\Phi^*\big]$$

$$+h(g_\alpha^+ + g_\alpha^-) - \int_{-h}^{h} z f_\alpha dz + \frac{1+\nu}{2(1-\nu)}\int_{-h}^{h}(h^2 - z^2)\,f_{3,\alpha}dz + R_{3+\alpha}\big[Q_{\alpha3};\gamma\big]. \qquad (7)$$

Systems (5–7) without the remainder terms R yield 2D systems of refined theories with control parameters γ. By choosing the concrete value of γ we can get any of the existing refined theories, while for other values of γ these mathematical models are new. E.g. by $\gamma = 0,1$ [1.5] coincides with Reissner's equation with respect to deflection in the linear case. Some systems (with arbitrary control parameters) were constructed in [3], Chaps. 2 and 3, for anisotropic dynamic piezoelectric and electrically conductive elastic plates of variable thickness.

Let us consider Equation (5), where the principal terms are

$$D\Delta^2 w, \quad D'\Delta[w,\varphi] = D'\big([\Delta w,\varphi] + [w,\Delta\varphi] + 2[\partial_\alpha w,\partial_\alpha\varphi]\big).$$

Using relations of form (1.9a)in [4], we obtain $\partial_{11}\varphi = \bar{\sigma}_{12}$, $\partial_{12}\varphi = -\bar{\sigma}_{12}$, $\partial_{22}\varphi = \bar{\sigma}_{11}$. Then the above expression can be rewritten as follows (see [4]):

$$D'\Delta[w,\varphi] = D'\big[\bar{\sigma}_{\alpha\beta}\partial_{\alpha\beta}\Delta w + \partial_{\alpha\beta}w\Delta\bar{\sigma}_{\alpha\beta} + 2\bar{\sigma}_{\beta\gamma,\alpha}\partial_{\beta\gamma}w_{,\alpha}\big]. \qquad (8)$$

Calculations and simple analysis of the symbolical determinant by these expressions show that the characteristic form of a system like (5) and (6) can be a positive or a negative number, zero or an arbitrary continuous function of x, y. Note that $ED' = 2(1+2\gamma)(1+\nu)D$. Therefore if $\{f\}$ denotes a physical dimension of f, then it is obvious that $\{\Delta^2 w\} = \{\Delta[w,\Phi/E]\}$. We recall that here E is Young's modulus of elasticity and ν is Poisson's ratio.

Thus, the first and the second summands in (8) define nonlinear wave processes for a static case. The structure of the third summand obviously corresponds to a 2D solution of the Kadomtsev-Petviashvili type.

A direct way of constructing one- and 2D solutions of Boussinesq, Burgers, Korteweg-de Vries, Kadomtsev-Petviashvili, Dorodnitsin equations and other systems describing turbulent flows, shock waves, one- and 2D solitons in fluids and continuum plasma physics is given in a substantial number of works (see, for instance, [5], [6]). For $\Gamma = 1$, in Equation (1) there appears a term like the one we have exhibited in (8). On the other hand, terms of the form $\Delta[u,\varphi]$ appear in Navier-Stokes type equations when $\Delta u \neq 0$. Thus we prove that the nonlinear phenomena observed in problems of solid mechanics can be detected in Navier-Stokes type equations, and vice versa.

Finally, we remark that the matter conservation laws are valid for KMR type systems (5–7); this conclusion follows from R. Kienzler, D. K. Bose [7] and from the fact that the KMR systems of the refined theories are equivalent to the proposed problems (1–3) (for $\Gamma = 1$) with the corresponding initial and boundary conditions since these KMR systems have approximation of at least second order (for details see [3], Chap. 2).

2 New Mathematical Models for Poroelastic Media

The aim of the second part of the paper is to create and justify new mathematical models for anisotropic, poroelastic media, and to apply them to a variety of dynamic and steady-state nonlinear problems for thin-walled structures.

The best-known model is M. Biot's theory for poroelastic media. If this theory is considered to be applicable in the isotropic case, then it should be pointed out that this theory has certain disadvantages for weak anisotropy. Neither Biot's implies the modern theory for elastic media considered by C. Truesdell, P. Ciarlet and other authors. We propose to develop a new mathematical model for anisotropic poroelastic media. As a special case, from this model we can derive in particular Biot's theory and also the modern nonlinear theory of elasticity.

In Biot's linear theory, the corresponding differential operators with respect to spatial variables have double degeneration since the symbolic determinants contain as a cofactor the symbolic minor corresponding to the *graddiv* operator (see [8], formula (14)). In Biot's nonlinear theory, the anisotropic property of media depends on a ratio of strain and stress tensors, but not on the character of media. It should be said that mathematical models presented in [4] are free of these disadvantages.

In this connnection, let us consider the problem of constructing a $3D$ model with three spatial variables for a poroelastic medium. We denote by Ω the domain in the 3D Euclidean space R^3. In the Cartesian coordinate system, a point is denoted by $x = (x_1, x_2, x_3)$, and a time interval by $(0, T)$. Thus, at each point of the mixture (macro-point) we consider the following averaged values of stress and strain tensors and displacement vector:

$$\boldsymbol{\sigma} = (\sigma_{11}, \sigma_{22}, \sigma_{33}, \sigma_{32}, \sigma_{31}, \sigma_{12})^T, \quad \boldsymbol{\varepsilon} = (\varepsilon_{11}, \varepsilon_{22}, \varepsilon_{33}, \varepsilon_{32}, \varepsilon_{31}, \varepsilon_{12})^T,$$

$$\boldsymbol{u} = (u_1, u_2, u_3)^T, \quad \boldsymbol{p} = (p_{11}, p_{22}, p_{33}, p_{32}, p_{31}, p_{12})^T,$$

$$\boldsymbol{\zeta} = (\zeta_{11}, \zeta_{22}, \zeta_{33}, \zeta_{32}, \zeta_{31}, \zeta_{12})^T, \quad \boldsymbol{w} = (w_1, w_2, w_3)^T, \tag{9}$$

$$\varepsilon_{ij} = \frac{1}{2}(u_{i,j} + u_{j,i} + u_{k,i}u_{k,j}), \quad \zeta_{ij} = \frac{1}{2}(w_{i,j} + w_{j,i} + w_{k,i}w_{k,j}).$$

Equilibrium equations for a mixture are written in the form

$$\partial_i(\sigma_{ij} + \sigma_{kj} \cdot u_{i,k}) = \partial_{tt}(\rho_1 u_i + \rho_2 w_i) + f_i,$$

$$\partial_j(p_{ij} + p_{kj} \cdot w_{i,k}) = \partial_{tt}(\rho_2 u_i + \rho_3 w_i) + \frac{\eta}{\mu}\partial_t w_i + \varphi_i, \quad (x,t) \in Q_T. \tag{10}$$

Here $f = (f_1, f_2, f_3)^T$, $\varphi = (\varphi_1, \varphi_2, \varphi_3)^T$, are the mass force vectors, ρ_i is the density, and η/μ is defined as in [8], [9].

As different from law (5.1) in [9], we define a law of Hooke's type as follows:

$$\boldsymbol{\sigma} = \boldsymbol{B}\varepsilon + \boldsymbol{C}\zeta, \tag{11}$$

$$\boldsymbol{p} = \boldsymbol{C}\varepsilon + \boldsymbol{M}\zeta, \tag{12}$$

where $B, A = B^{-1}$ are the 6×6 and symmetric matrices of rigidity and pliability, respectively; $C = \{c_{ij}\}_{6\times6}$, $M = \{m_{ij}\}_{6\times6}$ are also symmetric matrices.

It is assumed that at least one plane of elastic symmetry, which is parallel to Ox_1x_2 plane, at each point of the body passes, i.e. the matrices C and M contain at most 13 nonzero constants and

$$b_{i4} = b_{i5} = b_{46} = b_{56} = C_{i4} = C_{i5} = C_{46} = C_{56} = m_{i4} = m_{i5} = m_{46} = m_{56} = 0. \quad (13)$$

From (11) and (12) we have $\sigma_{ii} = B_i\varepsilon + C_i\zeta$, $p_{ii} = C_i\varepsilon + M_i\zeta$, $\sigma_{ij} = B_{9-(i+j)}\varepsilon + C_{9-(i+j)}\zeta$, $p_{ij} = C_{9-(i+j)}\varepsilon + M_{9-(i+j)}\zeta$, where B_i, C_i, M_i are the i-th rows of the respective matrices.

Let us now introduce the notation

$$\tau_{ij} = (\sigma_{ij}, p_{ij})^T, \quad \epsilon_{ij} = (\varepsilon_{ij}, \zeta_{ij})^T, \quad U_i = (u_{ij}, w_{ij})^T.$$

Using the above notation and (13), we rewrite the equilibrium equations (10) and relations (11), (12) as follows (see [10]):

$$\partial_j(\tau_{ij} + \tau_{kj} \otimes U_{i,k}) = \rho\partial_{tt}U_i + \rho_0\partial_t U_i + F_i, \quad (14)$$

$$\tau_{ii} = A_{i1}\epsilon_{11} + A_{i2}\epsilon_{22} + A_{i3}\epsilon_{33} + A_{l6}\varepsilon_{12},$$

$$\tau_{\alpha3} = A_{6-\alpha4}\epsilon_{32} + A_{6-\alpha5}\epsilon_{13}, \quad \tau_{12} = A_{61}\epsilon_{11} + A_{62}\epsilon_{22} + A_{63}\epsilon_{33} + A_{66}\epsilon_{12},$$

where

$$A_{mn} = \begin{pmatrix} b_{mn} & C_{mn} \\ C_{mn} & m_{mn} \end{pmatrix}, \quad \rho = \begin{pmatrix} \rho_1 & \rho_3 \\ \rho_3 & \rho_2 \end{pmatrix}, \quad \rho_0 = \begin{pmatrix} 0 & 0 \\ 0 & \eta/\mu \end{pmatrix}, \quad F_i = (f_i, \varphi_i)^T.$$

Here the symbol $\otimes$ denotes the operation $(a_1, a_2)^T \otimes (b_1, b_2)^T = (a_1b_1, a_2b_2)^T$.

Analogous 3D nonlinear models for anisotropic binary mixtures are presented in [10], which are a generalization of the previously known models for thin-wolled structure poroelastic and binary mixtures. The models constructed in [10] are not only of certain independent scientific interest, but also have such a form of a spatial model that allows us both to construct and to justify KMR type systems of DEs in stationary and nonstationary cases. Under the justification we mean the assumption of "physical soundness" of these models in the Truesdell-Ciarlet sense (for details see, for example, [11], Chap. 5, [12], Chap. 17). As is known, even in the case of an isotropic elastic plate of constant thickness the subject of justification was an unsolved problem. The point is that T. von Kármán, A. E. Love, S. Timoshenko, L. Landau & E. Lifshits and others considered a Saint-Venant-Beltrami compatibility condition as one of the equations of the corresponding system of differential equations. This fact was recently also confirmed by P. Podio-Guidugli [13]. In our models we give a correct equation, which is especially important for dynamic problems. The corresponding system in this case contains wave processes not only in the vertical, but also in the horizontal direction. The equation has the following form:

$$\left(\Delta^2 - \frac{1-\nu^2}{E}\rho\Delta\partial_{tt}\right)\varphi = -\frac{E}{2}\,[w,w] + \frac{\nu}{2}\left(\Delta - \frac{2\rho}{E}\partial_{tt}\right)(g_3^+ + g_3^-) + \frac{1+\nu}{2h}\,f_{\alpha,\alpha}. \quad (15)$$

The accuracy of the afore-mentioned mathematical models is confirmed due to the new term introduced by the author of this paper and describing the effect of a boundary layer. The existence of this term not only explains some paradoxes inherent in the 2D theory of elasticity (I. Babushka, S. Lukasievicz), but also plays an important role in studying, for example, process of crack and hole formation (for details see [3], Chap. 1, Item 3.3). Furthermore, note that in [10] the equations of form (15) are constructed with respect to certain stress tensor components by the differentiation and summation of two differential equations. Other equations of KMR type differing from equations of form (15) are equivalent to a system where the order of each equation is not higher than two.

3 Problems of Constructing Von Kármán Type Systems

Problems of constructing 2D models without using hypotheses of geometrical and physical character are topical and the interest shown in them for the past five decades is connected in the first place with the name of I.I. Vorovich.

Among the works dedicated to the construction and justification of the plate and shell theory a special mention should be made the monograph [11], where the problem of the physical soundness of the von Kármán system is studied. In particular, Ciarlet wrote:"The 2D von Kármán equations for non-linearly elastic plates, originally proposed by T. von Kármán in 1910 (see page lxiii), play an almost mythical role in applied mathematics. While they have been abundantly, and satisfactoriy, studied from the mathematical standpoint, as regards notably various questions of existence, regularity, and bifurcation of their solutions, their physical soundness has often been seriously questioned. Using the same method as in Chap. 4 we show in this chapter that the von Kármán equations may be given a full justification by means of the leading term of a formal asymptotic expansion (in terms of the thickness of the plate as the "small" parameter) of the exact 3D equations of nonlinear elasticity associated with a specific class of boundary conditions that characterizes the von Kármán plates" (see [11], Chap. 5, pp. 367–406, [12], Chap. 17, pp. 694–699).

We remind that for the justification of the von Kármán theory the basic restrictive assumption (used in [11], Chap. 5) are, for example, the following relations, artificial on the whole and typical of asymptotic methods:

$$u_\alpha^\varepsilon(x^\varepsilon) = \varepsilon^2 u_\alpha(\varepsilon)(x), \quad \sigma_{33}^\varepsilon(x^\varepsilon) = \varepsilon^4 \sigma_{33}(\varepsilon)(x), \quad g_3^\varepsilon(x^\varepsilon) = \varepsilon^4 g_3(x)$$

(for completeness, see [11], Chap. 5, pp. 367–406).

The same relations were also used in [14] for orthotropic-elastic plates.

Below, following [3], Chap. 1, we propose a direct method of constructing von Kármán equations and justifying their physical soundness. The terms in these equations have a concrete physical meaning sense and are as follows:

the averaged components of a displacement vector, bending and twisting moments, shearing forces, surface efforts and rotation of a normal. Further, von Kármán equations follow for a equilibrium elastic body when the principal vector and the principal moment vanish.

Let the initial spatial problem of elasticity for an anisotropic homogeneous elastic plate have form (1)–(4) ($\Gamma = 1$) with the corresponding boundary conditions (see, for example, [3], Chap. 1). Then we have

$$
\begin{aligned}
2h\Bigg[&\left(c_{\alpha\alpha}\partial_{\alpha\alpha} + \frac{3}{2}c_{\alpha 6}\partial_{12} + \frac{1}{2}c_{66}\partial_{3-\alpha\,3-\alpha}\right)\overline{u}_\alpha \\
&+\left(\frac{1}{2}c_{\alpha 6}\partial_{\alpha\alpha} + \left(c_{12}+\frac{1}{2}c_{66}\right)\partial_{12} + c_{3-\alpha 6}\partial_{3-\alpha\,3-\alpha}\right)\overline{u}_{3-\alpha}\Bigg] \\
&+h\Bigg[\left(c_{\alpha\alpha}\partial_\alpha + c_{\alpha 6}\partial_{3-\alpha\,3-\alpha}\right)\left(\overline{u}_{3,\alpha}\right)^2 + \left(c_{\alpha 6}\partial_\alpha + c_{66}\partial_{3-\alpha}\right)\overline{u}_{3,1}\overline{u}_{3,2} \\
&+\left(c_{12}\partial_\alpha + c_{3-\alpha 6}\partial_{3-\alpha}\right)\left(\overline{u}_{3,3-\alpha}\right)^2\Bigg]+b_{33}^{-1}\left(b_{\alpha 3}\partial_\alpha + b_{\alpha 6}\partial_{3-\alpha}\right)\int_{-h}^{h}\sigma_{33}dz=\overline{f}_\alpha. \quad (16)
\end{aligned}
$$

For the linear case, the system of Equations (16), where we neglect the remainder terms R, corresponds to the problem of defining a generalized plane stress-stain state. For the nonlinear case, system (16) immediately implies one of the fundamental equations of the von Kármán system, corresponding to the Airy function, if each equation is differentiated and the result is summed (for details see the discussion below).

3.1 The isotropic case

For the coefficients we obviously have $c_{\alpha\alpha} = \lambda^* + 2\mu$, $c_{66} = 2\mu$, $c_{12} = \lambda^*$, $c_{\alpha 6} = 0$, $\lambda^* = 2\lambda\mu(\lambda + 2\mu)^{-1}$, where λ and μ are the Lamé coefficients. Now system (16) can be rewritten as

$$
\begin{aligned}
(\lambda^*+2\mu)\partial_1\tau+\mu\partial_2\omega&=\frac{1}{2h}\,f_1+\mu\left(\partial_1(\overline{u}_{3,2})^2-\partial_2(u_{3,1}u_{3,2})\right)-\lambda_1(\sigma_{33,1},1), \\
(\lambda^*+2\mu)\partial_2\tau-\mu\partial_1\omega&=\frac{1}{2h}\,f_2+\mu\left(\partial_2(\overline{u}_{3,1})^2-\partial_1(u_{3,1}u_{3,2})\right)-\lambda_1(\sigma_{33,2},1),
\end{aligned}
\quad (17)
$$

where $\lambda_1 = \lambda/2h(\lambda+2\mu)$ and the functions $\tau = \overline{\varepsilon}_{\alpha\alpha}$, $\omega = \overline{u}_{1,2}-\overline{u}_{2,1}$ correspond to plane expansion and rotation.

By (17), the second equation with respect to the Airy function in the von Kármán system takes the form

$$
\begin{aligned}
(\lambda^* + 2\mu)\Delta\overline{\varepsilon}_{\alpha\alpha} &= (\lambda^* + 2\mu)\Delta\left(\frac{1}{2\mu} - \frac{1}{\mu(3\lambda + 2\mu)}\right)\left(\overline{\sigma}_{11} + \overline{\sigma}_{22}\right) \\
&= \mu\left(\partial_{11}\left(\overline{u}_{3,2}\right)^2 - 2\partial_{12}\left(\overline{u}_{3,1}\overline{u}_{3,2}\right) + \partial_{22}\left(\overline{u}_{3,1}\right)^2\right) + \frac{1}{2h}\overline{f}_{\alpha,\alpha} \\
&\quad +\frac{1}{2h}\left(\frac{\lambda(\lambda^* + 2\mu)}{2\mu(3\lambda + 2\mu)} - \frac{\lambda}{\lambda + 2\mu}\right)\int_{-h}^{h}\Delta\sigma_{33}dz,
\end{aligned}
$$

or

$$\Delta\bigl(\overline{\sigma}_{11} + \overline{\sigma}_{22}\bigr) = -\frac{E}{2}\bigl[\overline{u}_3, \overline{u}_3\bigr] + \frac{\nu}{2h}\int_{-h}^{h}\Delta\sigma_{33}dz + \frac{1+\nu}{2h}\,\overline{f}_{\alpha,\alpha}. \tag{18}$$

If we introduce the Airy function in the well-known way

$$\sigma_{\alpha\beta} = (-1)^{\alpha+\beta}\partial_{3-\alpha\,3-\beta}\Phi, \quad \overline{\sigma}_{\alpha\alpha} = \frac{1}{2h}\int_{-h}^{h}\partial_{3-\alpha\,3-\alpha}\Phi\,dz = \Delta\Phi^*, \tag{19}$$

then (18) implies the second equation of the von Kármán system

$$\Delta^2\Phi^* = -\frac{E}{2}\bigl[\overline{u}_3, \overline{u}_3\bigr] + \frac{\nu}{2}\Delta(g_3^+ + g_3^-) + \frac{1+\nu}{2h}\,\overline{f}_{\alpha,\alpha}, \tag{20}$$

where $[u, v]$ is the Monge-Ampere form.

3.2 The orthotropic case

Assuming $c_{\alpha 6} = 0$, from (16) obviously follows

$$2h\left[c_{\alpha\alpha}\partial_\alpha\overline{\varepsilon}_{\alpha\alpha} + (c_{12} + c_{66})\partial_\alpha\overline{\varepsilon}_{3-\alpha\,3-\alpha} + \frac{1}{2}(-1)^{3-\alpha}c_{66}\partial_{3-\alpha}\bigl(\overline{u}_{1,2} - \overline{u}_{2,1}\bigr)\right]$$

$$+hc_{66}\left[\partial_{3-\alpha}\bigl(\overline{u}_{3,1}\overline{u}_{3,2}\bigr) - \partial_\alpha(u_{3,2})^2\right] = \overline{f}_\alpha - b_{\alpha 3}b_{\alpha 3}^{-1}\int_{-h}^{h}\sigma_{33,\alpha}dz - R_\alpha^{AN}, \tag{21}$$

where

$$\overline{\varepsilon}_{\alpha\alpha} = \frac{1}{2h}\int_{-h}^{h}(u_{\alpha,\alpha} + u_{k,\alpha}u_{k,\alpha})dz.$$

If the coefficients b and c satisfy the condition of generalized transversality ([3], p. 27), i.e. if the relations $c_{11} = c_{22} = c_{12} + c_{66}$, $b_{13} = b_{23}$ are true, then (21) immediately implies

$$\begin{aligned}
c_{11}\partial_1\tau + \frac{1}{2}c_{66}\partial_2\omega &= \frac{1}{2h}\overline{f}_1 - b_{13}b_{33}^{-1}\frac{1}{2h}\int_{-h}^{h}\sigma_{33,1}dz \\
&\quad - hc_{66}\left[\partial_2\bigl(\overline{u}_{3,1}\overline{u}_{3,2}\bigr) - \partial_1\bigl(\overline{u}_{3,2}\bigr)^2\right] - R_1^{AN}, \\
c_{11}\partial_2\tau - \frac{1}{2}c_{66}\partial_1\omega &= \frac{1}{2h}\overline{f}_2 - b_{23}b_{33}^{-1}\frac{1}{2h}\int_{-h}^{h}\sigma_{33,2}dz \\
&\quad - hc_{66}\left[\partial_1\bigl(\overline{u}_{3,1}\overline{u}_{3,2}\bigr) - \partial_2\bigl(\overline{u}_{3,2}\bigr)^2\right] - R_2^{AN}.
\end{aligned} \tag{22}$$

Systems (16), (17), (20) and (21) can be combined into one equation that corresponds to the function Φ^* from the von Kármán equations. It is obvious that this procedure is true in the case of differentiability of the functions $\overline{u}_\alpha$. We remind that the latter functions are averaged with respect to the thickness of the plate in the horizontal components of the displacement vector.

The other fundamental equation of the von Kármán system corresponds to a bending problem. For clarity, we now give the corresponding relations when Ω_h is an isotropic elastic plate of constant thickness (for a more general case, where an elastic plate of variable thickness and with a finite displacement is anisotropic and nonhomogeneous, see [3], Chap. 1, and [22])

$$\frac{(1-\nu)D}{2}\Delta u_\alpha^* + \frac{(1+\nu)D}{2}\partial_\alpha \Delta u_{\beta,\beta}^* - \frac{3(1-\nu)D}{2h^2(1+2\gamma_\alpha)}\left(u_\alpha^* + \overline{u}_{3,\alpha}\right)$$
$$= f_\alpha^* + R_{\alpha+2}[u_\alpha], \tag{23}$$
$$\frac{3(1-\nu)D}{2h^2(1+2\gamma_3)}\left(\Delta \overline{u}_3 + u_{\alpha,\alpha}^*\right) = f_3^* + R_5[\overline{u}_3].$$

Here u_α^* are the rotations of the normals, R_{2+i} are remainder terms, D is cylindrical rigidity of the bending, and γ_α are arbitrary parameters.

Obviously, the Equations (16) (or (17–21)) and (23) without the remainder terms are a complete system of KMR type DEs with respect to the functions $\overline{u}_i(x,y)$ and $u_\alpha^*(x,y)$.

Note that nonlinear 2D models for Reissner type DEs with boundary layer effects for elastic plates were constructed for the first time in [3].

System (23) also has another equivalent form if as the unknown functions we choose the averaged deflection $\overline{u}_3$ and the shearing force Q_α (for details see [3], Chap. 1). Neglecting in (23) the remainder terms R for $\gamma = -0,5$, we obtain

$$D\Delta^2 w = (1+\Delta)(g_3^+ - g_3^-) + 2h[w,\varphi] + h(g_{\alpha,\alpha}^+ + g_{\alpha,\alpha}^-) - \int_{-h}^h z(f_{\alpha,\alpha} - f_3)dz, \tag{24}$$

where w and φ are the approximate values of the functions $\overline{u}_3$ and Φ^*. We remind that the function Φ^* has been introduced for convenience. At the same time as noted by Antman "this work [3] was the first that gave the von Kármán equations a rational position within the general theory of nonlinear elasticity" ([12], p. 699).

For the dynamic case, for clarity, we consider here the linear isotropic case with the right-hand side of the initial system having the form $\rho\partial_{tt}u_i + f_i(x,y,z,t)$ ($\rho =$const), where, f_i are the known functions and ρ is the matter density. Then, by virtue of the formulas of this subsection, the following equations follow immediately:

$$\frac{E}{2(1+\nu)}\Delta v_\alpha + \frac{E}{2(1-\nu^2)}\partial_\alpha v_{\beta,\beta} = \rho\partial_{tt}v_\alpha + \overline{f}_\alpha,$$
$$\frac{(1-\nu)D}{2}\Delta \omega_\alpha + \frac{(1+\nu)D}{2}\partial_\alpha \omega_{\beta,\beta}$$
$$-\frac{3(1-\nu)D}{2h^2(1+2\gamma_\alpha)}\left(\omega_\alpha + \omega,_\alpha\right) = \frac{2h^3}{3}\rho\partial_{tt}\omega_\alpha + f_\alpha^*, \tag{25}$$
$$\frac{3(1-\nu)D}{2h^2(1+2\gamma_3)}[\Delta \omega + \omega_{\alpha,\alpha}] = \frac{4h^3}{3}\rho\partial_{tt}\omega + \overline{f}_3.$$

In the nonlinear case, instead of (20) we obviously have (15), where the new term $\Delta\partial_{tt}\varphi$ describes transversal wave processes.

Note that the system of Equations (25) is a direct product of the equilibrium equations (17), while all the corresponding relations in [11], [15]–[16] and other papers are nothing else but the geometrical identity (compare with (1.4_1), $i = 1$, $k = l = 2$):

$$(-1)^{\alpha+\beta}\partial_{3-\alpha\,3-\beta}\varepsilon_{\alpha\beta} = -0,5[u_3, u_3].$$

In the linear case, mathematical models of type (17), (20), (21) and (22) are a convenient unified form for applying directly the methods developed in [17]–[20] and other works for all classical BVPs. Furthermore, models of type (17) and (22), which are systems of Cauchy-Riemann DEs, are natural and convenient expressions for the application of the two-stage method of least squares and the FEM, whereas models (2.21) and (2.22) in [21] contain, instead of two, eight scalar equations.

For completeness, it should be said that for the 3D case the values $e = \operatorname{div} u$, $\omega = \operatorname{rot} u$ were introduced by Love [18]. For the 2D linear case, the values $e = u_{a,a}$ and $\omega = u_{1,2} - u_{2,1}$ were considered by Biot in the late 30th of the 20th century.

References

1. C. A. J. Fletcher, *Computational Techniques for Fluid Dynamics* 2, Springer Verlag, 1988.
2. P. G. Ciarlet, *Mathematical Elasticity, Vol. 1: Three Dimensional Elasticity*, NH, 1993.
3. T. Vashakmadze, *The Theory of Anisotropic Elastic Plates*, Kluwer, 1999.
4. T. Vashakmadze, To the New Treatment for some Poro-elastic Thin-walled Structures, J. Georg. Geophysic. Soc., Issue A. Vol. 9A, 2005, 100–110.
5. M. J. Ablovitz, H. Segur, *Solitons and the Inverse Scattering Transforms*, SIAM, 1981.
6. A. Newel, *Solitons in Mathematics and Physics*, SIAM, 1985.
7. R. Kienzler, D. K. Bose, Matherial Conservation Laws Established within Consistent Plate Theory, IUTAM Symposium, 23–28 April, 2007, Tbilisi, Relation of Shell, Plate, Beam and 3D Models, Book of Abstracts, 36–37.
8. M. A. Biot, Generalized Theory of Acoustic Propagation in Porous Dissipative Media. J. Account. Soc. Am. v. 35, No 5, part 1, 1962, 1254–1264.
9. M. A. Biot, Nonlinear and Semilinear Rheology of Porous Solids. J. Geophys. Res. 78, 1973, 4924–4937.
10. T. Vashakmadze, R. Janjgava, To Construction and Justification of von Kármán-Reissner type Systems of Equations for Elastic Plates with Binary Mixtures, Proceed. Geor. Technical Univ., v. 447, N 1, 2003, 53–65.
11. P. G. Ciarlet, *Mathematical Elasticity, Vol. II: Theory of Plates*, NH, 1997.
12. S. S. Antman, *Nonlinear Problems in Elasticity*. Springer, II Ed., 2005.
13. P. Podio-Guidugli, Von Kármán Equations without Abdicating Reason, Kluwer, 2002, 1–27.

14. H. Begehr, R. Gilbert, C. O. Lo, The Two-Dimensional Orthotropic Plate. J. Elasticity, 26, 1991, 147–167.

15. T. Von Kármán, Festigkeitsprobleme im Maschinenbau, *in Encyclopadie der Mathematischen Wissenschaften*, IV/4C, 1910, 311–385.

16. A. A. Love, *A Treatise of Mathematical Theory of Elasticity*, IVth Ed., CUP, 1959.

17. N. Muskhelishvili, *Some Basic Problems of Mathematical Theory of Elasticity*, Noordhroff, 1963.

18. I. Vekua, *Generalized Analytic Functions.* Oxford-London-New York-Paris, 1962.

19. R. Lekhnitskii, *Anisotropic Plates*, Gordon & Beach, NY, 1968.

20. R. Gilbert, Wei Lin, Functional-Theoretic Solutions to Problems of Orthotropic Elasticity. J. Elasticity, 15, 1985, 143–154.

21. S.-Y. Yang, C. L. Chang, Analysis of Two-stage Least-squares FEM for Planar Elasticity Problems. Math. in Appl. Sci., 22, 1999, 713–732.

22. T. Vashakmadze, On Dynamical Problems for 3D Anisotropic Theory of Elasticity. Proceed. Javakhishvili Tbilisi St. Univer., Vol. 353 (22–23), 2003, 109–113.

List of Participants

Armenia

Aghalovyan, Lenser A.
Bagdoev, Alexander
Belubekyan, Vagharshak
Safaryan, Yuri
Sargsyan, Samvel

Austria

Mang, Herbert A.

Bulgaria

Gavrilova, Elena

Canada

Hansen, Jorn S.
Oguamanam, Donatus C.D.

Georgia

Avalishvili, Gia
Avalishvili, Mariam
Bantsuri, Revaz
Buchukuri, Tengiz
Chkadua, Otar
Chinchaladze, Natalia
Duduchava, Roland
Gordeziani, David
Jaiani, George

Kapanadze, Gogi
Khomasuridze, Nuri
Kipiani, Gela
Kukudzanov, Sergey
Meunargia, Tengiz
Natroshvili, David
Odishelidze, Nana
Shavlakadze, Nugzar
Tvalchrelidze, Avtandil
Vashakmadze, Tamaz
Zirakashvili, Natela

Germany

Altenbach, Holm
Begehr, Heinrich
Ebel, Adolf
Kienzler, Reinhold
Schlebusch, Rainer
Schulze, Bert-Wolfgang
Wendland, Wolfgang L.

Italy

Freddi, Lorenzo
Paroni, Roberto
Podio-Guidugli, Paolo
Ricci, Paolo Emilio

Netherlands
Dick H. Van Campen

Poland
Bojarski, Bogdan

Romania
Birsan, Mircea

Russia
Ershov, Yuri

Switzerland
Chipot, Michel

Turkey
Aksoy Ü.
Çelebi A. O.

UK
Mikhailov, Sergey

Ukraine
Makarov, V.L.